# Leibniz: Dissertation on Combinatorial Art

# Leibniz: Dissertation on Combinatorial Art

*Translated with introduction and commentary by*
MASSIMO MUGNAI, HAN VAN RULER,
AND MARTIN WILSON

OXFORD
UNIVERSITY PRESS

# OXFORD

UNIVERSITY PRESS

Great Clarendon Street, Oxford, OX2 6DP,
United Kingdom

Oxford University Press is a department of the University of Oxford.
It furthers the University's objective of excellence in research, scholarship,
and education by publishing worldwide. Oxford is a registered trade mark of
Oxford University Press in the UK and in certain other countries

Published in the United States of America by Oxford University Press
198 Madison Avenue, New York, NY 10016, United States of America

British Library Cataloguing in Publication Data
Data available

Library of Congress Control Number: 2019949395

ISBN 978-0-19-883795-4

Printed and bound by
CPI Group (UK) Ltd, Croydon, CR0 4YY

# Preface

History is bound to cherish the youthful work of a later genius, especially if its themes anticipate those of masterpieces to come. In the case of Leibniz's *Dissertatio de arte combinatoria* (1666), however, scholarship has been curiously reserved in dealing with this particular *travail de jeunesse*. Although studies by such eminent Leibniz specialists as Louis Couturat (1868–1914), Karl Dürr (1875–1928), and Eberhard Knobloch have added much to the interpretation of Leibniz's first publication in print, a critical edition of the *Dissertatio* has thus far been wanting, no doubt due to its arduous phrasing, its wide-ranging subject matter, as well as its complex overall theme.

All of these factors make Leibniz's *De arte combinatoria* (henceforth *DAC*) a text that is not easy to edit. Working on the present edition, moreover, the editors came across many further obscurities complicating an annotated rendition of the work. Its mix of languages and subject matters is only a minor hurdle in comparison to the strenuous way in which Leibniz at times gives expression to the ideas contained in it, which, for instance, may include descriptions in Latin of complex mathematical operations. Another problem encountered was to find the many texts by authors cited in the *DAC*, a difficulty that was enhanced by Leibniz's recurrent habit of quoting authors indirectly—and on occasion incorrectly.

The wealth of materials in *De arte combinatoria* and the dense argumentation it contains did not impede Martin Wilson to work on an English translation of the text over a period of many years, until he finally suggested to prepare it for publication in the wake of the appearance of Arnold Geulincx's *Ethics* in 2006, a project on which he and Han van Ruler had been working together.[1] Finding numerous puzzling details in Leibniz's text and a multitude of references unaccounted for in the extant secondary literature, it would take Martin Wilson and Han van Ruler some further years of textual and bibliographical investigations to come up with a reasonably reliable translation. At this point, Massimo Mugnai joined Martin

---

[1] Geulincx (2006).

and Han and the project began to take the shape of an edition aiming not only to satisfy the requirements of the inner circle of Leibniz's scholars but a larger spectrum of readers interested in Leibniz's philosophy and the origins of the combinatorial calculus in general. This implied not only the obvious task of preparing an intelligible translation of the *DAC*, but even that of explaining the many difficult passages of the text. As a corollary to this task, it became indispensable to find all writings and authors mentioned by Leibniz, some of them very obscure or almost unknown, and to equip the text with a rich apparatus of footnotes.

Thus, the annotated translation that lies before you is the product of a collective enterprise in which Martin Wilson has been the principal translator of Leibniz's text. Massimo Mugnai and Han van Ruler have revised and edited the translation to produce the text as it now stands, and jointly contributed to its annotation. Han has taken care of the largest part of the footnotes, whereas Massimo, besides preparing the Latin version of the *DAC* here included, has provided the main text with an Introduction, aiming to situate in its historical context this early work of Leibniz and to present the main results contained in it.

In the process of the collaborative work on the project, Martin Wilson was sadly found missing for a period, due to a stroke he suffered on 1 March, 2011. Though quickly regaining most of his colourful personality, Martin has not been able to fully recover. The editors would like to express their gratitude to Astley Nursing Home in Worcestershire, England, for the care they have given Martin ever since, as well as to Barry Sweeting, who has been acting as Martin's representative and became a good friend.

The editors are also grateful to Matthias Àrmgardt, Albrecht Heeffer, Franco Montanari, Jan Papy, Horst Pfefferl, Wolfgang Schlosser, and Andrew Weeks for their helpful comments on some questions of detail. Richard Arthur, Vincenzo De Risi, Enrico Giusti, Enrico Pasini, and Monica Ugaglia have read a first draft of the Introduction, finding mistakes and kindly suggesting amendments. A special word of thanks is due to Mariano Giaquinta and Eberhard Knobloch for their generous advice on many technical questions.

Working on the annotation of the *DAC* during years in which there has been a stupendous increase in the number of electronic sources available online, Han van Ruler was nonetheless pleased to be helped by the cooperative staff of a huge number of libraries still important for their treasures— Leiden University Library, the Herzog August Bibliothek Wolfenbüttel, the British Library, and the Bibliothèque Nationale de France in particular.

Besides expressing his debt to the staff of these and many other libraries, he also wishes to thank Atsuko Fukuoka, Iqbal Faridi, and Peter Guttner, as well as Patrick Masereel and Vincent de Keijzer for their hospitality in Wolfenbüttel, London, and Paris over the years.

Finally, the editors are grateful to Maria Rosa Antognazza and Paolo Mancosu for their constant encouragement and to Peter Momtchiloff of Oxford University Press, whose confidence and dedication have given an enormous spurt to the realization of this long-term project.

Massimo Mugnai, Han van Ruler
Firenze and Leiden

# On this Edition

The Latin text here presented of Leibniz's *Dissertatio de arte combinatoria* is faithful to the original publication of the *DAC* (Leipzig: Fick and Seubold, 1666), the full title of which reads:

> *Dissertatio De Arte Combinatoria, In qua Ex Arithmeticae fundamentis Complicationum ac Transpositionum Doctrina novis praeceptis extruitur, & usus ambarum per universum scientiarum orbem ostenditur; nova etiam Artis Meditandi, Seu Logicae Inventionis semina sparguntur. Praefixa est Synopsis totius Tractatus, & additamenti loco Demonstratio Existentiae Dei ad Mathematicam certitudinem exacta.*

In a new, ameliorated edition, the text of the *DAC* was reproduced by Carl Immanuel Gerhardt (1816–99) in volume IV of his edition of *Die philosophische Schriften von Gottfried Wilhelm Leibniz* (Berlin: Weidmann, 1880 / Hildesheim: Georg Olms, 1960), 27–104.

In the text below, however, the reader will find references in square brackets (in bold) to the 1666 edition and also square brackets (in italics) to the page numbers of the *Akademie*-edition of the *DAC*:

> Gottfried Wilhelm Leibniz, *Sämtliche Schriften und Briefe* herausgegeben von der Deutschen Akademie der Wissenschaften zu Berlin, Sechste Reihe, Erster Band, Darmstadt: Reichl, 1930 / Berlin Akademie Verlag, 163–230.

In the translation here presented, Leibniz's typography has been followed almost literally, even if the structure of individual sentences has been adapted in such ways as to yield a readable English text.

In dealing with typos, omissions, miscalculations, and mistaken references in the 1666 original, as well as with Leibniz's recurrent inclusion of Greek and German words into the Latin text, consistency in referring to these has not been an aim in itself. Obvious editorial interventions will be marked by the use of square brackets in the text of the translation.

In choosing case by case whether simply to copy the original or to make editorial interventions, it has been the editors' intention to offer an intuitively understandable account of what the author must have had in mind. Latin text, translation, and annotations will hopefully provide the reader with sufficient materials to do the same and reconstruct the mathematical and philosophical objectives Leibniz set himself when writing the *DAC*.

As far as the editors know, besides the present edition there is only one entire translation of the *DAC* into a modern language: G. W. Leibniz, *Dissertación de arte combinatoria*, translated by Manuel A. Correia, in G. W. Leibniz, *Obras filosóficas y científicas. Escritos Matemáticos*, (ed.) Mary Sol de Mora Charles (Granada: Editorial Comares), 2015, vol. 7B, 547–643. A previous translation into Catalan was edited, again by Manuel Correia, in 1992.

## Quotation marks

In what follows, double quotation marks have generally been used for quotations, and single quotation marks for highlighting (the use of) specific concepts or expressions. Double quotation marks, however, may occasionally also point out odd or unusual expressions.

In the footnotes, double quotation marks are also used for quotations in Latin, but quotations of Latin verse and references to (the use of) Latin concepts and expression are generally given in *italics*.

Finally, Leibniz himself uses double quotation marks to indicate the solutions he offers to the problems dealt with in the *DAC*—a custom that the editors have chosen to copy in the translation.

## References

Next to page number references (after ':') and references to volumes ('vol.', 'vols'), the reader will also find references to the signature marks ('sig.', 'sigg.'), which are found especially in the front matter of early-modern prints, as well as, where applicable, to columns rather than pages ('col.', 'cols').

## Abbreviations

A list of bibliographical abbreviations and abbreviated titles for reference works occurring in the notes to both the Introduction and the Translation of the *DAC*, may be found in sections A. (*Leibniz Works*) and D. (*Abbreviations*) of the Bibliography, pp. 287 and 301–2, below.

# Contents

# Introduction

## 1 The project for the 'characteristic art'

Leibniz was twenty years old when the *Dissertation on Combinatorial Art* (henceforth *DAC*) was first published. As he would later write: "the booklet was composed when the author was still young, in the year 1665, and was edited at Leipzig in the year 1666".[1] The work expanded the text of a dissertation he had defended earlier that same year at the Leipzig Faculty of Philosophy, aiming for a university position. This first embryonic form of the *Dissertation* was entitled *Disputatio Arithmetica de complexionibus* ('Arithmetical Disputation on Complexions') and contained the first part of the 1666 work, from *Cum Deo* ("With God's Help") through to the end of section 8 of the first problem.[2]

At the time, Leibniz was not particularly well trained in mathematics, as he was to admit later on the occasion of a re-edition of the *DAC*. When, indeed, the *DAC* was published again without the author's permission by the Frankfurt book dealer Cröker in 1690, this provoked a reaction on Leibniz's part, who wrote a short notice in the *Acta Eruditorum*.[3] As he writes there, the booklet was "not sufficiently polished" and no longer agreed with the ideas he held "at present" (1691). Moreover, the book left something to be desired:

> ... some problems concerning numbers could have been solved in a far better way and the solutions are susceptible of more accurate proofs. At the time, indeed, [the author] had only a vague idea of higher mathematics and thus, unaware of the discoveries of others and not sufficiently acquainted

---

[1] GP 4: 103. The original (see below, note 4) has '1668'.
[2] Cf. A VI, 1: 170–5. The *Arithmetical Disputation* included four more corollaries (A VI, 1: 228–30). The entire title of the *Arithmetical Disputation* was: *Disputatio Arithmetica de Complexionibus, Quam in Illustri Academia Lipsiensi Indultu Amplissimae Facultatis Philosophicae pro loco in ea obtinendo prima vice habebit* M. Gotifredus Guilielmus Leibnuzius Lipsiensis, J. U. Baccal. D. 7. Martii Anno 1666. H. L. Q. C. Lipsiae, Literis Sporelianis.
[3] Cf. Knobloch (1974: 412).

with mathematical analysis, he was able to produce something of his own on the basis of a hasty reflection ( ... )[4]

Despite this, however, Leibniz also notes that "the booklet" contains "many new meditations" that he does not regret, concerning the "art of discovery" (*ars inveniendi*) and the "excellent" idea of an alphabet of human thoughts.

The 'art of discovery' and the *alphabet of human thoughts* are two issues strongly connected with one another. They both pertain to the discipline of logic, for which Leibniz manifested a genuine interest from an early age onwards. According to an autobiographical passage included in a text probably written in the year 1682, Leibniz, when he was "not yet twelve", was filling "sheets of paper with wonderful meditations about logic, attempting to overcome the subtleties of the scholastics."[5] In the same vein, in a text composed during the years 1683–5, he writes:

> As a boy I learned logic, and having already developed the habit of digging more deeply into the reasons for what I was taught, I raised the following question with my teachers. Seeing that there are categories for the simple terms by which concepts are ordered, why should there not also be categories for complex terms, by which truths may be ordered? I was then unaware that geometricians do this very thing when they demonstrate and order propositions according to their dependence upon each other. It seemed to me however that this could be achieved universally if we first had the true categories for simple terms and if, to obtain these, we set up something new in the nature of an alphabet of thoughts, or a catalogue of the highest genera or of those we assume to be highest, such as $a, b, c, d, e, f,$ out of whose combination inferior concepts may be formed.[6]

Later on, the idea of building an 'alphabet of thoughts' evolved into a very complex project, which remained a central topic among Leibniz's philosophical interests until the end of his life: the composition of a 'universal characteristic', or 'characteristic art' (*ars characteristica*).

The Latin word *character* has the same meaning as the English words 'mark', 'sign'; and with the expression *ars characteristica*, Leibniz refers to

---

[4] Leibniz's notice was published in the *Acta Eruditorum* (Leipzig) of the year 1691 (pp. 63–4, cf. GP 4: 103–4) and has been recently republished in Leibniz, *Essais*, vol. III: 995–6.

[5] GP 7: 126. For "the age of twelve", see what G. H. Parkinson writes in *Logical Papers*, p. x, n. 4, where, relying on Fischer-Kabitz (1920: 712), he considers it an exaggeration on Leibniz's part.

[6] L: 229 (A VI, 4: 538).

a system of symbols denoting concepts and propositions, on which one may operate according to certain rules to discover new truths and to find proofs for the old ones. To construct the *characteristic*, two fundamental steps were needed: 1) performing an analysis of all human concepts, so that we may have a complete list of all simple concepts; 2) recombining the simple concepts, thus generating all possible complex concepts and all possible truths. Step (1) requires that we dispose of a general inventory of all known concepts and truths, and this motivated Leibniz to plan a further step: 3) the construction of a general encyclopedia of human knowledge. Leibniz's optimistic project is that of analysing each complex concept in its component parts, until we reach the simple, primitive ones. At the first stage of the project, clearly witnessed by the *DAC*, Leibniz takes for granted (in accordance with the Lullist tradition) that we human beings could attain the first, simple concepts corresponding to the properties of God. Later on, however, around 1686, Leibniz states that, even though we cannot reach the simple concepts, we may stop our analysis at a certain point and consider as simple, i.e. primitive, those concepts that we cannot further analyse. Thus, in this case the simple concepts are simple *for us*, not absolutely.[7]

Once the encyclopedia were constructed, however, the problem would still remain of finding the rules to perform the analysis and re-combination of the concepts; and it is for this reason that Leibniz adds another stage to the previous ones: 4) the constitution of the so-called 'general science' (*scientia generalis*).

Leibniz thinks of the general science in terms of a very abstract discipline, i.e. as a kind of 'meta-science', which was to investigate the various methods and procedures (deductive and inductive, logical and empirical) internal to each scientific field and displayed by people involved in their respective research activities. Of course, the study of logic would play a pre-eminent role in this general science and it is therefore no wonder that Leibniz, in the period from 1676 to 1690, developed a series of essays on the logical calculus. These essays were aimed at finding the general rules and principles governing any kind of logical inference, and may be considered as the first successful attempt, in Western culture, to develop an algebraic treatment of logic before George Boole's seminal work.[8]

---

[7] Cf., for instance, A VI, 4A: 742, 746, 964.

[8] Boole (1847). On Leibniz's logical writings and Boolean algebras, see Castañeda (1990); Swoyer (1994); Lenzen (2004); Mugnai (2010); Malink-Vasudevan (2016), and the debate between Mugnai and Malink-Vasudevan (2017).

Almost in the same period as his inquiries on logical calculus, Leibniz applied himself to the task of determining the main features of the so-called 'rational grammar' (*grammatica rationalis*), which, according to his intentions, should constitute a fundamental tool toward the construction of a 'universal language'.

In some texts, Leibniz seems to identify 'universal language' and 'characteristic art': he models the *characteristic* on the example of a natural language based on a grammar reduced to its essential elements. In general, however, Leibniz thinks of the construction of a universal language as a step toward the constitution of the universal characteristic, which, as we have seen, presupposes a large-scale program with lots of elements besides the universal language, including the encyclopedia, the general science, the logical calculus, etc.

If we compare what Leibniz writes in the *DAC* with the project for the characteristic art as it emerges in the years following his stay in Paris (March 1672–October 1676), it is not difficult to find in the 1666 work many seeds of the future developments. From the *DAC*, indeed, the project for the characteristic art inherits the following, fundamental ideas:

1) An analysis of concepts is needed to reach the first, simple concepts;
2) By means of a synthetic procedure, complementary to that of analysis, all possible complex concepts may be derived out of the simple ones;
3) The process of synthesis can be performed by combining the concepts, simple and complex, on the basis of the rules of logic;
4) If we assign to each simple concept a symbol of an alphabet, we may unambiguously represent in symbolic form any complex concept whatsoever;
5) Given that any elementary truth can be expressed combining two concepts corresponding, respectively, to a subject and a predicate, the process of combining the simple concepts will generate all possible truths, constituting the basis of the *art of discovery*;
6) If we employ *numbers* to represent concepts, we may make logical inferences simply operating on numbers, as in an arithmetical calculus, instead of operating directly on concepts.

To appreciate the novelty and originality of Leibniz's project, as it is articulated in the *DAC*, however, we have to dwell a little on the cultural background against which these ideas were conceived.

## 2 'Lullism' in the seventeenth century: art of memory and first steps towards the creation of an encyclopedia

In the first half of the fifteenth century, humanists strongly reacted against scholasticism in general and to scholastic logic in particular, which they tended to consider as a discipline emblematic of a philosophical attitude that was preoccupied primarily with frivolous topics and exclusively devoted to produce sophistries. With the rediscovery of the works of Euclid and the publication of texts by authors of late antiquity such as Heron of Alexandria and Pappus, the standard of rigorous reasoning ceased to be identified with Aristotle's *Organon*.

This does not mean, however, that, during the middle ages, Euclid's *Elements* were unknown. In the first half of the twelfth century Adelard of Bath translated into Latin an Arabic manuscript of the *Elements*. Adelard's edition was followed by several other translations made from Arabic sources and by the translation made by Gerardo da Cremona. In the thirteenth century, Campano da Novara prepared a text of Euclid's work, which was later employed by Luca Pacioli and Nicolò Tartaglia for their editions. Medieval commentaries on the *Elements*, however, with the exception of those belonging to the Arabic tradition, were usually quite poor. In the second half of the sixteenth century, the *Elements* were translated into modern languages.[9] In the same period, Greek manuscripts began to circulate in Europe and Bartolomeo Zamberti translated the *Elements* into Latin directly from the Greek (1505), thus opening a new era in the philological reception of the text. In 1533 Simon Grynaeus published the *editio princeps* of the Greek text. In this sense it seems proper to speak of a 'rediscovery' of Euclid in the Western World.[10]

Books on logic from the fifteenth and sixteenth century often expressed the need to develop old-fashioned logical matters according to the method followed by mathematicians. In the introduction to his treatise on logic (first published in 1492), Girolamo Savonarola (1452–98), for example, criticizes the obscurity that characterized the traditional way of teaching logic and states that he aims to offer a synthesis of the entire dialectic 'according to the usual practice of mathematicians'.[11] Thus, it is no wonder that Jacob

---

[9] Into Italian (1543), German (1562), French (1564), English (1570), Spanish (1576), and Dutch (1606).

[10] See De Risi (2016).

[11] Savonarola (1982: 3) complains that many people abstain from studying logic because of the intrinsic difficulties of the discipline, "the obscurity of Aristotle's books and the sophistries

Pelletier (1515–82), more than fifty years after Savonarola's logic, in his introduction to the Latin translation of the *Elements* (1557),[12] writes that "every proof which leads us to truth is geometrical in character", and that "we are not able to distinguish the true from the false if we have not previously been well acquainted with Euclid."[13] From the beginning of the sixteenth century onwards, even though in many cases logic continued to be taught along traditional lines, authors on the subject of logic start raising the issue of its relationships with mathematics.

In antiquity and in medieval times, logic and mathematics had been considered as two quite independent disciplines.[14] Even though it is probable that Aristotle elaborated his theory of syllogisms drawing inspiration from contemporary ideas in mathematics, and even though Stoic arguments were employed by later commentators to justify certain theorems in Euclid's *Elements*, the idea of reducing logic to a calculus seems to have been conceived for the first time in the period between the second half of the sixteenth and the first half of the seventeenth century.[15] Pierre de la Ramée (1515–72), for instance, claims that the Greek word for 'syllogism' properly means a calculus or a computatio and Thomas Hobbes in his *Computatio sive Logica* plainly identifies the act of thinking with that of performing an addition or a subtraction: "By ratiocination, I mean computation. Now, to compute is either to collect the sum of many things that are added together, or to know what remains when one thing is taken out of another. Ratiocination, therefore, is the same with addition and subtraction."[16] Clearly, Hobbes's idea that logical operations simply amount to adding or subtracting concepts, was too naïve to permit any interesting development in the direction of a logical calculus. Yet, the claim that reasoning is the same as computing was quite new and invited people to create stronger relationships between logic, understood as the discipline which dictates the canon for correct reasoning, and mathematics.

---

and obscurity" which afflict the language of (scholastic) logicians. To these he opposes the clarity and simplicity to which mathematicians are accustomed. This contraposition is typical of the humanist thinkers and, as far as I know, there is no trace of it in logic handbooks written before the fifteenth century.

[12] The translation concerned the first six books only.
[13] Pelletier (1557: 12).      [14] Cf. Muller (1981); Acerbi (2010: 15–86).
[15] On the relationships between logic and mathematics in the seventeenth century, see Mugnai (2010).
[16] Hobbes (1839a: 3). We find an analogous claim concerning syllogism in Gassendi (1658, vol. 1: 34).

Another important consequence of re-introducing Euclid's works in Western culture was that together with these works even some ancient commentaries on the *Elements* (like Pappus's commentary, for instance; fourth century AD), were to become widely known. Relying on Pappus, the astronomer and mathematician Christopher Clavius (1538–1612) accordingly contributed to an increasing interest in the distinction between *analysis* and *synthesis* among philosophers and mathematicians.[17]

Roughly speaking, given a certain mathematical problem, *analysis* was usually described as a method for going backwards from the problem to some general principles that were involved in it and which could constitute a basis for finding a solution. *Synthesis* was the opposite: from the principles to the solution. *Analysis* was intended as a kind of heuristic process, whereas *synthesis* was devised as a method for proving theorems starting from first principles and axioms. Thus, analysis and synthesis were viewed as two essential tools for the project of renewing logic both as an art of judgment (*ars iudicandi*) and as an art of discovery (*ars inveniendi*).

The processes of analysis and synthesis played an essential role in another project characterizing the cultural climate at the beginning of the seventeenth century: the project of the creation of a universal language. This project, in turn, was linked to the old tradition of so-called *Lullism*, from the medieval thinker Ramon Lull (1232–1316), who is credited as a founding father of combinatory and as a source for Leibniz's ideas on combinatorial art.[18]

In his *Ultimate General Art* (*Ars Magna*, 1305), Ramon Lull proposes a program the main features of which may be summarized as follows. Once you have the first concepts, which he believes are nine in number and correspond to the properties of God (Goodness, Power, Wisdom, Truth, Glory, etc.), Lull proposes the construction of a machinery composed by concentric discs of paper rotating around a rigid axle. Each disc (properly a circle) is divided into nine sectors and in each sector a letter is typed representing a concept. The outer disc is fixed and contains the letters representing the first concepts from the combination of which all others are generated. The inner circles, which contain letters of other general concepts, different from those corresponding to the divine properties, may each revolve in opposite direction with respect to the others. Moving the inner circles and keeping the outer one fixed, the 'artist', i.e. the expert or

---

[17] For the development of the concept of analysis in Western philosophy, see Beaney (2016).
[18] On Lull's art and logic, see Bonner (2007).

master of the art, could create all possible combinations of the letters printed in the sectors. With two circles, he might give rise to all possible combinations of *two* letters; with three circles to all possible combinations of *three* letters, etc. Lull's idea was that using this device systematically, one could reproduce all complex concepts of all created things.[19] Lull considered the *General Art* as a tool for producing *true knowledge* and believed that simply manipulating a few discs of paper and combining their inscribed letters one could discover new truths in every field of human knowledge. His aim was that of creating a general theory of the principles of all sciences: a kind of 'science of all sciences', the nature of which was not merely formal and which situated itself in a terrain intermediate between logic and metaphysics.[20]

Starting from the second half of the sixteenth century, a renewed interest in Lullism was largely motivated by the reaction against traditional ways of teaching (Aristotelian) scholastic logic. The latter seemed too formal and too remote from real life to be of any interest, whereas the Lullian *Art*, which appeared to be playing with concepts of *real* properties and was hunting for truths concerning real beings, was more promising in the eyes of people who attempted to give a practical orientation to philosophy.

## 3 The project for the creation of a universal language

As we have seen, Lull employed letters of the Latin alphabet to represent concepts, but other signs or characters could also be invented to denote the 'primitive concepts'. One can, for instance, devise some set of particularly simple signs, easy to be manipulated and remembered, and then generate, according to certain rules, the complex configurations corresponding to all complex concepts. In this way, each concept could be associated with a kind of 'picture', explicitly representing the degree of complexity of the concept itself. This latter is a typical example of what, at the beginning of the seventeenth century was considered a project for the construction of a *universal language*. The connection with the *Art* of Ramon Lull was not a necessary requisite for the project: many authors, however, were driven to associate the two because of contingent analogies, mainly grounded on the

---

[19] This is a quite simplified account of Lull's art: for a more accurate explanation, see Bonner (2007).
[20] Cf. Bonner (2007: 127) and Chapter 5 in particular.

recourse to the methods of analysis and synthesis. This is the case, for example, with Seth Ward (1617–89), Savilian Professor of astronomy at Oxford, who had an interest in Lull's theories and gave a very clear exposition of the project for a universal language:

> [ ... ] for all Discourses being resolved in sentences, those into words, words signifying either simple notions or being resolvable into simple notions, it is manifest, that if all the sorts of simple notions be found out, and have Symbols assigned to them, those will be extremely few in respect of the other, (which are indeed Characters of words, such as *Tullius Tiro's*)[21] the reason of their composition easily known, and the most compounded ones at once will be comprehended, and yet will represent to the very eye all the elements of their composition, and so deliver the natures of things: and exact discourses may be made demonstratively without any other pains than is used in the operations of specious Analytics.[22]

'Specious analytics' (*analytica speciosa*) is a name also employed by Leibniz in the *DAC* to designate the *algebra* introduced by François Viète (1540–1605); the term *speciosa*, a derivation of *species*, alluding to the possibility of dealing simultaneously with several things of the same kind. As Seth Ward stresses, the creation of a universal language requires first that we are able to decompose our complex concepts into their component parts. To execute this task, however, we need something like a register of all (or at least of the most important) complex concepts. Once this register was ready, it would contain the greatest part of human knowledge all at once. It is from this idea that the project for the construction of an encyclopedia took its origin, an encyclopedia that would contain the entire knowledge acquired by human beings in the course of the centuries. Thus, in the period considered, Lullism, the search for a universal language and the project for constructing an encyclopedia were three different, but fairly interconnected subjects, pursued by a considerable number of philosophers in Europe. Amongst these, Johann Heinrich Alsted (1588–1638), the celebrated author of a huge encyclopedia in seven volumes (Herborn 1630), and an author well-known to the young Leibniz, played a very influential role.[23]

---

[21] Marcus Tullius Tiro (first century BC) was first a slave and then a freeman of Cicero who invented the 'Tironian notes', the first Latin shorthand system.

[22] Ward (1654: 21).

[23] On Alsted's influence on the young Leibniz, see Loemker (1961); Mugnai (1973). For an exhaustive account of the relationships between Alsted and Leibniz on matters related to

As Alsted writes in his *Clavis Artis Lullianae et verae logicae* (*A Key for the Lullian Art and the True Logic*, 1609), Lullism was poorly represented in Germany at the time, whereas it was flourishing in Spain, France, and Italy. In Italy, in particular, many philosophers (here called *speculatores*, 'speculators') applied themselves intensively to the study of this *art*, becoming a kind of addict to it in the course of doing so.[24] Alsted calls Ramon Lull "a mathematician and a cabbalist"[25] and praises him for the use of letters of the alphabet to denote concepts, as well as for his recourse to mathematical symbols like circles. He considers the former as valid aids to memory and the latter as tools on which the teaching of any science is based.[26]

In comparison with other texts devoted to the same topic (Bruno's works included), Alsted's *Clavis* constitutes a rare example of clarity and conciseness: it gives a terse account of how Lull's art works and weds the Lullian combinatory with some features borrowed from the reformed logic proposed by Pierre de la Ramée (1515–72), thereby giving rise to a logical system mainly oriented at the development of topical arguments, i.e. arguments conceived with the purpose of ruling dialectical disputes.[27]

## 4 Atomism and 'immeation'

Compared with the works of authors belonging to the tradition of Lullism, the *DAC* differentiates itself first of all because of the emphasis put on the theory of syllogism, then for the attempt to extend the combinatorial calculus to any kind of matter. Finally, even though Leibniz's mathematical training was not very developed at the time, an intense mathematical flavour that can hardly be found among the followers of Ramon Lull emanates from the *DAC*.

According to the image of the world that we get reading the *DAC*, everything having a complex structure is a whole which can be decomposed into lesser wholes and these latter, in turn, into their least components parts

---

religion, see Antognazza-Hotson (1999); on Alsted's encyclopaedism and the 'second Reformation' in Germany, see Hotson (2000). On Lullism, the encyclopedic projects and the universal language, see Rossi (2006).

[24] Alsted (1609: 19).       [25] Alsted (1609: 20).
[26] Alsted (1609: 21). For a further development of the doctrine, the reader of the *Clavis* is invited to consult the commentaries to the *Art* written by Giordano Bruno.
[27] On the relationships between Ramus's doctrines and the Herborn encyclopedists (on Ramus's philosophy in general), see Ong (2004).

(atoms or some molecular aggregates of some sort). By a simple process of the combination of the parts, one may, in turn, re-compose the whole. The same procedure can be applied to non-material things as well: to concepts, for example, propositions, geometrical figures, and numbers. Thus, combinatory applied to mereology, the doctrine of wholes and parts, becomes a kind of all-purpose tool in Leibniz's hands; a tool that allows one to reveal the intimate secrets of nature:

> Since all things that exist, or can be thought, may be said to be made up of parts, either real or at least conceptual, whatever differs in kind must necessarily differ either in parts, and here lie the Applications of *Complexions*, or by a different situs, hence the application of *Dispositions*.[28]

The system of variations elaborated in the *DAC*, Leibniz observes, "leads the mind that yields to it almost through all infinity, and embraces at once the harmony of the world, the inner workings of things, and the series of forms."[29] Introducing the seventh specimen, devoted to the application of combinatory to geometrical figures, Leibniz emphatically writes:

> With these complications, not only can geometry be enriched by an infinite number of new Theorems, as every complication brings into being a new, compound figure, by the contemplation of whose properties we may devise new theorems and new demonstrations; but (if it is indeed true that great things are made up of little things, whether you call them atoms or molecules) we also have a unique way of penetrating into the arcana of nature. This is because the more one has perceived of the parts of a thing, the parts of its parts, and their shapes and arrangements, the more perfectly one can be said to know the thing. The structure of these figures is first to be investigated abstractly, in plane and solid geometry: after which, when you enter upon natural history and the question of being, that is, upon the question of the real constitution of bodies, the vast portals of Physics will stand open, and the character of the elements, the origin and mixture of the qualities, the origin of mixtures, the mixing of those mixtures, and everything that formerly lay hidden in darkness will be revealed.[30]

---

[28] *DAC*: 93 (A VI, 1: 177).     [29] *DAC*: 133 (A VI, 1: 187).
[30] *DAC*: 135 (A VI, 1: 187).

Later in life, Leibniz would come back to the fact that, for a brief period in his youth, he had favoured atomism. That he had done so, is quite in agreement with the parenthetical remark made in the passage above, which shows no prejudice against atoms as the least components of natural bodies.[31] Leibniz's atomism, however, is wedded in the *DAC* with a metaphysical doctrine centered on the notion of *immeation* or *perichoresis*, according to which everything is related with everything on the basis of the two relations of *similarity* and *dissimilarity*. 'Perichoresis' is the transliteration into English of a Greek word meaning 'circle' or the effect of revolving, which was employed in theology "to explain the relation of 'coherence' of the three persons of the Trinity".[32] The Latin word 'immeatio' was coined to translate 'perichoresis' and according to Johann Heinrich Bisterfeld, Leibniz's main source on this issue, it was meant to designate "the varied concourse, combination and complication of relations".[33] In the *DAC*, Leibniz praises Bisterfeld's work:

> I shall finally and briefly summon everything that can be culled from the metaphysical doctrine of the relations of Being to Being in order to form Places out of the different kinds of relation, and appropriate maxims characteristic of each of them, out of the theorems. I believe that I have seen this done, in more than just the summary fashion of encyclopaedists, by Johann Heinrich Bisterfeld in his *Universal Lamp, or Epitome of the Art of Meditating*, published in Leiden, 1657, which is entirely based on what he calls immeation and *perichoresis*, that is to say, the universal likeness of all things in all things, as well as the universal unlikeness of all things to all things, the principles of which are Relations. Anyone who reads that book will gain an increasing understanding of how to apply the art of complication.[34]

Thus, Leibniz deeply modifies the account of the world usually associated with atomistic doctrine, which was based on the aggregation and disaggregation of atoms governed by mere chance, without any trace of intrinsic

---

[31] On Leibniz's atomism, see Garber (2009: 62–70, 81–2); Arthur (2016).

[32] Antognazza (1999: 50).

[33] Bisterfeld (1657a; 185–6) (Antognazza's translation; Cf. Antognazza (1999: 50)). As Maria Rosa Antognazza remarks, Bisterfeld, in his recognition of the universal analogy with Trinity, "was certainly not alone. Rather he was working within a long tradition rooted in Augustinian thought." Ideas similar to those of Bisterfeld are to be found in Alsted and in "that other and still more illustrious student of Alsted: Jan Amos Comenius." (Antognazza 1999: 52).

[34] *DAC*: 70.

finalism. To this view he substitutes the picture of a world composed of beings connected by a net of reciprocal *relations* that generate a mutual union and communion analogous to that of the three persons in the holy Trinity. This picture was clearly meant to avoid the materialistic and atheistic consequences implicit in genuine atomistic theories as, for example, the view displayed by Lucretius in his poem *On the nature of things*. It is not difficult to see the existence of a latent conflict between Leibniz's acceptance of atomism on the one hand, and his agreement to a metaphysics motivated by religious issues and influenced by a philosophy largely inspired by doctrines of Neo-platonic origins on the other.[35] As soon as the conflict came to light, in the years immediately following the edition of the *DAC*, Leibniz decided to abandon the atomistic theory, but in a certain sense he never stopped flirting with it: when he attempts to explain to his contemporaries his *hypothesis of the monads*, he uses the expression 'formal atom' to characterize what he properly intends to denote with the word 'monad'.[36] As formerly in the *DAC*, in his mature thought Leibniz attempts to reconcile an atomistic perspective with an anti-materialistic metaphysics.

## 5  Themes from the DAC in Leibniz's mature philosophy

Besides the presence of some ideas that will constitute a substantive part of the project for the *characteristic art* later on, and besides an inclination to mix atomism with a strongly anti-materialistic perspective, we find in the *DAC* other items clearly recognizable in Leibniz's mature works as well. Among them, the following are particularly worth mentioning:

1) The notion of *blind thought*;
2) The distinction between propositions that are eternally true and propositions the truth of which is founded "not on essence but on existence";
3) A detailed and systematic treatment of the theory of the syllogism.

Leibniz calls 'blind thought' (*cogitatio caeca*) any thought that is developed essentially by means of words or other symbols. Such a thought is *blind*

---

[35] On the influence of Neoplatonic doctrines on Leibniz's thought, see Mercer (2000).
[36] Cf., for instance, what Leibniz writes in the *New System of the Nature of Substances and Their Communication* (1695): "So, in order to get to these real unities I had to have recourse to a formal atom, what might be called a real and animated point, or to an atom of substance, which must contain some kind of form or activity in order to make a complete being" (*Philosophical Texts*: 143).

because by employing words or any other kind of symbols, we *do not see* our own thoughts, which remain concealed behind a curtain of material signs, spoken or written or whatever.

The category of *blind thought* plays a prominent role in Leibniz's mature philosophy. Even early on during his stay in Paris, Leibniz emphasizes the fact that human beings cannot grasp directly by an act of intuition any thought beyond a certain degree of complexity.[37] In other words, we are forced to take recourse to symbols to formulate and to express concepts or thoughts sufficiently complex. We need a language, or at least something like a language, not only to communicate, but even to develop our own thoughts.[38]

The idea that some kind of material tool is necessary for thinking is pervasive in Leibniz's philosophy and strictly connected with the opinion according to which, with the only exception of God, all spiritual beings (even the angels) are associated with a body (even though a very 'thin' one). According to the Platonic tradition, Leibniz thinks that this 'bodily part' prevents any kind of spiritual beings from being directly acquainted with ideas and thoughts in their 'purity', as it were.

The distinction between propositions that are eternally true and propositions the truth of which is founded "not on essence but on existence" foreshadows the distinction between 'truths of reason' and 'truths of fact' as it is expressed, for example, in the *New Essays* (1702).[39] In the *DAC*, the propositions "that are eternally true" are synonymous with "theorems", i.e. "propositions that hold not because of the will of God, but by their very nature", and are contrasted with "singular propositions" like "Augustus was emperor of the Romans" and with "observations", i.e. "universal propositions whose truth is founded not on essence but on existence".[40] In a slightly different form, we find the same distinction in the so-called *Monadology*:

> There are two kinds of truths, those of *reasoning* and those of fact. Truths of reasoning are necessary and their opposite is impossible, while those of fact are contingent and their opposite is possible. When a truth is necessary one can find its reason through analysis, resolving it into ever more simple ideas and truths until one reaches primitives.[41]

---

[37] Cf., for instance, A VI, 3: 462.

[38] This does not exclude the possibility that, in certain peculiar cases, we may have direct acquaintance with some ideas or truths, see Mugnai (2014).

[39] Cf., for instance, *NE*: 446.     [40] *DAC*: 177 (A VI, 1: 199).

[41] *Monadology*, section 33 (transl. Rescher 1991: 120).

In the *New Essays*, Leibniz explicitly considers the distinction between propositions of reason and propositions of fact as 'his own',[42] and in the same vein as in the *DAC*, he writes that propositions of fact "can also become general, in a way; but that is by induction or observation, so that what we have is only a multitude of similar facts [ ... ]". General propositions of reason, instead, "are necessary".[43]

Finally, the *DAC* emphasizes Leibniz's precocious interest in the theory of syllogism. Even though Leibniz in 1686 had worked out a logical calculus, which was supposed to include the traditional syllogistic theory, he never stopped investigating the logical mechanisms governing the syllogistic doctrine of Aristotelian origin. It is quite remarkable that late in his life, only one year before his death, Leibniz devoted himself to writing an essay on the syllogistic figures, aiming to give a new rigorous foundation to traditional syllogistic theory.[44] In the *New Essays* we find an apology for the syllogism:

> It must be admitted that the scholastic syllogistic form is not much employed in the world, and that if anyone tried to use it seriously the result would be prolixity and confusion. And yet – would you believe it – I hold that the invention of the syllogistic form is one of the finest, and indeed one of the most important, to have been made by the human mind. It is a kind of universal mathematics whose importance is too little known. It can be said to include an *art of infallibility*, provided that one knows how to use it and gets the chance to do so – which sometimes one does not.[45]

Many years before the *New Essays*, when he began to write the first texts on logical calculus, Leibniz elaborated a very general notion of *logically valid* arguments, which includes the traditional syllogisms as a particular case. This point, too, is clearly stated in the *New Essays*:

> But it must be grasped that by 'formal arguments' I mean not only the scholastic manner of arguing, which they use in the colleges, but also any

---

[42] "The distinction you draw appears to amount to mine, between 'propositions of fact' and 'propositions of reason'." (*NE*: 446).

[43] *NE*: 446.

[44] This essay has recently been published by Wolfgang Lenzen in *Schriften* (2019: 428–43). A partial edition of it appeared in the *Leibniz Review*, vol. 20, 2010: 117–35.

[45] *NE*: 478.

reasoning in which the conclusion is reached by virtue of the form, with no need for anything to be added. So: a sorites, some other sequence of syllogisms in which repetition is avoided, even a well drawn-up statement of accounts, an algebraic calculation, an infinitesimal analysis – I shall count all of these as formal arguments, more or less, because in each of them the form of reasoning has been demonstrated in advance so that one is sure of not going wrong with it.[46]

We have thus, with a 'broad brush', given a picture of the cultural background against which Leibniz conceived the *DAC*; it is time now to enter into some details concerning the main topics discussed in this work.

## 6 The first two problems and their solutions

Leibniz begins the *Dissertation* with the clear statement that his investigation properly belongs to *metaphysics*.[47] Metaphysics, indeed, "is concerned first with Being, and then with the affections of Being";[48] and it is on these latter that the doctrine of combinations rests. Leibniz does not say explicitly what he means by the words 'being' (Latin: *ens*) and 'affection' (*affectio, modus*), but tacitly assumes some familiarity on the part of the reader with these terms and writes that just as affections of bodies (colour, weight, and figure) are not themselves bodies, so the affections of a being are not beings.

According to the scholastic doctrine (inspired by Aristotle's *Categories*), Leibniz considers the totality of the affections as divided into three main categories: *quality, quantity*, and *relation*. A *quality* is an "absolute affection (or mode)" of a being; a *quantity* is a relative affection "which holds between a thing and one of its parts"; a *relation* is an affection, which holds between a thing and some other thing. If metaphysics is understood as the discipline that investigates "what is common to all kind of Beings", then *quality*,

---

[46] *Ibidem.*

[47] As Eberhard Knobloch observes: "Leibniz [...] spoke of the "ars combinatoria", a philosophical term to which he attached different meanings in the course of his life, but which always embraced more than what we call combinatorics today and was not even restricted to mathematical problems." (Knobloch 1974: 410)

[48] *DAC*: 69 (A VI, 1: 170).

*quantity*, and *relation*, because they concern the most general affections of all things, belong to metaphysics as well.

After introducing these three categories, Leibniz recognizes two kinds of relations: *unity* and *agreement*. We have a relation of *unity*, when several things are put together to form a *whole*: in this case, the things forming the whole are called *parts*. Leibniz seems to have quite a general idea of a whole, as something that obtains by simply juxtaposing or adding things together: it does not necessarily presuppose some kind of intrinsic link among the parts.[49] Moreover, Leibniz does not limit himself to considering *material* wholes only: on the contrary, he admits any kind of wholes, even *immaterial ones,* such as a piece of music or a poetical work. He is quite reticent about the relation of *agreement*, probably because it does not play an important role in the art of combinations. In later texts, written after the edition of the *DAC,* the relation of *agreement* (Lat. *convenientia*) seems to be not much different from *similarity under some respect.* Thus, for example, two 'things' agree if they have properties (colour or shape) that are similar or if they share a similar structure.[50]

Given a whole, it can be divided into lesser parts and these parts can be rearranged in different ways. Thus, for instance, let a whole consist of *A, B,* and *C.* The lesser wholes "are its parts *AB, BC,* and *AC;* and the disposition of the smallest parts, or what are taken as the smallest parts (that is, Unities), can be varied both in relation to each other and to the whole".[51] Leibniz calls *variation* the change of relations (including change of order) amongst the parts of a given whole (what we now call *permutation*); and *complexion*, the "union of a lesser whole within a greater whole" (what we now call *combination*). Loosely speaking, a *complexion* is the grouping of the elements of some domain of things (abstract or material), according to a given number: thus, for example, given a domain of four elements, we have six *complexions* of *two* elements out of them. To indicate that a domain (a whole) made of x elements must be decomposed into complexions each of *n* elements, Leibniz uses the notation '$x^n$' (where 'x' is the number of the things belonging to the intended domain). The disposition of the parts in relation to each other and to the whole is called *situs*.

---

[49] Around 1686, Leibniz will assume that a part should be 'homogeneous' to its whole; see, for instance, A VI, 4B: 1671 (discussion with Fardella (1690)): "because a part is always homogeneous with the whole".

[50] Cf. A VI, 4A: 336: "*A relation of agreement* is either of similarity or dissimilarity."

[51] *DAC:* 73 (A VI, 1: 171).

After introducing the main notions on which the combinatorial art should rest, Leibniz proposes a list of problems preceded by a short preamble, which explains how he intends to develop his investigation:

> There are three things that have to be considered: *Problems, Theorems,* and *Applications.* To some of the problems I have appended an application and, where it seemed worthwhile, theorems. Moreover, to some of the problems I have added an explanation of the solution.[52]

Thus, Leibniz considers a totality of twelve problems and a variety of applications, ranging from the theory of syllogism to some puzzles typical of recreational mathematics as, for instance, how many ways a given number of persons could sit down at one table in some order or other. Problems 1–3 concern combinations, whereas problems 4–12 are centred on permutations. The chapters are of different length and some of them quite disproportionate, when compared to others. As Knobloch remarks:

> The longest chapter (problem 2) alone takes up 28 pages of the Academy edition, while the shortest fills a mere 2 1/2 lines. On the other hand, the final paragraph following the formulation of the 12th problem refers to all of the last 6 problems, whereas the second chapter includes part of the first.[53]

Among the applications, the most interesting ones from a philosophical point of view are those concerning logic and the syllogism, and it is on these that we will concentrate in these introductory remarks.

The first problem (PROBLEM I) tackled by Leibniz is:[54]

(I) *given a number with its corresponding exponent, to find the complexions.*

Leibniz offers two solutions: one concerning the general case, when the exponent is any number whatsoever, the other involving the particular case with exponent 2. The first solution sounds quite cumbersome:

---

[52] Cf. *DAC*: 79 (A VI, 1: 173).   [53] Knobloch (1974: 412).

[54] To distinguish the main problems into which the *DAC* is articulated from the problems internal to the various applications, we write in capitals the first.

Let the number of complexions of the preceding exponent with the preceding number be added to the number of complexions of the given exponent with the preceding number; the result will be the required number of complexions.[55]

To understand Leibniz's words, let us put the number of *complexions* = 'C'; the number of 'things' to be combined = '$n$ '; and the number of the exponent = $k$. In this case, '$n^k$ ' denotes all complexions of $k$ *elements out of $n$ things*. If we suppose $n = 5$ and $k = 3$, then '$5^3$' denotes all complexions (combinations) of 3 elements that can be made out of 5 elements. Leibniz's proposed solution corresponds to the formula:

$$(F1) \ C\binom{n}{k} = C\binom{n-1}{k-1} + C\binom{n-1}{k}$$

As this formula shows, the total number of complexions of $k$ elements out of $n$ 'things' is determined by adding the number of complexions of $k-1$ elements out of $n-1$ things *to* the number of complexions of $n$ elements out of $n-1$ things. To stick to our example, to calculate the number of complexions of three elements out of five things, we must: a) calculate the number of complexions of two elements out of four things; b) calculate the number of complexions of three elements out of four things; and, finally; c) add together these two numbers. In this case, however, to determine the number of complexions of $k$ elements out of $n$ things, we need to know the number of complexions of $n-1$ things: as Leibniz remarks, to find the complexions, "the complexions of the preceding number are a prerequisite".[56]

After having expressed in words the content of (F1), Leibniz proceeds to construct a table devised as a tool to find all possible complexions of $k$ elements out of $n$ 'things', for $k$ and $n$ comprised between the numbers 1 and 12.[57] It is to this table that Leibniz constantly refers, when he proposes the solutions to the combinatorial problems formulated in the *DAC*.

---

[55] Cf. *DAC*: 81 (A VI, 1: 174).

[56] *DAC*: 83 (A VI, 1: 174). The total number of complexions can be found in a more general and direct way applying the formula (unknown to Leibniz at the time of the *DAC*):

$$C\binom{n}{k} = \frac{n!}{k!(k-1)!}.$$

[57] As a matter of fact, however, the column with the number of 'things' starts with '0', even though, as Knobloch (1974: 413) remarks, Leibniz "explicitly excludes combinations of no elements since he gives the number of all possible combinations as $2^n - 1$".

The sub-case of Problem I, with 2 as exponent can be formulated as follows:

(I, 1)   Given a set with $n$ 'things', find all possible complexions (without repetitions) of 2 elements out of them.

To solve this problem, Leibniz appeals to a method well known among 'Arithmeticians':

> Thus, it is clear that the number of com2nations is composed of the terms of an arithmetical progression, whose difference is 1, and which are numbered inclusively from 1 to the number one less than the number of things; that is, from all numbers less than the number of things added together. But, as Arithmeticians teach us, such a series of numbers can be added together very concisely in this way, by multiplying the highest number in the series by the next greater number, the product of which divided by 2 gives the required total [...].[58]

In this case too, Leibniz employs words to explain how to solve the problem: "Multiply the Number of things by the next smaller number, and this product divided by 2 will be what is required".[59] This simply paraphrases the formula:

$$(F2) \; C\left(\begin{matrix} n \\ k=2 \end{matrix}\right) = \frac{n(n-1)}{2}$$

The second problem (PROBLEM II) tackled by Leibniz is:

(II)   *given a number, to find simply the complexions,*

i.e. given a set of things, to determine all possible combinations (without repetitions) of these things. Leibniz states the correct solution, which can be summarized by the formula:

$$(F3) \; \sum_{k=1}^{n} \binom{n}{k} = 2^k - 1,$$

---

[58] *DAC*: 89 (A VI, 1: 176).     [59] *Ibidem*.

where '$n$' = the number of things belonging to a given set. Leibniz confesses, however, of being quite uncertain about the reasons, which justify this solution:

> The *Proof*, or *reason why*, is difficult to conceive of; or if you can conceive of it, difficult to explain.[60]

## 7 The 'application' (usus) to traditional syllogistic

After having presented the first two problems and their solutions, Leibniz illustrates several 'applications' (Lat.: *Usus*). Of these, the *sixth* one concerns the theory of syllogism and the *tenth* the part of the logic of (categorical) propositions that properly belongs to the logic of discovery (*logica inventionis*). Let us first take a look at the theory of syllogism.

Leibniz does not define what a *syllogism* is and simply assumes that the reader is well acquainted with it. From a logical point of view, his treatment of syllogistic is far from being rigorous: he presents, for instance, the notion of *mood* before introducing that of *figure*. The latter appears only after the moods *Darii, Datisi*, etc. have been discussed, whereas these presuppose the different kinds of figures. At various moments, Leibniz also offers rules for distinguishing valid syllogisms from invalid ones, without showing any concern for a systematic exposition. Particularly revealing of a general lack of rigour, moreover, is the last paragraph of Chapter 24, where Leibniz is forced to make a step back and to give a quick account of some syllogistic rules that are useful to understand not only what follows, but even some of the issues in the preceding chapters. Putting together the scattered remarks about the general features of a syllogism, however, it is possible to reconstruct a coherent account of traditional syllogistic from the *DAC*.

In accordance with scholastic tradition, Leibniz takes for granted that any atomic proposition is composed of a subject, a predicate, and a copula (the verb 'to be'). Concerning the copula, he says that it is twofold: 'is' (*est*) and 'is not' (*non-est*), and that it is part of the predicate:

> Accordingly, just as there are two primitive signs used by Algebraists and Analysts, + and −, so there are, as it were, two copulae, *is* and *is not*: with

---

[60] *DAC*: 91 (A VI, 1: 176).

the former the mind puts things together, with the latter it takes them apart. In this sense, *is* is not strictly a copula, but part of a predicate, and the copulae are two: one that is named *not*, and another that is unnamed, but included in the *is* whenever *not* is not added to it.[61]

Since the copula is part of the predicate, each atomic proposition can be considered as obtained by combining two terms, a subject and a predicate, respectively. Leibniz assumes that a syllogism is an argument made of three propositions in which only three terms are involved.[62] This implies that each term figures twice in a syllogism.

To denote the terms, Leibniz employs the Greek letter 'μ', the Latin letter 'M', and the German letter '𝔐', which correspond to the more common 'P', 'S', and 'M' of the logical tradition (and which will be used in the present context, instead of the symbols chosen by Leibniz).

Two propositions are called the *premises* and the third the *conclusion* of the syllogism. To determine the combinations giving rise to the propositions that may enter a syllogism, Leibniz gives the following rules:

(1) It is not permitted to combine the same term with itself (there is no proposition like, for instance, 'SS');
(2) S and P are combined only in the conclusion, and in such a way that S is always placed first;
(3) In the first premise, M and P are combined, in the second M and S.

As in any standard account of syllogistic, 'P' refers to the so-called 'major term', which is the predicate of the conclusion; 'S' refers to the 'minor term', which is the subject of the conclusion; and 'M' refers to the *middle term*, which does not figure in the conclusion, because it disappears after having united, in virtue of its mediation, the minor and major term. Leibniz calls 'major premise' the premise that contains the major term and 'minor premise' that which contains the minor term; and denies that a simple change of the order of the premises produces a change in the *figure* of the syllogism.

Figures are determined, as usual, by the position of the middle term in the premises:

It is therefore clear that the variety of the figures arises from the position of the middle term within the premises. When it comes first in the major premise alone, and second in the minor premise, the figure is the Aristotelian First Figure. When it comes second in both the major and the minor premise, the figure is the Aristotelian Second Figure. When it comes first in both premises, the figure is the Aristotelian Third Figure; and when it comes second in the Major premise alone, and first in the Minor premise, it is the Fourth Figure of Galen [ ... ][63]

Thus, we have the following four syllogistic figures:

| I | II | III | IV |
|---|---|---|---|
| MP SM | PM SM | MP MS | PM MS |
| SP | SP | SP | SP |

In the quoted passage, Leibniz distinguishes the three Aristotelian figures from the fourth, which, according to the tradition, he attributes to Galen. But, what is more interesting, arguing in favour of the fourth figure he observes that if one adopts the Aristotelian way of expressing predication, the fourth figure is easily changed into the first, and vice versa. Leibniz, indeed, remarks that a universal affirmative proposition like 'Every α is β', is expressed by Aristotle as 'β inheres in every α' and that if one adopts this way of speaking, the fourth figure takes the form:

M inheres in every P; S inheres in every M

from which the conclusion follows:

Every S is P

or, transposing the premises and adopting Aristotle's way of speaking:

3')   P inheres in every S

---

[63] Leibniz, indeed, does not hesitate to accept the fourth figure: "To the enemies of the Fourth Figure I make this one objection for the time being, that the Fourth Figure is as good as even the First" (*DAC*: 119 (A VI, 1: 183)).

The transposition of the premises is needed, if the subject of 3') has to be the minor term. Leibniz remarks that the same can be done with the other syllogistic figures, i.e. that they may be reduced to the corresponding Aristotelian form, and he shows himself to be quite proud for having discovered this technique.

Leibniz's remarks reveal a clear awareness, not very usual at the time, of the differences between Aristotle's own idea of a syllogism and that held by the syllogistic theory prevailing in the sixteenth century. In 1584, Giulio Pace (Julius Pacius) (1550–1635) published a new Latin translation of the *Organon* containing the Greek text side-by-side with some explanatory remarks on the margins, which quickly became a reference book for people interested in Aristotle's logic. It is quite probable that Leibniz had some acquaintance with Aristotle's logical doctrines through Pace's edition.

Having established what a syllogism is, Leibniz gives a definition of 'syllogistic mood': A *mood* is *a disposition or form of a syllogism taking account of both quantity and quality* of the propositions involved.[64]

With regard to *quality*, a proposition may be either affirmative (A) or negative (N). With respect to *quantity*, Leibniz distinguishes 4 different kinds of propositions:

Universal (U): 'Every S is P';
Particular (P) 'Some S are P';
Indefinite (J): 'S is P';
Singular (S): 'Socrates (or: 'This man') is a philosopher'.

Usually, as far as quantity is concerned, the traditional Aristotelian syllogistic recognizes only universal and particular propositions. In the *Prior Analytics*, Aristotle considers indefinite propositions too, i.e. propositions in subject-predicate form in which the quantity of the subject is not explicitly expressed, but he says that they have to be considered equivalent to the particular ones. The point here is that in ancient Greek, and in Latin as well, propositions corresponding to English sentences like 'man is animal' were permitted and used without sounding particularly odd. Aristotle's reason for considering an indefinite proposition as logically equivalent to the particular and not to the universal one is that, the quantity of its subject not being

---

[64] As Dürr (1949: 9) remarks, Leibniz doesn't employ the word *modus* (*mood*) to designate only valid syllogisms.

mentioned explicitly, it is safer, for drawing inferences, to assume a less binding commitment than the one implied by a universal proposition.

## 8 Singular propositions and some syllogistic rules

Another interesting aspect of Leibniz's list of the different kinds of propositions, in comparison with the traditional (mainly scholastic) treatment of syllogistic, is the inclusion of singular propositions. This is a novelty, although one could hardly attribute it to Leibniz in particular, since it may easily be found in other seventeenth-century texts on logic. Thus, for example, John Wallis (1616–1703) shows genuine interest in singular sentences in his *Logic* and Honoré Fabri (1608–88), an author with whom Leibniz was to become well acquainted from 1670 onwards, devotes particular attention to singular propositions in his lectures on logic, attempting even to construct an Aristotelian square of oppositions for them.[65] Some years later, this same attempt will be made by Girolamo Saccheri (1667–1733) in the *Logica demonstrativa*, in which there is a chapter specifically entitled 'On singular propositions'.[66]

Singular propositions play an ambiguous role in traditional syllogistic: on the one hand they motivate a classical analysis of the elementary propositions in terms of subject and predicate; on the other, they are forced into the scheme of the four categorical propositions through their assimilation either to universal or to particular propositions. In a certain sense, they impose their structure on all the other propositions, but at the same time they are accepted amongst these latter only at the expense of being considered 'particular cases' of universal or particular propositions (or of both). Thus, in the seventeenth century, people interested in singular propositions had to settle the question of how to interpret these propositions, whether as particular or as universal, with the precise aim of employing them in arguments having syllogistic form. The great majority of authors argued that singular propositions might be considered as cases of universal ones: since there is only one Socrates, 'Socrates is a philosopher' has to be considered as equivalent to 'Every Socrates is a philosopher'. This opinion was shared, for example, by John Wallis.[67] In the *DAC*, Leibniz also states that singular propositions are equivalent to the universal ones. To justify this

---

[65] Cf. Wallis (1687); Liber II: 125; Fabri (1646); *Controversia XXIX*: 67.
[66] Saccheri (1701: 105–9).        [67] Wallis (1687); Liber II: 125.

claim, he refers to Johann Raue's analysis of singular sentences containing a proper name in the role of the subject.[68] Given a sentence like *Socrates is a son of Sophroniscus*, Raue reduces it to: *Whoever is Socrates, is a son of Sophroniscus* and, as Leibniz remarks, this is equivalent, from the logical point of view, to saying: *Every Socrates is a son of Sophroniscus*, even though Socrates is Sophroniscus' only child. Leibniz gives two reasons in favour of this choice: 1) the expression 'Every Socrates' refers to the *man* Socrates, not to the *name* 'Socrates'; 2) the word 'Every' does not imply "a multitude but a grouping of individuals", i.e. it has 'distributive character' and does not necessarily presuppose the existence of more than one individual. Therefore, Leibniz may conclude that "every singular Proposition should be taken in a Syllogism to stand, so far as mood is concerned, for a Universal proposition".[69]

Having determined the four types of quantities, Leibniz starts calculating the number of all possible moods and amongst these, that of the *useful* ones. The notion of a *useful*, as opposed to that of a *useless* mood, is abruptly introduced in Chapter 18 of the *DAC*, without any explanation. It is only later on, in Chapter 20 that Leibniz clarifies that, *as regards quantity*, useful moods are those that obey the two rules:

R1) from pure particulars nothing follows;
R2) the conclusion cannot surpass any of the premises in quantity.[70]

Before introducing these rules, however, Leibniz argues that each proposition entering the syllogism must belong to one of the four types above, and calls *equal* the syllogism made of three propositions of the same quantity. Clearly, *equal syllogisms* may be only four: U, U, U; P, P, P; J, J, J; S, S, S. All the syllogisms that are not equal are *unequal*, and are of two sorts: a) *unequal in part*, when they have two propositions of the same and the third of a different quantity; b) *wholly unequal*, when all their propositions differ in quantity. Speaking of the four syllogisms that are *equal*, Leibniz

---

[68] Cf. Raue (1638: 160–8).

[69] *DAC*: 115 (A VI, 1: 183). Later on, in *Some Logical Difficulties*, written in the first years of the eighteenth century, Leibniz will state that a singular proposition "is equivalent to a universal and a particular [...]. For *some Apostle Peter* and *every Apostle Peter* coincide, since the term is singular." (GP 7: 211). In the *New Essays*, however, he remains faithful to the position of the *DAC*: "But it is as well to notice that singular propositions are counted, so far as their form goes, among universal ones. For although there is indeed only a single Apostle Peter, it can still be said that anyone who was the Apostle Peter denied his Master." (*NE*: 485)

[70] *DAC*: 107 (A, VI, 1: 180). Even Giulio Pace, in his commentary to the *Organon*, distinguished *useful* from *useless moods*. Cf. Pacius (1597: 133ff).

remarks that only two, the first and the fourth, are *useful*.[71] Then, Leibniz proceeds to calculate the number of all combinations, useful and useless, of the syllogisms *unequal in part*, which amounts to thirty-six.[72] This number, however, if we apply R1 and R2 above, reduces to the half, i.e. eighteen.

To determine the number of all syllogisms unequal in whole, Leibniz first remarks that, because the propositions composing a syllogism are three and the possible quantities that each of these may assume are four, this amounts to knowing how many collections of three things can be made out of four things. The answer is: four. Because the three propositions in each syllogism may vary their order, giving rise to six different combinations, we have that 4×6 = 24 is the number of all syllogisms unequal in whole. Of these, only twelve are *useful*. Thus, the number of all possible moods, useful and useless according to quantity, is 64 (= 4 + 36 + 24), whereas the number of the useful ones (employing R1 and R2 to discriminate them) is 32 (= 2 + 18 + 12).

Whereas, as regards quantity, Leibniz groups syllogisms into *equal* and *unequal*, with respect to quality he differentiates *similar* from *dissimilar* syllogisms, and observes that, qualities being only two (Affirmative and Negative) and each syllogism being composed of three propositions, it is impossible to have syllogisms dissimilar in whole. *Similar syllogisms* are of two kinds only: A, A, A and N, N, N. The *dissimilar syllogisms* are of two kinds as well: A, A, N and N, N, A, but they may be varied in six different ways. Thus, the number of all syllogisms, *similar* and *dissimilar* is 2 + 6 = 8. Of these, applying the two following rules governing quality:

R3)  from purely negative propositions nothing follows;
R4)  the conclusion follows the part inferior in quality,[73]

---

[71] The mood S, S, S is *useful*, because Leibniz considers it equivalent to U, U, U, and this latter is *useful*. P, P, P and J, J, J are *useless* because from particular and indefinite premises no conclusion can logically follow.

[72] To calculate this number, Leibniz proceeds as follows. Given that a syllogism is composed of three propositions, and that two of them are of the same quantity, we have first to calculate the number of possible combinations of two quantities out of the given four (U, P, J, S), which is six. Then, because in case of syllogisms unequal in part, one proposition is taken twice and the other is maintained fixed, the number of all possible combinations is 6×2 = 12. If we take into account the order of the three propositions of the syllogism, the final number of all possible variations is 12×3 = 36.

[73] This rule presupposes that a negative proposition is inferior with respect to an affirmative one and states that a mood with a negative premise can be useful only if the conclusion is negative as well.

we have only three that are useful (A, A, A; N, A, N; A, N, A).[74]

Because moods take into account *quantity* and *quality* at the same time, the total number of *useful* and *useless* moods is 64×8 = 512.[75] To obtain the only useful ones, Leibniz first multiplies the thirty-two useful moods according to quantity by the three moods useful according to quality, getting ninety-six. Then, from ninety-six he subtracts all moods that are contained in the mood *Frisesmo*, which are eight, the reason being that "even though *Frisesmo* is a mood that in a certain manner holds good in itself, it is not, however, in any figure".[76] The final verdict is that the number of all useful moods is 96−8 = 88.

## 9 A classification of syllogistic moods

As we have seen, Leibniz calls 'mood' the set constituted by three categorical propositions in subject-predicate form with their quantity and quality specified. Using, as usual, the letters 'A' and 'N' to denote quality and the letters: 'U', 'P', 'J', 'S' to denote quantity, Leibniz, for example, represents as 'UA, SN, UA' a mood with, respectively, the first premise universal affirmative, the second singular negative, and the conclusion universal affirmative. This mood is *useless*, because it violates rule R4: the second premise being negative, the conclusion needs to be negative too. Examples of *useful moods*, instead, are: UA, UA, UA; UA, PN, PN; UA, SN, SN. Then, Leibniz distinguishes the mood 'in itself', as it were, from the mood associated with a particular syllogistic figure, which he calls 'figured mood'. Given that all moods (useful and useless) are 512, if each of them is disposed in each of the four figures, the number of all figured moods (useful and useless) amounts to 2.048.

If we are interested, however, in the *useful figured moods* only, we cannot simply multiply the number of *useful moods* with that of the four syllogistic

---

[74] As Couturat remarks, Leibniz does not mention another important rule of traditional syllogistic, concerning the quality of propositions: from affirmative premises no negative conclusion follows (see Couturat (1901: 3, n. 4); Dürr (1949: 13)).

[75] Leibniz could have followed a less complicated method to calculate the total number of moods (useful and useless). Given that each proposition of a syllogism may be of one of two different qualities (A and N), and that in a syllogism there are three propositions, $2^3$ = 8 is the number of all moods to which a syllogism may give rise according to quality. Therefore, sixty-four being the number of syllogistic moods according to quantity, the number of all syllogistic moods, useful and useless according to quantity and quality is 64×8 = 512.

[76] *DAC*: 107 (A VI 1: 180).

figures, because, as Leibniz remarks, "not all of the 88 useful moods find a place in each figure".[77] In other words: each figure has some structural constraints preventing that some useful mood could be put in it. These constraints, as stated by Leibniz, are:

R5) the Major Premise of the first and second figure must be always universal (U);

R6) the Minor Premise of the first and third figure must be always affirmative (A);

R7) in the second figure the Conclusion must be always negative (N);

R8) in the third figure the Conclusion must be always particular (P);

R9) in the fourth figure there are three conditions: a) the Conclusion cannot be a universal affirmative (UA); b) the Major Premise is never particular negative (PN); c) if the minor is negative (N), then the major has to be a universal affirmative (UA).

Rules R5 to R9 correspond to the usual prescriptions that, from the Middle Ages until the late nineteenth century (and beyond), one may easily find in any introductory text to syllogistic. Leibniz uses them to partition all useful moods into eight classes characterized first by *simple general* and then by *figured general moods*. Each class is composed of eight moods and each mood is expressed in the usual way by three couples of letters, specifying quantity and quality of the propositions that enter the mood. Leibniz calls 'simple special' each mood belonging to a given class: 'simple' because it is not figured, and 'special' because it is an instance, or species, of a more general mood, which Leibniz denotes, according to the scholastic use, by means of the vowels A, E, I, O. An example may help to better understand what Leibniz means.

Consider the following eight moods (each item is divided by a vertical stroke from its immediate successor):

UA,UA,UA|SA,SA,SA|UA,UA,SA|UA,SA,UA|SA,UA,UA|SA,SA,UA|SA,UA, SA|UA,SA,SA.

If we want to construct a *valid* syllogism with one of these moods, we are forced to employ the first figure only: 'putting' these moods in any other

---

[77] *DAC*: 123 (A VI, 1: 184).

figure, indeed, would amount to producing an *invalid* syllogism.[78] Now, if we compare Leibniz's way of writing a mood with that usually employed in traditional syllogistic, an easy translation is at hand: 'UA' denotes a universal affirmative proposition, which traditionally is expressed by means of the letter 'A'; 'UN' denotes a universal negative proposition, which is expressed by means of the letter 'E'. 'SA' and 'SN' denote two *singular* propositions, respectively, affirmative and negative: but because Leibniz states that a singular proposition is logically equivalent to a universal one, 'SA' and 'SN' are in their turn each equivalent to a universal proposition, and thus they may be expressed as 'A' or 'E' as well. Hence, it is not difficult to recognize that in the traditional syllogistic jargon, *all* the eight moods above are represented by the three vowels: 'AAA'; and, according to Leibniz, 'AAA' is precisely the 'simple general mood' which subordinates under itself all the eight 'simple special moods' above. Once associated with the first figure, the 'simple general mood' AAA becomes the *figured general mood* traditionally known as *Barbara* under which, as Leibniz writes, 'any of the simple special Moods falls'.[79]

Taking recourse to a notation that Leibniz will employ in later years, we may represent the different types of moods as follows:

1) 'AAA' = *general special mood*;
2) 'UA,UA,UA' = *simple special mood*;
3) $\underline{A(MP); A(SM)}$ = *general figured mood* (first figure).[80]
       A(SP)
4) $\underline{UA(MP); UA(SM)}$ = *special figured mood*.[81]
        UA(SP)

---

[78] Even though Leibniz does not give an explicit definition of *validity* of an argument, it is not difficult to infer from what he says in the *DAC* that he considers *valid* a syllogism when, for all possible substitutions of meaningful terms at the place of the letters 'S', 'M', 'P', if it comes out that the premises are true, then it does not happen that the conclusion is false.

[79] *DAC*: 125 (A VI, 1: 185).

[80] Instead of the 'classical' notation of scholastic origins, which puts the vowels *a, e, i, o* between subject ('S') and predicate ('P') to denote quality and quantity of a categorical proposition, Leibniz puts the vowels (in capital letters) in front of the couple of letters playing the role of subject and predicate. Thus, the traditional universal affirmative 'SaP' becomes, according to Leibniz's notation: 'ASP'. Cf. A VI, 4: 498–501. To avoid ambiguity we have added brackets: 'A(SP)'.

[81] These are not explicitly mentioned by Leibniz (Dürr (1949: 10)). A *special figured mood* originates from a *general figured mood* simply specifying the *quantity* of the propositions. Thus

UA(MP); JA(SM)

     JA(SP)

is another *special figured mood* derived from:

A(MP); A(SM)

     A(SP).

# 10  Useful and useless, valid and invalid moods. A general method for finding a counterexample

As Leibniz remarks, it may happen that different syllogistic figures give rise to valid syllogisms, when applied to the same class of simple special moods (or to the same general mood). This is the case, for instance, with the simple special moods characterized by the *general special mood* 'EIO':

UN,JA,JN|UN,PA,PN|UN,PA,JN|UN,JA,PN|SN,JA,JN|SN,PA,PN|SN,PA,JN| SN,JA,PN.

If we apply to each of these moods each figure in turn, beginning from the first one on the left, we have the four *figured general moods*: *Ferio* (I), *Festino* (II), *Ferison* (III), *Fresismo* (IV), which are all valid and under which all the above simple special moods fall. It is worth noting that, independently of their belonging to different figures, these four moods share the three vowels: E, I, O, i.e. the same vowels which denote the general special mood characterizing the eight simple special moods. Leibniz illustrates by means of a table all the relationships that subsist between the different kinds of moods and the syllogistic figures. Introducing the table, he is very proud for having revealed "the secret harmony of the figures" which, as he writes "all have the same number of moods".[82]

The table is composed of twelve rows, each individuated by a simple general mood and containing eight simple special moods. On the right of each of the first eleven rows are listed the traditional *figured general moods* which, in one or more of the four figures, give rise to a valid syllogism, with quantity and quality disposed as prescribed by the eight special moods belonging to the row. The twelfth row is individuated by the simple general mood 'JEO' and contains eight special moods associated with the figured mood *Frisesmo*, which does not belong to any of the four canonical figures and which therefore is not valid.[83] According to the rule (R5) above, indeed, a syllogism belonging to the first or to the second figure, must have the major premise always universal; and according to (R6) the minor premise of a syllogism of first or third figure must be affirmative: thus, these two rules prevent *Frisesmo* from belonging to any of the first three figures. Obviously, the possibility remains that *Frisesmo* would belong to the

---

[82] *DAC*: 123 (A VI, 1: 184).     [83] *DAC*: 127 (A VI, 1: 185–6).

fourth figure, but (R9) fully justifies the exclusion of this mood. In this case, however, Leibniz prefers to abstain from employing a merely syntactic rule and rather suggests a semantic method, i.e. a strategy that takes into account the *matter*, not only the form, of a syllogism. Leibniz's method consists in finding three terms that, substituted for the terms of a given syllogism, make the premises of the latter true and the conclusion false. If we find these terms, then we have found what we now call a 'counterexample' to the syllogism; and the existence of a counterexample shows that the syllogism is *invalid*. As Leibniz remarks:

> The normal use of 'instances' is to confirm inferences, but sometimes that is not a very reliable procedure, though there is a way of choosing instances which would not come out true if the inference were not valid.[84]

Aristotle in the *First Analytics* systematically applies the method of the counterexample, but Leibniz seems to ignore this and he states proudly:

> And here, in passing, I shall offer some useful advice (which is to some extent supported by this example) on what, so to speak, constitutes Proof, or the art of examining a proposed mood, and wherever it does not follow by virtue of form but in virtue of matter, immediately finding a counter-example, something I do not recall reading that Logicians have ever done.[85]

Leibniz's pride, however, is justified insofar as he attempts to find a general procedure for constructing a counterexample for each invalid mood that does not violate the classical rules mentioned above.[86] It is probably true that he is the first to have conceived this as a project in itself, even though the advice he gives about the construction of a counterexample is not so easy to understand:

> [ ... ] for UA, choose a proposition whose matter does not permit of simple conversion, for example, this proposition, *All men are animals*, rather than *All men are rational animals*; and the more remote the genus you choose, the better it will serve. For UN, choose a proposition in which species are denied of each other as near neighbours under the same proximate genus,

---

[84] *NE*: 481.     [85] *DAC*: 127 (A VI, 1: 186).
[86] As Marko Malink has suggested to me (private communication), to determine the procedure for constructing a counterexample, Leibniz seems to have drawn inspiration from *Bramantip*.

for example, man and brute; and which is not convertible by contraposition into UA, that is, neither its subject nor its predicate is an infinite term.[87] For PA and JA, always choose a proposition that is not the subaltern of any UA, but in which the species is affirmed particularly of the widest possible genus. For JN and PN, choose a proposition that is not the subaltern of any UN, of which neither term is infinite, and in which a species is denied of the most remote genus.[88]

Once the mood *Frisesmo* has been proved invalid, looking at the table constructed by Leibniz we see that the total number of all *figured general moods* is twenty-four, and that each syllogistic figure has exactly six valid moods. Using the same scholastic names employed by Leibniz, we list these moods with their corresponding figures as follows:

---

[87] An 'infinite term' is a term with the sign of negation prefixed, as, 'non-man', 'non-animal', etc. The *converse by contraposition* of a categorical proposition is an "immediate inference in which from a given proposition another proposition is inferred having for its subject the contradictory of the original predicate" (Keynes 1906: 134). Thus, given the sentence 'All S is P', according to Keynes, it has two contrapositives: *No not-P is S* and *All not-P is not-S*. Leibniz here rightly remarks that a *universal negative* proposition (UN) cannot be converted by contraposition in a *universal affirmative* (UA), with its subject and predicate changed into *infinite terms*.

[88] *DAC*: 127 (A VI, 1: 186). Because this text is quite concise and not easy to understand, the following reconstruction is largely conjectural.

Consider the four syllogistic figures and, in each figure, all possible combinations of the four letters A, E, I, O. For simplicity's sake, take into account the two premises only, instead of the entire syllogism. Thus, the number of all combinations in each figure is sixteen ($4^2$), i.e. there are sixteen possible two-premises syllogisms in each figure. From these syllogisms subtract, first, those that violate the rules mentioned above and, second, all combinations corresponding to the valid moods belonging to each figure (because they, obviously, cannot be refuted). To the remaining moods we may apply Leibniz's rules for creating a counterexample, with the further assumption that no set corresponding to each of the three terms contained in the premises is completely disjoint from both the sets corresponding to the other two terms. Thus, for instance, consider the mood 'I, E, O', in the first figure: *Some M is P, No S is M; therefore Some S is not P* and assume that the set corresponding to M intersects the set corresponding to P and that the sets corresponding to S and M are fully disjoint, with S a proper subset of P. According to this model, the premises are true and the conclusion false: we have found a counterexample to the mood. Leibniz's rules, however, seem to impose that S and M are disjoint and both *proper subsets* of the same set P, whereas, in our model, M and P simply intersect one another. This requires a little adjustment of Leibniz's rules: the condition that in the UA proposition species must be "denied of each other", for instance, seems to be not strictly necessary.

To determine his rules for a counterexample, Leibniz does not appeal to any general principle, thus it is quite probable that he reached them checking each figure in turn (employing a kind of *British Museum algorithm*). If this is the case, he may have taken advantage from the use of diagrams that he could have found in Sturm (I owe this remark to David Rabouin).

I figure: *Barbara, Celarent, Darii, Ferio; Barbari, Celaro;*
II figure: *Cesare, Camestres, Festino, Baroco, Cesaro, Camestros;*
III figure: *Darapti, Felapton, Disamis, Datisi, Bocardo, Ferison;*
IV figure: *Fapesmo, Fresismo, Ditabis, Baralip, Celanto, Colanto.*

In the 1691 paper in the *Acta Eruditorum*, in which he expresses his irritation for the re-edition of the *DAC*, Leibniz suggests some corrections to this list:

> The determination of the moods belonging to the syllogistic figures was mistaken as well. Even though, indeed, it is very true that in each of the four figures we may find the same number of useful moods, i.e. six (which has been remarked here for the first time), a mood in the fourth figure has been wrongly disposed. Thus, in place of OAO [ ... ], which does not exist, we must substitute the mood AEE [ ... ].[89]

This means that in the fourth figure we have to cancel the mood *Colanto* and put *Calerent* in its place.[90]

As we have seen, in the *DAC* Leibniz is very proud of having revealed "the secret harmony" of the syllogistic figures, which all have "the same number of moods".[91] In the passage just quoted, from the *Acta Eruditorum*, written about twenty-five years after the *DAC*, Leibniz reveals the same attitude; and more than forty years later, in a letter to Louis Bourguet, he shows that he has not changed opinion:

> Exactly as arithmetic and geometry, the logic of syllogisms is truly demonstrative. When I was young, I proved not only that there really are four figures, which is easy to prove, but even that each figure has neither more nor less than six useful moods, whereas usually the first and the second figures are credited with only four moods and the fourth with five. I even proved that the second and third figure derive immediately from the first, without making recourse to the conversions which, in their turn, are proved by means either of the second or of the third figure. And

---

[89] GP 7: 104.
[90] As Couturat (1901: 7, fn) observes, in a letter to Vincent Placcius dated 16 November 1686, Leibniz recognizes his mistake concerning the fourth figure. Here, he writes that all the moods of the fourth figure properly are: AEE, AAI, EAO, [IAI], EIO, AEO, with *Calerent* at the place of *Colanto* (in the original, Leibniz forgets to mention 'IAI', i.e. *Ditabis*).
[91] *DAC*: 123 (A VI, 1: 184).

I proved also that the fourth figure belongs to a lower level and that it is in need either of the second or of the third figure or (which amounts to the same) of the conversions.[92]

## 11  The influence of Wirth (Hospinianus), Pierre de la Ramée, and Thomasius

The part of the DAC on traditional syllogistic heavily rests on the work of Johannes Wirth (Johannes Hospinianus, 1515–75), with respect to which Leibniz openly recognizes his debt, and of which he gives a short account. The idea of grouping moods into the two categories of special and general moods, for example, stems from Wirth. Leibniz adds to these categories those of simple and figured mood, giving rise to the mixed categories of simple special, simple general and special and general figured moods. Moreover, it is from Wirth that Leibniz takes the idea of considering as new distinct moods those obtained applying *subalternation* to the conclusion of the traditional ones: *Barbara, Camestres, Celarent,* and *Cesare.* Last but not least, Wirth's combinatorial approach to the theory of syllogism was surely in keeping with Leibniz's philosophical insights. Leibniz, however, does not accept Wirth's method entirely. He refuses, for instance, to accept the idea that singular propositions have to be considered logically equivalent to particular ones:

> In these speculations of Wirth there is something to be applauded, something to be desired. I applaud the discovery of new moods: *Barbari, Camestros, Celaro, Cesaro;*[93] I applaud what he correctly noted, that the moods which have attracted the traditional names, such as *Darii,* etc, relate to the moods subsumed under them in the same way as a genus relates to a species ( . . . ). But I cannot equally approve of the fact that he equated

---

[92] GP 3: 569. What Leibniz writes here about the fourth figure is not so evident from the DAC, and it is probably the result of some afterthought posterior to 1666. Cf. Letter to Cornelius Dietrich Koch (GP 7: 478): "Years ago I found that the valid moods of each figure are in number of six and that they can be neither more nor less." And GP 7: 519 (Letter to Gabriel Wagner, 1696).

[93] These moods, however, are not so 'new' as Leibniz thinks: *Barbari,* for instance, was quoted by Domingo De Soto in his commentary to Peter of Spain's *Summulae* (De Soto 1554: 125b), whereas other authors, like, for example, Ludovico Carbone (Carbo 1597: 148), were well aware that by means of subalternation, all syllogisms concluding with a universal proposition may conclude with the corresponding particular.

Singulars with particulars, a circumstance that distorted all his arguments, and resulted in fewer useful moods than should be the case, as will presently become clear.[94]

In general, Leibniz gets his results in a more direct and simpler way than Wirth, who "proceeds otherwise, and in a roundabout manner".[95]

Besides Wirth, other sources of Leibniz's systematic reconstruction of traditional syllogistic are Pierre de la Ramée and Jakob Thomasius. In Leibniz's words, Thomasius, relying on Ramus, has shown that 'the conversion of a proposition can be demonstrated through a syllogism by adding to it an identical proposition'.[96] According to the traditional doctrine, a *conversion* is an inference rule that authorizes the passage from a categorical sentence $p$ with subject S and predicate P to a categorical sentence $p'$ with subject P and predicate S. There are two types of conversion: a) the so-called conversion by limitation (*per accidens*); b) simple conversion. Examples of *simple conversion* are:

| (UN) | (PA) |
|---|---|
| No men are stones | Some philosophers are piano players |
| No stones are men | Some piano players are philosophers |

An example of *conversion by limitation* is:

(UA)
All men are animals
Some animals are men.

That UN sentences can be simply converted is quite easy to understand if we consider that, in this case, the predicate is excluded from all the extension of the subject, and vice versa; whereas in the case of PA sentences, because the word 'some' refers to the individuals 'falling' under both subject *and* predicate, it is obvious that subject and predicate may change their places the one with the other. Conversion by limitation, instead, is justified by the fact that in UA sentences the predicate may refer to more individuals, besides those denoted by the subject (for example, converting without limitation a sentence like 'All men are animals', which is true, amounts to state 'All animals are men', which is false).

---

[94]  *DAC*: 113 (A VI, 1: 182).      [95]  *DAC*: 107 (A VI, 1: 180).
[96]  *DAC*: 130 (A VI, 1: 186–7).

For what concerns PN sentences of the general form 'Some S are not P', they were usually considered *not convertible* by traditional syllogistic. It is not difficult to see why, if we consider a sentence like 'Some animals are not vertebrate': converting it, we obtain the false statement 'Some vertebrate are not animals'. In more general terms: if A is a proper subset of the set B (i.e. if B contains all the As and something more), then it is true that some members of B do not belong to A, but it is certainly false that some members of A do not belong to B.

In the *DAC*, however, Leibniz takes for granted that the conversion rule may be applied to PN propositions as well; and on the basis of the method proposed by his teacher Thomasius, he attempts to prove that conversion (usually considered an immediate inference), can be justified on firm logical ground by means of a syllogism with an identical premise.[97] One may use *Baralip*, indeed, a syllogism of the fourth figure, to justify the inference from 'All A are C' to 'Some C are A':

(*Baralip*)
All A are C; All C are C
Some C are A.

To validate the inference from 'Some A are C' to 'Some C are A', Leibniz employs *Disamis* (III figure); whereas *Cesare* (II figure) is used to justify the inference from 'No A are C' to 'No C are A':

(*Disamis*)                     (*Cesare*)
Some A are C; All C are C       No A are C; All A are A
      Some C are A                    No C are A

Unfortunately, as we have said, Leibniz tries to prove the rule of conversion even in the case of particular negative propositions, i.e. in the case of the inference from 'Some A are not C' to 'Some C are not A'. Leibniz proposes two different proofs, the first in *Bocardo* (third figure) and the second using *Colanto* (fourth figure):

(*Bocardo*)                         (*Colanto*)
Some A are not C; All A are A       Some A are not C; All C are C
      Some C are not A                    Some C are not A

---

[97] Leibniz's proofs are easy to follow and we limit ourselves to reproducing some of them here, without commenting them.

Yet, Leibniz's presumed 'proof' is flawed by three mistakes: 1) in the original text, Leibniz wrongly assumes that the first syllogism is in the mood *Baroco* (II figure), not in *Bocardo*; 2) the mood *Colanto* does not exist at all, as Leibniz himself will make clear in the 1691 paper on *Acta Eruditorum*; 3) conversion, as we have remarked above, cannot be applied to particular negative propositions.[98]

## 12 Using numbers to encode concepts (and propositions)

After having discussed the application of the combinatorial art to the traditional doctrine of syllogisms, Leibniz introduces the seventh application, devoted to "the complications of geometrical figures". From this part of the *DAC*, a flavour emanates that may remind the reader of essays on geometry of a later date, the period of Leibniz's maturity. It is quite evident, however, that the investigation here displayed is based on a rather narrow knowledge of geometry and that any further development would require completely new grounds.[99]

The eighth application concerns "the forming of cases among jurists", whereas the ninth is more general, consisting in finding all "subaltern genera and species", once "the species of a division are given". To illustrate the ninth application, Leibniz discusses the different forms of government (Oligarchy, Monarchy, etc.) that have been recognized within Western culture since Aristotle.[100]

The tenth application concludes the part of the logic of invention about terms (concepts), and presents two general problems that belong to the theory of propositions.[101] Since the propositions that Leibniz considers are the four categorical propositions of traditional syllogistic, the new investigation can be viewed as a closer examination of fundamental issues concerning the theory of syllogism. Leibniz introduces the tenth application as follows:

> It is time to pass from Divisions to Propositions, the second part of the Logic of Invention. A proposition is composed of a subject and a predicate,

---

[98] As Couturat remarks (1901: 10–11), and as Leibniz himself recognizes in paragraph 32 of the *DAC*, if we introduce negative terms ('not-stone', 'not-animal', etc.) then inferences as the following are perfectly legitimate: "Some men are not-stones; therefore some not-stones are men". In this case, however, the proposition "Some men are not-stones" is a particular *affirmative* with negated predicate.

[99] *DAC*: 135–9 (A VI, 1: 187–9).     [100] *DAC*: 145–8 (A VI, 1: 189–91).

[101] *DAC*: 149 ss. (A VI, 1: 192).

so that all propositions are com2nations. Hence, in the inventive Logic of propositions the problem is to solve the following: 1. *Given a subject, to find predicates*. 2. *Given a predicate, to find subjects; and in each case both affirmative and negative*.[102]

To solve both problems, which, as we shall see, will give rise to other 'sub-problems', we need first of all to display a systematic analysis of concepts. Each concept must be analysed into its component formal parts, i.e. of each concept we must lay down a definition and then a definition of each concept entering the definition, and so on, right down to simple parts. Simple parts are those that cannot be defined any further. Quoting Aristotle, Leibniz observes that: "One does not always look for a definition".[103] Thus, the first terms are not definable and are "better understood not by means of a definition, but by means of an analogy".[104]

The second step consists in putting all the primitive concepts in one class and employing symbols to designate them. Leibniz remarks that it "will be most convenient if they are numbers".[105] Once the first class of concepts has been laid down, a second class can be formed, simply giving rise to com-binations each made of two concepts belonging to the first class. A third class, accordingly, can be formed with triples of first concepts, and so on: concepts "that are composed of the same number of primitive terms should be included in the same class".[106]

Because complex terms are the result of combining two or more simple concepts and because simple concepts are designated by means of natural numbers, complex concepts can be considered as the arithmetical product of its composing numbers. What is remarkable here is Leibniz's awareness of the fact that, once one has employed numbers to designate concepts, one may 'forget' the concepts and simply operate with numbers and with their recognized properties.

Leibniz proposes even to employ fractions to facilitate the individuation of concepts in a given class. The denominator of each fraction should represent the class to which the corresponding concept belongs, whereas the numerator might designate the proper place of the concept in the class. Leibniz's own example is the following. Consider the four simple concepts corresponding to the numbers '3, 6, 7, 9'; because they are simple, they all

---

[102] *DAC*: 149 (A VI, 1: 192).   [103] Cf. Aristotle, *Metaph.* IX 6, 1048a: 36–7.
[104] *DAC*: 159 (A VI, 1: 195).   [105] *DAC*: 161 (A VI, 1: 195).
[106] *Ibidem*.

belong to the first class. Now, consider the second class made by the combinations of two simple concepts, in this order:

(1) 3.6; (2) 3.7; (3) 3.9; (4) 6.7; (5) 6.9; (6) 7.9.

Suppose, further, that the third class, made of triples of numbers has been determined. Leibniz suggests that a fraction as, for instance, 1/2 should designate the first term ('3.6') of the second class, according to the order just specified. Analogously, '5/2' will designate the fifth couple of the same sequence; whereas '4/3' and '5/3', designate respectively, the fourth and the fifth triple in the third class, etc.[107]

To the same complex concept different numerical expressions may correspond, but the possibility of reaching the simple concepts helps to avoid ambiguity. Thus, to continue Leibniz's example, suppose that the numbers 3, 6, 9 designate simple concepts. In this case, the same complex concept may be represented in four different ways: as '3.6.9', displaying all the simple elements composing it, or as: '1/2.9', '3/2.6', '5/2.3'. The fraction '1/2' designates the first couple of numbers ('3.6') belonging to the second class of combinations (according to the order above); and this couple, together with '9', designates the complex concept '3.6.9' as well. In their turn, '3/2' and '5/2' correspond, respectively, to the third and fifth couple and each, together with '6' and '3', designates the same complex concept ('3.6.9').

The analysis of a complex concept, as proposed by Leibniz, is analogous to the decomposition of a number in its prime factors. Commenting on Leibniz's *DAC*, Louis Couturat uses an example that explains the reasons of the analogy.[108] Consider the number 210 and suppose that it is the number of a complex concept. Its prime factors are: 2, 3, 5, 7 and they may be considered as the numbers of the simple concepts composing the complex concept corresponding to 210. Thus, we have a first class of unary combinations, and three other classes corresponding, respectively, to combinations of two, three and four prime numbers:

1)  (2), (3), (5), (7);
2)  (2.3), (2.5), (2.7), (3.5), (3.9), (5.7)
3)  (2.3.5), (2.3.7), (2.5.7), (3.5.7)
4)  (2.3.5.7)

---

[107] *DAC*: 161–3 (A VI, 1: 195).    [108] Couturat (1901: 41, fn 1).

Each of these combinations corresponds to a divisor of 210. The combination of the fourth class, for instance, corresponds to 210 itself; the second combination of the third class corresponds to 42; the third of the second class corresponds to 14, etc. If we interpret the numbers as concepts, each combination corresponds to a concept composing the complex concept symbolized by 210. This means that we may think of the divisors of a given number as the *predicates* which, in a true sentence, compose the concept corresponding to that number. Thus, for example, if in the case of the concept corresponding to 210, we predicate the combination of concepts of the fourth class, we obtain an identity (it is, indeed, the *definition* of the concept). As Leibniz remarks:

> The predicates of a given subject are all its primitive terms, and all derivative terms nearer to the primitive terms, whose own primitive terms are all in the given subject.[109]

## 13  Finding all possible subjects and predicates

We now have all the elements needed to solve the first problem of the tenth application (*usus*).

Problem 1): *Find all the possible predicates P, which can be truly stated of a given subject S in a universal affirmative proposition (UA).*

Solution 1): *Given the concept S belonging to the class k of combinations, the number of all true UA propositions having S as a subject (i.e. the number of all predicates that can be combined with S, giving rise to a true UA proposition) is $2^k - 1$.*

As Leibniz explains, this solution presupposes a distinction between two kinds of truths and, in particular, concerns *theorems* and *necessary truths* only:

> I must attend finally to the fact that this whole art of complication is directed towards theorems, or propositions that are eternally true, that is, propositions that hold not because of the will of GOD, but by their very nature. But all singular propositions are, as it were, either *historic*, for example, *Augustus was Emperor of the Romans*, or *observations*, that is,

---

[109] *DAC*: 163 (A VI, 1: 195).

universal propositions, but universal propositions whose truth is founded not on essence but on existence; for they are all true as if by chance, but in fact by the will of GOD. For example, *All men in Europe of mature years have cognisance of GOD*. No demonstration of such things can be granted, only induction; except that sometimes an observation can be demonstrated from another observation through the medium of a Theorem.[110]

As usual, Leibniz presents only in words the solution to this problem. Moreover, the solution proposed is flawed by a rather odd type of miscalculation.[111]

The second problem that Leibniz tackles concerns the particular affirmative propositions (PA):

Problem 2): *find the number of all particular affirmative propositions with a given subject S belonging to the class k of combinations.*

Leibniz's solution is the following:

[ ... ] find predicates UA of the given term in the manner just explained; and subjects UA [ ... ] Then add both numbers, because from a proposition UA proposition PA can be derived either simply by conversion or by subalternation.[112]

In other words:

Solution 2): *First calculate the number of all true UA propositions that can be formed with subject S (see preceding problem) and then add it to the number of all true UA propositions that can be formed with S as a predicate. From the result subtract 1 (the identical proposition 'All S are S').*

To solve this problem, we need to know how to find the number of UA true propositions with the given concept S as predicate. Leibniz, however, postpones the discussion of the UA propositions to that of the PA propositions; and when introducing the solution to the second problem, he explicitly refers to the following:

Problem 3): *Find all the possible subjects S, which can be truly stated of a given predicate P, in a universal affirmative proposition (UA).*

---

[110] *DAC*: 177 (A VI, 1, p. 199).    [111] Cf. *DAC*: 165, note 120.
[112] *DAC*: 165 (A VI, 1: 196).

Leibniz sees this problem as a particular case of a more general one, which in its turn splits up into two parts:

*Given a head and a number of things, (3.1) find all combinations overall, or (3.2) all combinations of a particular class, containing the given head.*

The *head* is a combination consisting of one or more items common to all combinations that can be obtained out of a given number of 'things'.[113] Given a domain of five elements (for instance, the numbers from 1 to 5), the couple (1, 2) is the *head* of the complexions (1, 2, 3), (1, 2, 4), (1, 2, 5), (1, 2, 4, 5), which are composed of, respectively, three and four items belonging to the domain. Thus, Leibniz turns the problem of finding all possible subjects of a given predicate in a UA proposition into a problem concerning the 'head of a complexion'.

Leibniz attempts to solve problem (3.1) with the help of the table ℵ. He recognizes that, if $n$ is the 'number of things' and $k$ a given head, to get the number of all subsets of elements of $n$ containing $k$, we need first to subtract $k$ from $n$. Then, we have to find the number $n-k$ in the first row of the table and go the end of the column headed with it: the last number of the column offers the desired solution. This number can be characterized by means of the expression: $2^{n-k}$. Leibniz, however, actually stops at the penultimate number (not at the last one) of the column, thus showing that he thinks of the solution of the problem in terms of the expression $2^{n-k} - 1$. That Leibniz considers this as the solution is shown by the numerical examples that he chooses:

> Moreover, in the total number of complexions of 5 things (which is 31) *a* is found 15 times (the total number of complexions of 4 things), *ab* 7 times (the total number of complexions of 3 things)[114]

As a matter of fact, the total number of complexions of four things is sixteen, not fifteen, as Leibniz claims, and among the complexions of four things the couple *ab* occurs eight times, not seven.[115]

We meet the same situation with the second, more specific, problem (3.2). Here we are asked to find all occurrences of a given head, which are contained in the subsets of a particular class of combinations. Sticking to Leibniz's example, let us suppose that there are five things. The

[113] Cf. Iommi Amunategui (2015); Knobloch (1973: 30, 1974: 414).
[114] *DAC:* 169 (A VI, 1: 197).        [115] Cf. Knobloch (1973: 31).

combinations (complexions) that can be made out of them are distributed into five classes: the first class with five single elements, the second with ten couples, etc. We are requested to find how many times a single element, a couple or a term occurs in a given class. In this case too, Leibniz employs the table ℵ and starts subtracting the number of the elements composing the head (k) from the total number of elements (n). Once the number corresponding to n–k is determined, he further suggests subtracting k from the number corresponding to the class (let us call it 'c'). The last step is that of determining the position of n–k and c–k in the table in order to find the number that corresponds to the solution of the problem. All this amounts to calculating the values of the expression $\binom{n-k}{c-k}$ and then applying the formula:

$$\binom{n}{k} = \frac{n!}{(n-k)!k!}$$

Thus, if, for instance, we want to know, given five things: a, b, c, d, e, how many times the couple (a, b) occurs among the elements of class 3, we have:

$$\binom{5-2}{3-2} = \binom{3}{1} = \frac{3!}{(3-1)!1!} = \frac{6}{2} = 3.$$

The couple (a, b) occurs three times among the complexions of class 3. In this case, Leibniz's proposed result is correct.

Coming back now to the logical problem of finding all subjects of a given predicate in a universal affirmative proposition (Problem 3), we see that Leibniz again distinguishes two cases:

3(a)   the set of all primitive concepts of a given concept P is a proper subset of the set of concepts constituting the concept S;

3(b)   S and P are composed of the same set of primitive concepts.

In case 3(a), the concepts eligible for the role of subject in a true UA proposition having P as a predicate belong to the classes containing all primitive concepts of P (i.e. to the classes following the class of P). In case 3(b) the predicate is only a different description of the subject.

If we consider the first concepts of P as the *head* of the complexion represented by P, Problem 3(a) becomes that of finding all the complexions containing the given *head*. So, we have:

Solution 3(a): *If k is the number of the first concepts of a given concept P and n the number of the first concepts in general, then the number of all* *possible subjects S, which can be combined with P giving rise to true UA propositions, is $2^{n-k}-1$.*

The total number of combinations is subtracted by 1, to avoid the case of the identical proposition, i.e. of the proposition with the concept P as a subject and as a predicate as well. In this circumstance Leibniz gives the correct solution of the problem, even though, as we have seen, his solution of the general case involving the heads, is wrong.

Leibniz calculates the number of subjects, employing the table ℵ placed at the beginning of the *Dissertation*. As we have seen, the table has been built according to principles that permit a correct calculation of the number of sought combinations. Besides the correctness of the result, however, Leibniz's procedure is lacking of generality. The recourse to the table is cumbersome and, for numbers greater than those explicitly considered, it involves calculations that may be quite complex and very tedious to be performed.

Once Problem 3(a) has been solved, it is easy to solve Problem 3(b):

Solution 3(b): *If k is the number of the first concepts of a given concept P and n the number of the first concepts in general, then the number of all possible subjects S, which can be combined with P, giving rise to true UA propositions including the identical one, is $2^{n-k}$.*

Assuming as above that the number of simple concepts is five (*a, b, c, d, e*) and that the couple (*a, b*) corresponds to the simple concepts composing the predicate P, Leibniz subtracts 2 from 5 and concludes that the number of all possible subjects S, which can be combined with P, giving rise to true UA propositions is 9, i.e. $2^3-1 + 2$. This result, however is wrong, the correct one being 8, i.e. $2^3-1 + 1$. Leibniz's mistake, as Couturat argued, is probably due to the fact that he considers (*a, b*) and (*b, a*) as *two* different subjects, in the case of the identical proposition having as P the concept characterized by the couple (*a, b*).[116]

As regards the particular affirmative propositions, Leibniz discusses the problem:

Problem 4: *Find all possible subjects S, which can be truly stated of a given predicate P, in a particular affirmative proposition (PA).*

Leibniz's solution is:

Solution 4: *the number of all possible subjects S, which can be truly stated of a given predicate P in a particular affirmative proposition (PA), is the same*

---

[116] Couturat (1901: 44, n. 1).

*as that of the predicates that can be truly stated of the given concept P assumed as the subject.*

The reason for this solution is grounded on the fact that every particular affirmative proposition can simply be converted.

Until now, only affirmative propositions have been taken into consideration; as Karl Dürr remarked, Leibniz's treatment of the negative ones is "very short and sketchy".[117] In Leibniz's own words:

> Negative Predicates and Subjects can be found in this way: from certain given primitive Terms regarded as the Number of things, calculate all the terms, derivative as well as primitive, as the complexions in total. For example, if there are five primitive terms, the total number of complexions will be 31. From this result subtract all the universal affirmative predicates and universal affirmative narrower subjects: the result will be all the negative predicates. With subjects the reverse holds.[118] The Particular negatives can be calculated from the universals, just as we calculated PA from UA above.[119]

Thus, we have:

Problem 5: *Find all possible predicates P, which combined with a given subject S give rise to UN true propositions.*

Problem 6: *Find all possible Subjects S, which combined with a given predicate P, give rise to UN true propositions.*

According to Leibniz, the solutions are:

Solution 5: *First calculate the number of all possible combinations (complexions) of simple concepts. Then, from this number subtract the number of all true UA propositions that have the given concept S as a subject and the number of all UA propositions having S as a predicate.*

Solution 6: *First calculate the number of all possible combinations (complexions) of simple concepts. Then, from this number subtract the number of all true UA propositions that have a given concept P as a subject and the number of all UA propositions having P as a predicate. (That is, the number of true UN propositions, in which a given concept enters as a*

---

[117] Dürr (1949: 25).
[118] That is: the number of UN propositions in which a given concept C plays the role of predicate, is the same as the number of UN propositions in which C plays the role of subject.
[119] *DAC*: 173 (A VI, 1: 198).

*predicate is equal to that of UN propositions in which this same concept enters as a subject.)*

## 14  Finding the number of all valid syllogistic arguments having a given proposition as conclusion

After having determined the number of subjects and predicates in the four kinds of categorical propositions of traditional syllogistic, Leibniz tackles the problem of finding the number of all valid syllogistic arguments having a given proposition as conclusion. This amounts to finding all middle terms that, combined with the subject A and the predicate B of the given proposition, yield all possible valid syllogisms (the number of middle terms is the same as that of valid syllogisms).

According to Leibniz, a middle term:

> will be a predicate of the subject and a subject of the predicate, that is, a term containing A and contained by B. And a term is said to contain a term if all the primitive terms of the latter are in the former.[120]

If we suppose that $n$ = the number of the primitive terms of A, $k$ = the number of the primitive terms of B and $n > k$, we may say, following Leibniz's suggestion, that A *contains* B.[121] Suppose, further, that the proposition, which has been chosen as the conclusion of the syllogistic argument, has concept A as subject and concept B as predicate (it is a UA proposition of the form 'All A are B'). Then, the number of concepts, which may play the role of middle terms in this syllogistic argument corresponds to $2^{n-k} - 2$ (excluding the case that A and B belong to the same class, i.e. when A = B). In other words, the number of middle terms is the same as that of the combinations ('complexions'), which can be made out of $n$-$k$ items, minus the two combinations with identical subject and predicate.

---

[120]  *DAC*: 173 (A VI, 1: 198).
[121]  This being the case, the containment-relation between middle term, subject and predicate has to be considered according to the following order: the subject (A) contains the middle term, which, in its turn, contains the predicate (B).

Leibniz's next step consists in discussing how to find all valid syllogistic arguments having a given UN proposition as conclusion. He proposes a quite simple solution of this problem:

> Moreover, we shall easily find whether a predicate is denied of a subject if, when each term is resolved into its primitive terms, it is clear that neither contains the other. A negative can, however, also be proved like this: find all the predicates of the subject. Since the predicate must be denied of all of them, they will altogether be middle terms in proving the negative. Find all the subjects of the predicate: since all of them must be denied of the subject, they too will altogether be middle terms in proving the negative. With each of these calculated, we shall have the number of middle terms in proving the negative.[122]

Clearly, the UN proposition can easily be proved if its subject 'S', and its predicate 'P', once analysed into their simple components, do not have any concept in common. Because, assuming that S is composed of the four simple concepts *a, b, c, e* and P of the two simple concepts *a, d*, even though S and P have only *one* simple concept in common, the PA proposition 'Some S is P' is true and, for the so-called 'Aristotelian square', the UN 'No S is P' is false.[123] Therefore, it is safer to assume that S and P do not have first concepts in common.

To find all middle terms, which can be employed to form valid syllogisms with the given UN proposition as conclusion, Leibniz gives the following rule. First, find the number of, respectively, all possible predicates of S and of all possible subjects of P; then, add together these two numbers: this is the sought solution. It is quite obvious that Leibniz is here thinking of a first figure syllogism:

MP; SM
_____
SP.

In this case, the middle term M is the *predicate* of S in the minor and the *subject* of P in the major premise. If P can be universally denied of all the concepts playing the role of predicates, we may build so many valid syllogisms having the given UN proposition as conclusion, as there are predicates of S. Thus, the problem becomes that of finding all the predicates

---

[122] *DAC*: 177 (A VI, 1: 198).    [123] Couturat's example (Couturat 1901: 47).

of a given subject S in a UA proposition—a problem, which, as we have seen, has been solved by Leibniz.[124]

## 15  Some problems concerning variations (permutations). The 'head' of a set of variations

In addition to those examined so far, Leibniz presents other applications of Problems I and II (better to say, of the methods employed to solve Problem I and II).[125] Thus, for instance, he proposes an encoding of some concepts of Euclidean geometry assumed as primitive, aiming to derive from them all possible complex concepts. As a corollary of this application, Leibniz proposes the creation of "a Universal Notation, that is, intelligible to any reader, no matter what his language."[126] And this leads him to plan the construction of a *universal language* that he presents quite emphatically as follows:

> But once the Tables, or predicaments, of my art of complication have been constituted, new vistas will open. For the primitive terms from whose complication all the others are constituted are to be designated by symbols, and these symbols will be a kind of alphabet. And it will be desirable for the symbols to be as natural as possible ( ... ). If all this is correctly and skilfully accomplished, we shall have a universal script that will be as easy as it is general, readable without the aid of a dictionary, and allowing fundamental knowledge of everything to be absorbed at once. Accordingly, every such script should be made up of geometrical figures, and of pictures, such as the Egyptians used in ancient times and the Chinese use today, except that their pictures are not reduced to a fixed Alphabet, or to letters, and as a result impose an incredible strain on the memory, whereas exactly the contrary happens with my proposed script.[127]

PROBLEM III reads:

III): "Given a number of classes, and of things in the classes, to find the complexions of the classes".[128]

---

[124] Cf. *DAC*: 165–6 (A VI, 1: 196).

[125] Remember that these two Problems were the following: PROBLEM I: "Given a number and exponent to find the complexions" (*DAC*: 81; A, VI, 1: 174); PROBLEM II: "Given a number, to find simply the complexions" (*DAC*: 91; A VI, 1: 176).

[126] *DAC*: 183 (A VI, 1: 201).        [127] *DAC*: 187–8 (A VI, 1: 202).

[128] *DAC*: 196 (A VI, 1: 204).

From PROBLEM IV onwards, *variations* enter into play and with Problems from VII to X, variations are considered in connection with the notion of a 'head' of a variation. We have encountered this notion above, when dealing with the problem of finding all possible subjects of a given predicate in a universal affirmative proposition (UA). The notion emerges again in PROBLEM VII, where Leibniz discusses some aspects of the calculus of *variations* (*permutations*).

In the case of *complexions*, a 'head' is a complexion consisting of items common to all combinations that can be obtained out of a given number of elements. Analogously, Leibniz calls 'head' the items (or item), which are invariant in respect to a set of permutations (*variations*). From a mathematical point of view, this is one of the most original parts of the *DAC*: Leibniz discusses problems that nobody had raised before him.[129]

When dealing with *variations*, Leibniz distinguishes various kinds of 'heads'.[130] A head "may consist either of one thing or of many". In the case of one thing, either it has no relation at all with the other things to be varied, or there is one or more things *homogeneous* with it. Leibniz employs the word *monadic* here, to characterize the case of an absolute isolation, i.e. of a head with no homogeneous elements. A head is *multipliable* if its elements can permute.[131] Concerning the notion of *homogeneity*, the list of definitions at the very beginning of *DAC* indicates that "a *homogeneous thing* is something that is equally disposable in a given place, without affecting the head".[132] We may think of *homogeneity* as a property that allows substitutions of one or more items in the place of others.

If the head consists of only one element, then either it is *monadic* or outside it there are 'things' homogeneous with it.

If the head consists of several things, Leibniz distinguishes two further cases:

a) the head contains mutually homogeneous things;
b) the head does not contain homogeneous things.

If a) is the case, then either $a_1$) outside the head there are things homogeneous with the things contained in it, or $a_2$) there are not.

If b) is the case, then either $b_1$) outside the head there are things homogeneous with the things contained in it, or $b_2$) there are not.

---

[129] Cf. Knobloch (1973: 47–53).
[130] For this part of the *DAC*, I rely completely on Knobloch (1973: 47–50).
[131] Cf. *DAC*: 79 (A VI, 1: 173).     [132] *Ibidem.*

Assuming that there are $n$ 'things', of which $k$ (for $1 \leq k < n$) are in the head, Leibniz discusses the following cases.[133]

1. The head is *not multipliable*.

If outside the head there are no elements homogeneous with it, the number of permutations amounts to

$$(n - k)!$$

Leibniz calls this result 'A'.

If outside the head there are homogeneous elements, supposing with Leibniz that the head is composed of only one element and that there is only one thing $(a_1)$ homogeneous with it, the number of permutations is

$$(n - k)!(a_1 + 1)$$

2. The head is *multipliable*.

If outside the head there are no elements homogeneous with the elements of the head, then, putting $(n-k)! = A$ (as Leibniz does), the number of permutations is

$$A(k!)$$

If inside the head there are no homogeneous elements and, moreover, outside the head there are elements which are homogeneous with elements belonging to the head, then, on the hypothesis that an element $a_1$ belonging to the head is homogeneous with another element, putting, as Leibniz does, $A(k!) = B$, the number of permutations is

$$B(a_1 + 1).$$

As Knobloch observes, if other elements of the head, $a_2, a_3, \ldots, a_m$ respectively, have elements homogeneous with them, the number of permutations is

$$B(a_1 + 1)(a_2 + 1)\ldots(a_k + 1)$$

---

[133] As Leibniz emphasizes, the elements outside the head can be "discontiguous": the head can be "scattered among them". (See *DAC*: 261 (A, VI, 1: 220) and Knobloch 1973: 48).

If some homogeneous elements belong to the head and some are outside, assuming that the homogeneous elements inside the head determine a subset $s$ of $s_1$ homogeneous elements and that they are homogeneous with one element $a_1$ outside the head, the number of permutations is

$$B\binom{s_1 + a_1}{s_1}.$$

In case of several homogeneous subsets inside and outside the head, the number of permutations becomes

$$B\binom{s_1 + a_1}{s_1}\binom{s_2 + a_2}{s_2}....\binom{s_k + a_k}{s_k}.$$

Since his period of intense study of mathematics in Paris under the guidance of Christiaan Huygens, Leibniz continued to be strongly interested in combinatorial art, but with the exception of a short essay on probability (1690), he never published any further contributions to the field. The situation of Leibniz's mathematical inquiries concerning the latter is clearly resumed, once more, by Eberhard Knobloch:

> Contrary to his plans, Leibniz never published any further mathematical contributions to the Ars Combinatoria except for a short essay on the theory of probability [Leibniz 1690], but hundreds of mostly uncollated manuscripts among the more than 7300 pages of mathematical material he left behind bear witness of his numerous studies in this field. Disregarding the studies exclusively concerned with the theory of numbers or with algebraic problems, the relevant notes may roughly be grouped under five headings: 1. Combinatorial theory in a narrower sense (basic combinatorial operations). 2. Symmetric functions (together with the theory of equations). 3. Partitions (a part of additive theory of numbers). 4. Determinants (elimination of unknowns in systems of linear equations and equations of a higher degree). 5. The theory of probability and related fields (theory of games, calculation of rents and interest).[134]

Assuming that the reader by now has some familiarity with Leibniz's terminology, we list the remaining problems discussed in the DAC, starting with Problem IV (the reader interested in the solutions proposed by Leibniz is kindly referred to the main text of the DAC in the present edition and the footnotes added there):

[134] Knobloch (1974: 410–11).

PROBLEM IV: "Given a number of things, to find the variations of order".[135]

PROBLEM V: "Given a number of things, to find the variation of a purely relative situs, or neighbourood".[136]

PROBLEM VI: "Given a number of things to be varied, of which one or some are repeated, to find the variation of order".[137]

PROBLEM VII: "Given a head, to find the variations".[138]

PROBLEM VIII: "To find the variations in common with another given head".[139]

PROBLEM IX: "To find the heads that have common variations".[140]

PROBLEM X: "To find the heads of useful or useless variations".[141]

PROBLEM XI: "To find the useless variations".[142]

PROBLEM XII: "To find the useful variations".[143]

According to Leibniz, the notions of 'useful' and 'useless variation' mentioned in the last three problems imply a reference to the subject matter of the things to be varied. Thus, for instance, in the case of the traditional doctrine of the four elements (air, earth, water, and fire), it is the intrinsic nature of both fire and water to determine that they cannot be combined together, giving rise to a useless variation if they are combined. Clearly, here, a *variation* is not considered as something merely formal, involving only the order or disposition of the elements to be varied, but implies a more substantive relationship, like that of a 'real' combination or connection.

## 16  Atomism and mereology

Of the several threads linking the *DAC* to Leibniz's mature philosophy, one of the most robust is the project for the constitution of a general theory based on the notions of *part* and *whole*. When Leibniz wrote the *DAC*, he was clearly under the strong influence of an atomistic point of view. Of classical atomism, however, he seems to consider with diffidence the implicit materialistic ontology and to accept the very general claim that everything (material and spiritual) is generated through a combination of certain

---

[135] *DAC*: 222 (A VI, 1: 211).      [136] *DAC*: 247 (A VI, 1: 217).
[137] *DAC*: 249 (A VI, 1: 217).      [138] *DAC*: 260 (A VI, 1: 219).
[139] *DAC*: 264 (A VI, 1: 221).      [140] *Ibidem.*
[141] *Ibidem.*      [142] *DAC*: 266 (A VI, 1: 222).
[143] *Ibidem.*

minimal parts. Thus, whereas the first elements of the material world are very small 'atom-like' parts, the first elements of the 'mental world' are the simple concepts from the combination of which every complex concept and every true proposition originates.

In the *DAC*, however, Leibniz's main interest is to compute the number of possible combinations (complexions) and permutations (variations) of the parts in a whole. It is only later, after his stay in Paris, that Leibniz begins to elaborate the notion of a *composition* of the parts of a whole. In a short text, written in the years 1685–6, for example, he writes:

> If, as soon as several things are put together, we understand that one thing immediately originates, then those things are called 'parts' and the latter 'whole'. And it is not necessary that they all exist at the same time or place, but it is sufficient that they are considered at the same time. Thus, from all Roman emperors we compose simultaneously one aggregate.[144]

In a series of essays on logical calculus written around 1690, *composition* becomes a more precise operation: the *real addition*. Leibniz calls this kind of addition *real* in order to distinguish it from an arithmetical addition. Whereas in the case of an arithmetical sum we have 'A + A = 2A', the real addition is *idempotent*, i.e. it obeys the law according to which 'A + A = A'.[145] *Real addition*, however, like arithmetical addition, is *commutative* and *associative*. Thus, we may summarize as follows the main properties of *real addition*:

1) $A + A = A$ (idempotence);
2) $A + B = B + A$ (commutativity);[146]
3) $(A + B) + C = A + (B + C)$ (associativity).[147]

As Leibniz clearly states in the second postulate of an essay devoted to the calculus of *real addition*, this latter, as the general operation of *composition* mentioned above, is *non-restricted*:

> Any plurality of things, such as A and B, can be taken together to compose one thing, $A \oplus B$, or, L.[148]

---

[144] A VI, 4A: 627.   [145] Cf. *Logical Papers*: 132 (A VI, 4A: 834).   [146] *Ibidem.*
[147] Even though Leibniz does not mention explicitly the associative property, he employs it systematically.
[148] *Logical Papers*: 132 (translation slightly modified) (A VI, 4A: 834).

Further on, in the same essay, Leibniz again emphasizes the non-restricted nature of real addition:

[...] our general construction depends upon the second postulate, in which is contained the proposition that any thing can be compounded with any thing. Thus, God, soul, body, point and heat compose an aggregate of these five things.[149]

When Leibniz writes the essays on *real addition*, what he calls 'my hypothesis of the monads' is already defined, at least in its fundamentals. In these essays, the monad is compared to the geometrical point and the body to a segment: as the point is *in* the segment without being *part* of it, thus a monad is *in* the body, without being a *part* of it. A segment, instead, is in a line (or in a bigger segment) and is *part* of it. This motivates Leibniz's distinction between two mereological approaches, one based on the relation of *containment* and the other based on *parthood*. Leibniz, however, is not fully satisfied with this theory and attempts to find something that can play the role of a principle of composition, besides real addition. Once he has converted the atoms of the classical atomistic doctrines into 'spiritual atoms' or 'soul-like' individual substances, he attributes to them the same role that the soul (as the form of the body) has in the Aristotelian philosophy. It is here that the idea originates of a monad dominating a cluster of other monads, which play the role of a body.

Leibniz's variety of hylomorphism, however, is quite peculiar: in the last analysis, matter and form, i.e. the body and the dominant monad, are made of the same ingredients. Whereas the dominant monad is a simple, individual substance associated with a body, the latter is an *aggregate* of simple substances or monads. Thus, the problem arises what is the principle or the main cause responsible for the aggregation of the monads constituting a body. Clearly, if we have to take seriously the analogy between points and monads, the juxtaposition or mere addition of simple monads cannot generate a body (as the addition of a plurality of points does not generate a segment or a line).

A possible answer could be that an aggregate is constituted as soon as a plurality of monads is ruled or 'governed' by another monad, the 'dominating'

---

[149] *Logical Papers*: 139 (AVI, 4A: 842).

one. But then, another problem arises: how can simple substances, which are not extended, constitute a body, something that is extended in space?

What causes Leibniz the greatest trouble in these issues is clearly the strange mixture of a general 'compositional' account of reality inherited from the primitive atomistic view presented in the *DAC*, and the anti-materialistic notion of an atom understood as a soul-like 'individual substance'.

An atomistic attitude, or at least a general sympathy for atomism (even of the 'classical' variety), emerges at many places in Leibniz's writings. Of this, we have clear evidence in the form of Leibniz's continued interest in Lucretius, an interest spanning his entire philosophical career. A long passage from *On the Nature of Things*, for example, is quoted in the *DAC*.[150] Later on, in the *Theodicy* (III, 321), published more than forty years after the *DAC*, we find another long quote from the same work. As strange as it may appear, in the *Theodicy* Leibniz quotes from the Italian translation of Lucretius's poem, which circulated only as a manuscript until 1717, when it was published for the first time in London. This reveals, on Leibniz's part, something very near to a persistent fascination for a doctrine of which he was supposed to be a fierce adversary, because of its consequences in the field of religion.

---

[150] *DAC*: 243–5 (A VI, 1: 216)

# DISSERTATIO

## De

# ARTE COMBI-
# NATORIA,

*In qua*

Ex Arithmeticae fundamentis *Complicationum* ac *Transpositionum*
Doctrina novis praeceptis exstruitur, et usus ambarum per uni-
versum scientiarum orbem ostenditur; nova etiam
Artis Meditandi,

*Seu*

## Logicae Inventionis semina
sparguntur.

*Praefixa est Synopsis totius Tractatus, et additamenti loco*
Demonstratio

### EXISTENTIAE DEI,
ad Mathematicam certitudi
nem exacta.

AUTORE

## GOTTFREDO GUILIELMO
## LEIBNÜZIO Lipsensi,
Phil. Magist. et J. U. Baccal.

---

*LIPSIAE,*
Apud Joh. Simon. Fickium et Joh.
Polycarp. Seuboldum
*in Platea Nicolaea,*
Literis Spörelianis.
A. M. D C. L X V I.

# A DISSERTATION ON
# COMBINATORIAL
# ART,

*In which,*
starting from the foundations of Arithmetic, a theory of *Complications*
and *Permutations* is constructed out of new rules, their use across the
entire field of the sciences is demonstrated, and new seeds of the
art of Heuristic,

*or*

## The Logic of Discovery,
*are sown.*

*Prefixed to which is a Synopsis of the whole Treatise,
and by way of addition,* a Proof of the
EXISTENCE OF GOD,
laid out with Mathematical
certitude,

By
## GOTTFRIED WILHELM
## LEIBNIZ of Leipzig,
M.Phil, LL.B.

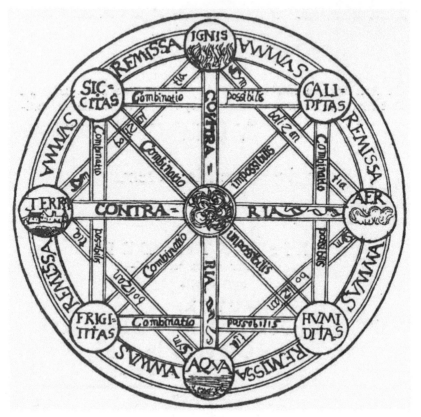

[167]

*VIRO*
SUMMO, MAGNIFICO, MAXIME
REVERENDO
*DNO*

# MARTINO GEIERO,

S. Stae. Theol. Doct. Serenissimi Electoris Saxoniae Supremo
Concionatori Aulico, Supremi Dresdensis Consistorii
Assessori, et Consiliario Ecclesiastico.
Theologo Incomparabili:

*Suo verò, praeter susceptionis beneficium, Patrono et Maecenati*
*maximo; rationem studiorum suorum constare voluit*

## AU[C]TOR.

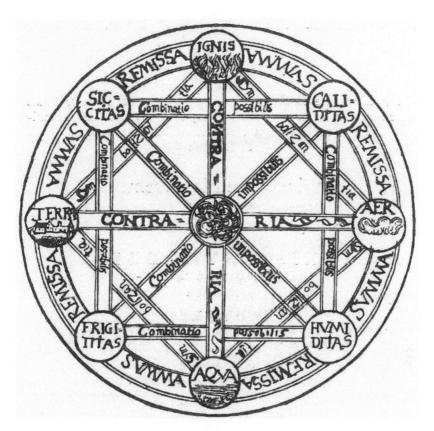

TO THE MOST DISTINGUISHED,
MAGNIFICENT, AND MOST VENERABLE
GENTLEMAN, HERR

# MARTIN GEIER,

Doctor of Theology of Sacred Scripture, Principal
Court Preacher to his Serene Highness the Elector
of Saxony, Assessor in the Upper Consistory of Dresden,
Ecclesiastical Councillor, Incomparable Theologian,[1]

*and indeed, besides the favour of having adopted him*, his Patron
and greatest Maecenas, **THE AUTHOR** has wished to commend
the account of his studies.

---

[1] The Leipzig theologian and university professor of Hebrew Martin Geier (1614–80) was
one of Leibniz's three godparents. Cf. Antognazza (2009: 27) and DBE 3: 602.

[*168*]  Synopsis Dissertationis
De
ARTE COMBINATORIA.

Sedes Doctrinae istius Arithmetica. Hujus origo. Complexiones autem sunt Arithmeticae purae, situs figuratae. *Definitiones* novorum terminorum. Quid aliis debeamus. Problema I. Dato numero et exponente Complexiones et in specie Combinationes invenire. Probl. II. Dato numero complexiones simpliciter invenire. Horum usus (1) in divisionis inveniendis speciebus: v. g. Mandati, Elementorum, Numeri, Registrorum Organi Musici, Modorum Syllogismi categorici, qui in universum sunt 512 juxta Hospinianum, utiles 88 juxta nos. Novi Modi figurarum ex Hospiniano: Barbari, Celaro, Cesaro, Camestros, et nostri Figurae IVtae Galenicae: Fresismo, Ditabis, Celanto, Colanto. Sturmii modi novi ex terminis infinitis, Daropti. Demonstratio Conversionum. De Complicationibus Figurarum in Geometria, congruis, hiantibus, texturis. Ars casus formandi in Jurisprudentia. Theologia autem quasi species est Jurisprudentiae, de Jure nempe Publico in Republica DEI super homines. (2) In inveniendis datarum specierum generibus subalternis, de modo probandi sufficientiam datae divisionis. (3) Usus in inveniendis propositionibus et argumentis. De arte Combinatoria Lullii, Athanasii Kircheri, nostra, de qua sequentia: Duae sunt copulae in propositionibus: *Revera*, et *Non*, seu + et −. De formandis praedicamentis artis Com2natoriae. Invenire: Dato definito vel termino; definitiones, vel terminos aequipollentes: Dato subjecto praedicata in propositione UA, item PA, item N. Numerum Classium, Numerum Terminorum in Classibus: Dato capite complexiones: Dato praedicato subjecta in propositione UA, PA, et N. Datis duobus terminis in propositione necessaria UA et UN argumenta seu medios terminos invenire. De Locis Topicis, seu modo efficiendi et probandi propositiones contingentes. Specimen mirabile Praedicamentorum artis Com2natoriae ex Geometria. Porisma de Scriptura Universali cuicunque legenti cujuscunque linguae perito intelligibili. Dni de Breissac specimen artis Com2natoriae seu meditandi in re bellica, cujus beneficio omnia consideratione digna Imperatori in mentem veniant. De Usu rotarum concentricarum Chartacearum in arte hac. Serae hac arte constructae sine

# Synopsis

Arithmetic the Place of this Doctrine. Its origin. Complexions belong to pure, dispositions to figurate Arithmetic. *Definitions* of new terms. What I owe to others. *Problem I*, given a number and exponent, to find the Complexions, and in particular the Combinations. *Problem II*, given a number, to find the overall Complexions. *Their applications.* (1) To finding species of divisions of, for example, a Mandate, Elements, Numbers, Registers of an Organ, and Moods of categorical Syllogism, of which there are altogether 512 according to Wirth, 88 useful according to me. New Moods of the figures according to Wirth: *Barbari, Celaro, Cesaro, Camestros*; and of the Fourth Galenic Figure, according to me: *Fresismo, Ditabis, Celanto, Colanto.* The new moods of Sturm using infinite terms. *Daropti.* The demonstration of Conversions. On the Complications of Geometrical Figures, congruent, hiant, and textured. The Art of forming cases in Jurisprudence. Theology, indeed, as a kind of species of Jurisprudence, that is, of Civil Law in the Republic of God over mankind. (2) To finding subaltern genera of given species, and of a method of proving the sufficiency of a given division. (3) The application to finding propositions and arguments. On the Combinatorial art of Lull, Athanasius Kircher, and myself, on which I make the following observations: there are two copulae in any proposition, *Really* and *Not*, i.e. + and –. On the formation of the predicaments of Com2natorial Art. Given something defined or a term, to find definitions or equivalent terms. Given a subject, to find predicates in propositions of the forms UA, PA, and N. To find the Number of Classes, and the Number of Terms in the Classes. Given a head, to find the complexions. Given a predicate, to find subjects in Propositions of the forms UA, PA, and N. Given two terms in a necessary proposition of the forms UA and UN, to find arguments, or middle terms. On Topical Places, or how to make and prove contingent propositions. A notable example of the Predicaments of com2natorial art taken from Geometry. A corollary concerning a Universal Script, intelligible to any reader who has learned any language. An example of Seigneur de Breissac's combinatorial art, or art of thinking about matters of war, for the improvement of which everything worthy of consideration by a commander may be brought to mind. On the use of concentric paper rings in this art. Locks constructed by the use of this art that may be opened without

clavibus aperiendae, Mahl–Schlosser, Mixturae Colorum. Probl. III. Dato numero Classium et rerum in singulis, complexiones classium invenire. Divisionem in divisionem ducere, de vulgari Conscientiae divisione. Numerus sectarum de summo Bono e Varrone apud Augustinum. Ejus Examen. In dato gradu Consanguinitatis numerus (1.) cognationum juxta *l. 1 et 3 D. de Grad. et Aff. (2)* personarum juxta *l. 10 D. eod.* [169] singulari artificio inventus. Probl. IV. Dato numero rerum variationes ordinis invenire. Uti hospitum in mensa 6 Drexelio, 7 Harsdörffero, 12 Henischio. Versus Protei, v. g. Bauhusii, Lansii, Ebelii, Rìccioli, Harsdörfferi: Variationes literarum Alphabeti, comparatarum Atomis; Tesserae Grammaticae. Probl. V. Dato numero rerum variationem vicinitatis invenire. Locus honoratissimus in rotundo. Circulus Syllogisticus. Probl. VI. Dato numero rerum variandarum, quarum aliqua vel aliquae repetuntur, variationem ordinis invenire. Hexametrorum species 76, Hexametri 26, quorum sequens antecedentem litera excedit, Publilii Porphyrii Optatiani: quis ille. Diphtongi *ae* Scriptura. Probl. VII. Reperire dato capite variationes. Probl. VIII. Variationes alteri dato capiti communes. Probl. IX. Capita variationes communes habentia. Probl. X. Capita variationum utilium et inutilium. Probl. XI. Variationes inutiles. Probl. XII. Utiles. Optatiani Proteus versus, J. C. Scaligeri (Virgilii Casualis), Bauhusii (Ovidii Casualis), Kleppisii (praxis computandi Variationes inutiles et utiles), Caroli a *Goldstein*, Reimeri. CL. Daumii 4, quorum ultimi duo plusquam Protei. Additamentum: Demonstratio Existentiae DEI.

keys, or *Mahl-Schlösser*.[2] The mixing of colours. *Problem III*, given a number of classes, and a number of things in each class, to find the complexions of the classes. To multiply one division by another. On a common division of Conscience. The number of schools of thought regarding the Highest Good according to Augustine's presentation of Varro. An examination of it. In a given degree of consanguinity, the number (1) of cognations according to Gaius, *Institutions of Civil Law*, Books 1 and 3, and (2) of persons according to Paulus, *Digest*, Book 10, *Of Degrees and* Affinities, found by means of a peculiar artifice. *Problem IV*, given a number of things, to find the variations of order, such as of guests at table, 6 in Drexel, 7 in Harsdörffer, and 12 in Henisch. Protean Verses, for example, of Bauhuysen, Lansen, Ebel, Riccioli, and Harsdörffer. The Variations of the letters of the Alphabet, which are compared to Atoms; Grammatical Blocks. *Problem V*, given a number of things, to find the variation of a neighborhood. The place of honour at a round table. The Syllogistic Circle. *Problem VI*, given a number of things to be varied, of which one or some are repeated, to find the variation of order. 76 species of Hexameters. 26 Hexameters, in which each exceeds the one before by a letter, composed by Publilius Porphyrius Optatianus. Who he was. The Writing of the Diphthong *ae*. *Problem VII*, given a head, to find the variations. *Problem VIII*, to find the Variations in common with another given head. *Problem IX*, to find heads that have common variations. *Problem X*, to find the heads of useful and useless variations. *Problem XI*, to find the useless variations. *Problem XII*, to find the useful variations. A Protean verse of Optatian, of J.C. Scaliger, one by chance of Virgil, of Bauhuysen, one by chance of Ovid,[3] of Kleppis (an exercise in calculating useless and useful variations), of Carolus à *Goldstein*, of Reimer, and 4 verses of the celebrated Daum, of which the last two are more than Protean. In addition, a Proof of the Existence of GOD.[4]

---

[2] *Mahl-Schlösser* is the German word for 'combination locks', singular *Mahl-Schloß* or *Mahlschloss*.

[3] The expression *Casualis*, here translated as "by chance", refers to the fact that the further examples from Virgil and Ovid were randomly chosen by Leibniz himself.

[4] All references to names and concepts occurring in the *Synopsis* may be found in editorial footnotes to the relevant passages in the main text.

## DEMONSTRATIO
# EXISTENTIAE DEI.

### *Praecognita*:

1. Definitio 1: *Deus* est Substantia incorporea infinitae virtutis.
2. def. 2. *Substantiam* autem voco, quicquid movet aut movetur.
3. def. 3. *Virtus infinita est Potentia principalis movendi infinitum*. Virtus enim idem est quod potentia principalis, hinc dicimus Causas secundas operari in *virtute* primae.
4. Postulatum. *Liceat quotcunque res simul sumere, et tanquam unum totum supponere*. Si quis praefractus hoc neget, ostendo. Conceptus partium est, ut sint Entia plura, de quibus omnibus si quid intelligi potest, quoniam semper omnes nominare vel incommodum vel impossibile est, excogitatur unum nomen, quod in ratiocinationem pro omnibus partibus adhibitum compendii sermonis causa, appellatur *Totum*. Cumque datis quotcunque rebus, etiam infinitis, intelligi possit, quod de omnibus verum est; quia omnes particulatim enumerare infinito demum tempore *possibile* est, licebit unum nomen in rationes ponere loco omnium: quod ipsum erit *Totum*.
5. Axioma 1. Si quid movetur, datur aliud movens.
6. Ax. 2. Omne corpus movens movetur.
7. Ax. 3. Motis omnibus partibus movetur totum.
8. Ax. 4. Cujuscunque corporis infinitae sunt partes, seu ut vulgo loquuntur, Continuum est divisibile in infinitum.
9. Observatio. Aliquod corpus movetur.

[*170*]                           Ἔκθεσις

(1) Corpus A movetur *per praecog.* 9. (2) E. datur aliud movens *per 5*. (3) et vel incorporeum, (4) quod infinitae virtutis est (*per 3*. (5) quia A ab eo motum habet infinitas partes *per 8)* (6) et Substantia *per 2*. (7) E. DEUS *per 1*, q. e. d.; (8) vel Corpus, (9) quod dicamus B, (10) id ipsumet movetur *per 6* (11) et recurret quod de corpore A demonstravimus,

# A PROOF OF THE EXISTENCE OF GOD

## *Preliminary Knowledge*

1. Definition 1. *God* is an incorporeal Substance of infinite virtue.
2. Definition 2. I call a *substance*, indeed, anything that moves or is moved.
3. Definition 3. *Infinite Virtue is the sovereign Power of moving the infinite.* For virtue is the same as sovereign power. Hence we say that secondary Causes operate by *virtue* of the prime Cause.
4. Postulate. *We may consider any number of things at the same time, and suppose them to be as one whole.* In case any refractory person should deny this, I prove it as follows. The concept of *parts* is that they are a plurality of Beings for which, if we can understand something concerning all of them, we invent a single name, called the *Whole*, to be used for the sake of brevity in reasoning to represent all the parts, because it is inconvenient or impossible to name them all. And given a certain number of things, even an infinite number, whenever we can understand what is true of all of them, then, because it is *possible* to enumerate all of them individually only in an infinite time, it will be permissible in argument to set one name to represent them all; and that name will be the *Whole*.
5. Axiom 1. If something is moved, there must be something else moving it.
6. Axiom 2. Every moving body is moved.
7. Axiom 3. When all parts are moved, a whole is moved.
8. Axiom 4. The parts of any body are infinite in number, or as is commonly said, a Continuum is infinitely divisible.
9. Empirical observation. Some body is moved.

## Proof

1) *By 9,* a body A is moved. 2) Therefore, *by 5,* there exists something moving it. 3) This is either something incorporeal, 4) which, *by 3,* is of infinite virtue, 5) because, *by 8,* the body A moved by it has infinitely many parts, 6) and *by 2,* is a Substance. 7) Therefore, *by 1,* GOD exists, Q.E.D.[5] 8) Or it is a Body, 9) which we may call B. 10) And, *by 6,* it is itself moved. 11) And there recurs what I have demonstrated of body A, so that there will either exist an incorporeal mover,

---

[5] Initialism for *Quod erat demonstrandum,* i.e. 'What was to be demonstrated'.

(12) atque ita vel aliquando dabitur incorporeum movens, nempe ut in A ostendimus *ab ἐκθ. 1, ad 7* DEUS, q. e. d.; (13) vel in omne infinitum existent corpora continue se moventia, (14) ea omnia simul, velut unum totum liceat appellare C *per 4.* (I5) Cumque hujus omnes partes moveantur *per ἐκθ. 13,* (I6.) movebitur ipsum *per 6.* (I7) ab alio *per 5* (I8.) incorporeo (quia omnia corpora in infinitum retro, jam comprehendimus in C *per ἐκθ. 14,* nos autem requirimus aliud a C *per ἐκθ. 17*) (I9) infinitae virtutis *(per 3,* quia quod ab eo movetur, nempe C, est infinitum *per ἐκθ. 13 + 14*) (20) Substantia *per 2.* (21) Ergo DEO *per 1.* Datur igitur DEUS. Q. E. D.

## [1] CUM DEO!

1. Metaphysica, ut altissime ordiar, agit tum de Ente, tum de Entis affectionibus: ut autem corporis naturalis affectiones non sunt corpora, ita Entis affectiones non sunt Entia.

2. Est autem Entis affectio (seu Modus), alia absoluta quae dicitur *Qualitas,* alia respectiva, eaque vel rei ad partem suam, si habet, *Quantitas*; vel rei ad aliam rem *Relatio,* etsi accuratius loquendo, supponendo partem quasi a toto diversam etiam quantitas rei ad partem relatio est.

3. Manifestum igitur neque Qualitatem neque Quantitatem neque Relationem Entia esse: Earum vero tractationem in actu signato ad Metaphysicam pertinere.

4. Porro omnis Relatio aut est *Unio* aut *Convenientia.* In unione autem Res inter quas haec relatio est dicuntur *partes,* sumtae cum unione, *Totum.*

12) namely, as I demonstrated *by the Proof* concerning A *in steps 1) to 7)*, GOD,
Q. E. D. 13) Or there exist to all infinity bodies continually moving one another.
14) *By 4*, it is permissible to speak of them together as C, as if they are a single
whole. 15) And since, *by 13) of the Proof*, all its parts are moved, 16) it will, *by 6*,
itself be moved, 17) *by 5*, by something else, 18) incorporeal, because *by step 14)
of the Proof* we are now considering as included in C all the bodies back to
infinity, and *by step 17) of the Proof* something other than C is required, 19) *by 3*,
of infinite virtue, because what is moved by it, namely C, is infinite *by steps 13)
and 14) of the Proof*, 20) *by 2*, a Substance, 21) and therefore, *by 1*, GOD.

Therefore God exists, Q. E. D.

## WITH GOD'S HELP!

1. Metaphysics, to begin at the top, is concerned first with Being, and then
   with the affections of Being; and just as the affections of a natural body
   are not themselves bodies, so the affections of a Being are not themselves
   Beings.[6]

2. There is one absolute affection (or Mode) of Being which is called
   *Quality*; and another, relative affection, which is called *Quantity* if it
   holds between a thing and one of its parts, or *Relation* if it holds between
   a thing and some other thing; although, to be more precise, if we suppose
   a part to be somehow distinct from the whole, even quantity is a relation
   between the thing and the part.

3. It is therefore clear that neither Quality, nor Quantity, nor Relation are
   Beings; but that treating of them explicitly[7] belongs to Metaphysics.

4. Furthermore, every relation is either a *Union* or an *Agreement*. In a
   union, indeed, the Things between which there is this relation are called

---

[6] In Leibniz's days, the discipline of metaphysics often combined a discussion of 'being' and
its 'affections' (*metaphysica universalis*) with a discussion of God, angels, and other forms of
spiritual being, such as heavenly intelligences and the human mind (*metaphysica specialis*). The
distinction was ultimately based on the combination of topics occurring in Aristotle's *Metaphysics*, but it was also criticized on account of the fact that the subjects of the *metaphysica
specialis* were thought not to fit the science of *ens in genere*, or 'being in general', in terms of
which the subject of metaphysics was defined. Cf. *Chauvin*: 402–3. Leibniz also disregards the
*metaphysica specialis* here and mentions only the kind of things treated in the *metaphysica
universalis*, or *ontologia* ('ontology'), such as the Aristotelian categories (quantity, quality,
relation, etc.), ultimately in order to present his own science of complexions as a science of
numbers, wholes, and parts, and thus as a derivative of metaphysics.

[7] The phrase *in actu signato* is a technical expression that refers to the deliberate character of
a designated act.

Hoc contingit quoties plura simul tanquam *Unum* supponimus. *Unum* autem esse intelligitur quicquid uno actu intellectus, s. simul, cogitamus, *v. g.* quemadmodum numerum aliquem quantumlibet magnum, saepe *Caeca* quadam *cogitatione* simul apprehendimus, cyphras nempe in charta legendo cui explicate intuendo ne Mathusalae quidem aetas suffectura sit.

5. Abstractum autem ab uno est *Unitas,* ipsumque totum abstractum ex unitatibus, seu totalitas dicitur *Numerus. Quantitas* igitur est Numerus partium. Hinc manifestum in reipsa Quantitatem et Numerum coincidere. Illam tamen interdum quasi extrinsece, relatione seu Ratione ad aliud, in subsidium nempe quamdiu numerus partium cognitus non est, exponi.

[*171*] 6. Et haec origo est ingeniosae Analyticae Speciosae, quam excoluit inprimis *Cartesius,* postea in praecepta collegere *Franc. Schottenius,* et *Erasmius Bartholinus,* hic *elementis Matheseos universalis,* ut vocat. [2] Est igitur *Analysis* doctrina de Rationibus et Proportionibus, seu Quantitate non Exposita; *Arithmetica* de Quantitate exposita, seu Numeris: falso autem Scholastici credidere Numerum ex sola divisione continui oriri nec ad incorporea applicari posse. Est enim numerus quasi figura quaedam

*parts*, and taken together with the union, a *Whole*. This occurs as often as we suppose a plurality of things at the same time to be *One*. *One*, instead, means something that we think of in a single intellectual act, that is, at the same time; as, for example, when by reading numerals on a piece of paper we grasp in a kind of *blind thought*[8] some very large number that all the years of Methuselah would not suffice to count up explicitly.

5. A *Unity*, indeed, is something abstracted from one; and the whole itself, or totality, when abstracted from all its unities, is called a *Number*. Quantity, therefore, is the number of the parts. Hence, it is clear that in the thing itself Quantity and Number coincide. But when the number of parts is not known, Quantity is sometimes exhibited as it were extrinsically, in relation or in *proportion* to another, that is, with the aid of an auxiliary.

6. And this is the origin of the Specious Analytics,[9] which *Descartes* was the first to work out, and afterwards *Frans van Schooten* and *Erasmus Bartholin* reduced to principles, the latter in what he calls *the elements of a universal Mathesis*.[10] *Analysis* is therefore a theory concerned with Ratios and Proportions, that is, with Implicit Quantity. It is *Arithmetic* that is concerned with Explicit Quantity, or Numbers. Moreover, the Scholastics were wrong to believe that number arises solely from division

---

[8] The notion of 'blind thought' (*cogitatio caeca*) would acquire a certain significance in Leibniz's mature writings, rendered in French as *pensée sourde* i.e. 'deaf thought'. A blind thought occurs whenever one performs mental operations using a sign (a word, e.g. or a symbol) without at the same time taking account of the idea that the sign represents. As Donald Rutherford has put it, with "blind" or "symbolic" thought, "the use and exchange of signs takes the place of a conscious apprehension of ideas". Cf. Rutherford (1995: 244).

[9] *Analytica Speciosa* refers to the first elementary form of Algebra, the word *species* in this context being a name for any letter of the alphabet employed in place of numbers or geometrical magnitudes in an algebraic calculation. The first systematically to employ letters to perform algebraic calculations was the French mathematician François Viète (1540–1603), who also distinguished *logistica speciosa*, where letters were used, from *logistica numerosa*, where calculations were directly made with numbers. Cf. Viète (1591).

[10] Descartes's *Geometrie*, which provided the foundations of analytic geometry, was first published in 1637 as one of the scientific treatises following the *Discourse on Method*. The idea of a universal mathesis is discussed in the preliminary *Discourse* itself, AT VI: 19–20. The term *mathesis universalis* primarily occurs in the *Regulae ad directionem ingenii*, AT X: 377–9, where Descartes introduces the idea of "a general science, which explains all the points that can be raised concerning order and measure irrespective of the subject-matter ( . . . )." Cf. AT X: 378; translation from CSM I: 19. Frans van Schooten the Younger (1615–60), a Leiden professor of mathematics, and tutor to Christiaan Huygens, published a Latin edition of Descartes's *Geometrie* with an elaborate commentary of his own: Schooten (1649). Cf. DDP II: 895–7. The second, 1659–61, edition in two volumes became "the standard mathematical work of the period." Cf. J. E. Hofmann, 'Schooten, Frans van', in DSB XII: 205–7; quotation from p. 206. Rasmus Bartholin (1625–98), Danish scientist and professor of mathematics and medicine at Copenhagen, studied mathematics with Frans van Schooten at Leiden and contributed to Van Schooten (1651). Cf. A. Rupert Hall, 'Bartholin, Erasmus', in DSB I: 481–2.

incorporea orta ex Unione Entium quorumcunque, v. g. DEI, Angeli, Hominis, Motus, qui simul sunt Quatuor.

7. Cum igitur Numerus sit quiddam Universalissimum, merito ad Metaphysicam pertinet. Si Metaphysicam accipias pro doctrina eorum quae omni entium generi sunt communia. Mathesis enim (ut nunc nomen illud accipitur) accurate loquendo non est una disciplina, sed ex variis disciplinis decerptae particulae quantitatem subjecti in unaquaque tractantes, quae in unum propter cognationem merito coaluerunt. Nam uti Arithmetica atque Analysis agunt de Quantitate Entium; ita Geometria de Quantitate corporum, aut spatii quod corporibus coextensum est. Politicam vero disciplinarum in professiones divisionem, quae commoditatem docendi potius, quam ordinem naturae secuta est, absit ut convellamus.

8. Caeterum Totum ipsum (et ita Numerus vel Totalitas) discerpi in partes tanquam minora tota potest, id fundamentum est *Complexionum*, dummodo intelligas dari in ipsis diversis minoribus totis partes communes, v. g. Totum sit A. B. C., erunt minora Tota, partes illius, AB. BC. AC: Et ipsa minimarum partium, seu pro minimis suppositarum (nempe Unitatum) dispositio inter se et cum toto, quae appellatur situs, potest variari.

9. Ita oriuntur duo *Variationum* genera: *Complexionis* et *Situs*. Et tum *Complexio* tum *Situs* ad Metaphysicam pertinet, nempe ad doctrinam de Toto et partibus, si in se spectentur: Si vero intueamur *Variabilitatem*, id est Quantitatem variationis, ad numeros et Arithmeticam deveniendum est. Complexionis autem doctrinam magis ad Arithmeticam puram, situs ad figuratam pertinere crediderim, sic enim unitates lineam efficere intelliguntur. Quanquam hic obiter notare volo, unitates vel per modum lineae rectae vel circuli aut alterius lineae li[3]nearumve in se redeuntium aut figuram claudentium disponi posse, priori modo in situ absoluto seu partium cum toto, *Ordine*; posteriori in situ relato seu partium ad partes, *Vicinitate*, quae quomodo differant infra dicemus def. 4 et 5. Haec prooemii loco sufficiant, ut qua in disciplina materiae hujus sedes sit, fiat manifestum.

of a continuum, and that it could not be applied to incorporeal things. For a number is, as it were, a kind of incorporeal figure arising from the Union of any kind of Beings, for example, of GOD, Angel, Man, and Motion, which considered together are Four.

7. Since, therefore, Number is something of the greatest universality, it rightly belongs to Metaphysics, if Metaphysics is understood as a theory of what is common to all kinds of beings. For a Mathesis (as the name is currently understood) is, to be precise, not one discipline but particulars gathered from a variety of disciplines, which deal with the quantity of the subject belonging to each of them and which consequently coalesced into one because of their cognate nature. And just as Arithmetic and Analysis treat of Quantity of Beings, so Geometry treats of Quantity of bodies, or of the space that is coextensive with bodies. (Heaven forbid that I should break up the social division of disciplines into professions, which follow rather from the convenience of teaching them than from the order of nature.)

8. Finally, a Whole itself (and thus Number, or Totality) can be divided into parts like lesser wholes; and that is the principle of *Complexions*, granted that these diverse lesser wholes have some parts in common. For example, let a Whole consist of A, B, and C. The lesser Wholes are its parts AB, BC, and AC; and the disposition of the smallest parts, or what are taken as the smallest parts (that is, Unities), can be varied both in relation to each other and to the whole, and this disposition is called a *situs*.

9. Thus, there arise two kinds of *Variation: Complexion* and *Situs*. And a *Complexion* and a *Situs*, if considered in themselves, belong equally to Metaphysics, namely, to the theory of a whole and its parts. If we look at *Variability*, that is, at the quantity of variation, we must turn to numbers and Arithmetic.[11] But I would maintain that the theory of Complexion belongs rather to pure Arithmetic, while the doctrine of situs belongs to figured Arithmetic, for in the latter the unities are conceived of as constituting a line. However, I want to note here in passing that unities can be arranged either in the mode of a straight line, or in the mode of a circle or some other closed line, or lines which outline a figure. In the first mode, they are in an absolute situs, or *Order* of parts within a whole; in the second mode, in a relative situs, or *Neighbourhood* of parts to parts. As to how these differ, I shall speak of this below in Definitions 4 and 5. This suffices by way of introduction to make clear the discipline in which the subject matter resides.

---

[11] On the notion of *variation* in Leibniz, see Knobloch (2012).

# DEFINITIONES

1. *Variatio* h. l. est mutatio relationis. Mutatio enim alia substantiae est alia quantitatis alia qualitatis; alia nihil in re mutat, sed solum respectum, situm, conjunctionem cum alio aliquo.

[172] 2. *Variabilitas* est ipsa quantitas omnium Variationum. Termini enim potentiarum in abstracto sumti quantitatem earum denotant, ita enim in Mechanicis frequenter Ioquuntur, potentias machinarum duarum duplas esse invicem.

3. *Situs* est localitas partium.

4. Situs est vel absolutus vel relatus: ille partium cum toto, hic partium ad partes. In illo spectatur numerus locorum et distantia ab initio et fine, in hoc neque initium neque finis intelligitur, sed spectatur tantum distantia partis a data parte. Hinc ille exprimitur linea aut lineis figuram non claudentibus neque in se redeuntibus, et optime linea recta; hic linea aut lineis figuram claudentibus, et optime circulo. In illo prioritatis et posterioritatis ratio habetur maxima, in hoc nulla. Illum igitur optime *Ordinem* dixeris;

5. Hunc *vicinitatem*, illum dispositionem, hunc compositionem. Igitur ratione ordinis differunt situs sequentes: abcd. bcda. cdab. dabc. At in Vicinitate nulla variatio sed unus situs esse intelligitur, hic nempe:

b      Unde festivissimus Taubmannus, cum Decanus Facultatis philo-

a   c    sophicae esset, dicitur Witebergae in publico programmate

d      seriem candidatorum Magisterii circulari dispositione complexus, ne avidi lectores intelligerent, quis suillum locum teneret. [4]

# DEFINITIONS

1. *Variation* here means change of relation. For change may involve
   substance, quantity, or quality; still another kind changes nothing in
   the thing, but only in its respect, its situation, or its conjunction with
   something else.
2. *Variability* is nothing other than the quantity of all the Variations. For
   the terms of powers, considered in the abstract, denote their quantity;
   thus in Mechanics, for instance, it is frequently said that the powers of
   two machines are each other's double.[12]
3. A *Situs* is the location of parts.
4. A Situs is either absolute or relative: the former is that of parts with respect
   to a whole, the latter that of parts with respect to parts. In the former we
   observe the number of places and their distance from the beginning and
   the end, while in the latter neither beginning nor end is discerned,
   and we observe only the distance of a part from a given part. Hence,
   the former is expressed by a line, or lines, not enclosing a figure or
   forming a closed curve, and best of all by a straight line; the latter by a
   line, or lines, enclosing a figure, and best of all by a circle. In the former
   the order of priority and posteriority are held to be of the greatest
   importance, in the latter of no importance. Therefore, the former is
   best called an *Order*;
5. The latter a *neighbourhood*, the former a disposition, the latter a
   composition. The following examples of situs therefore differ by
   reason of order: abcd, bcda, cdab, dabc. But in a Neighbourhood
   there is no variation, and only a single situs discerned, namely this:

   b      Hence, it is said that the excellently witty Taubmann, when he
   a   c  was Dean of the Faculty of Philosophy, used in the public
   d      programme    at    Wittenberg    to    arrange    the    names
          of candidates for the Master's Degree in a circle, in case too
   eager readers should make an inference about who occupied the least
   honourable place.[13]

---

[12] Interpreting variability as the capability of undergoing variations, Leibniz here justifies
that the abstract term 'variability' can be employed to denote quantities of variation.
[13] Friedrich Taubmann (1565–1613), a Latin poet, philologist, and professor at Wittemberg,
was known for his witticisms later collected as *Taubmanniana*, first published in 1702 and
reprinted many times after.

6. Variabilitatem ordinis intelligemus fere, quando ponemus *Variationes* καθ'ἐξοχήν *v. g. Res 4 possunt transponi modis 24.*

7. Variabilitatem complexionis dicimus *Complexiones.* v. g. *Res 4 modis diversis 15 invicem conjungi possunt.*

8. Numerum rerum variandarum dicemus simpliciter, *Numerum*, v. g. *4 in casu proposito.*

9. *Complexio,* est Unio minoris Totius in majori, uti in prooemio declaravimus.

10. Ut autem certa Complexio determinetur, majus totum dividendum est in partes aequales suppositas ut minimas (id est quae nunc quidem non ulterius dividantur), ex quibus componitur et quarum variatione variatur Complexio seu Totum minus; quia igitur totum ipsum minus, majus minusve est, prout plures partes una vice ingrediuntur; numerum simul ac semel conjungendarum partium, seu unitatum, dicemus *Exponentem*, exemplo progressionis geometricae, v.g. sit totum ABCD. Si Tota minora constare debent ex 2 partibus, v. g. AB, AC, AD, BC, BD, CD, exponens erit 2 sin ex tribus, v. g. ABC, ABD, ACD, BCD, exponens erit 3.

11. Dato Exponente Complexiones ita scribemus: si exponens est 2, *Com2-nationem* (combinationem) si 3, *Con3nationem* (conternationem) si 4, *Con4nationem*, etc. [*173*]

12. *Complexiones simpliciter* sunt omnes complexiones omnium Exponentium computatae, v. g. 15 (de 4 Numero) quae componuntur ex 4 (Unione), 6 (com2natione), 4 (con3natione) 1 (Con4natione).

13. *Variatio utilis* (*inutilis*) est quae propter materiam subjectam locum habere non potest; v. g. 4 Elementa com2nari possunt 6 Mahl sed duae com2nationes sunt inutiles, nempe quibus contrariae Ignis, aqua; aer, terra com2nantur.

6. I shall understand Variability of Order to be, roughly, when we take *Variations proper*.[14] For example, *4 Things can be transposed in 24 ways*.

7. I call Variability of complexions simply, Complexions. For example, *4 Things can be joined together in 15 diverse ways*.

8. The number of things to be varied I shall call simply the *Number*. For example, *4 in the case just mentioned*.

9. A *Complexion* is a Union of a lesser Whole within a greater Whole, as I stated in the introduction.

10. In order to determine a certain Complexion, the greater Whole must be divided into equal parts, supposed to be the smallest possible (that is, they cannot be further divided), out of which the Complexion or lesser Whole is composed, and from which it is varied by means of variation. Therefore, because it is a lesser Whole, it is greater or less according to how many parts enter into it at a time. The number of parts or unities to be joined together and at one time I shall call the *Exponent*, after the model of a geometric progression. For example, let the whole be ABCD. If the lesser Wholes must be composed of 2 parts, for instance, AB, AC, AD, BC, BD, CD, the exponent will be 2. But if the lesser Wholes must be composed of three parts, for instance, ABC, ABD, ACD, BCD, the exponent will be 3.

11. Given an Exponent, I shall describe the Complexions as follows: if the exponent is 2, as a *Com2nation* (combination); if 3, as a *Con3nation* (conternation); if 4, as a *Con4nation*, etc.[15]

12. *Overall Complexions* are all the complexions calculated with respect to all Exponents. For example, out of 4 things in Number there are 15, composed of 4 by Union, 6 by com2nation, 4 by con3nation, and 1 by Con4nation.

13. A *useful* (*useless*) *Variation* is one that [can or] cannot occur because of the nature of its subject matter. For example, the 4 Elements can be com2ned in 6 ways,[16] but 2 of these com2nations are useless, namely, those in which the contraries Fire and Water, and Air and Earth, are com2ned.[17]

---

[14] The Greek expression means 'par excellence'. 'Variations *par excellence*' are simple variations, what we now call 'permutations', or variations with respect to order, a type Leibniz here distinguishes from the variability of complexions, which is dealt with in Definition 7. On the notion of variation in Leibniz, see Knobloch (2012).

[15] The same way of writing was employed earlier by Marin Mersenne (see Knobloch 1973: 23, 1974: 412).

[16] Leibniz here uses the German term *Mahl* ('times', 'ways') in order to refer to the number of times or the plurality of ways in which the said comb2nation can occur. The term reoccurs at numerous other places in the text.

[17] Leibniz here refers to the ancient theory of the 'four elements': fire, air, water, and earth, from which everything in the natural world was thought to originate or out of which it was thought to be composed. The theory has its origin in a range of presocratic sources, and was systematically discussed by Empedocles, as well as by Aristotle (see below, note 38), who added a fifth element to the previous four. The theory of the four elements (and the fifth, the *quintessentia*) was further developed in the Middle Ages and the Renaissance period.

14. *Classis rerum* est Totum minus, constans ex rebus convenientibus in certo tertio, tanquam partibus; sic tamen ut reliquae classes contineant res contradistinctas; v. g. infra [5] probl. 3 ubi de classibus opinionum circa summum Bonum ex B. Augustino agemus.

15. *Caput Variationis* est positio certarum partium; *Forma variationis*, omnium, quae in pluribus variationibus obtinet, v. infr. probl. 7.

16. *Variationes communes* sunt in quibus plura capita concurrunt, v. infr. probl. 8 et 9.

17. *Res homogenea* est quae est aeque dato loco ponibilis salvo capite. *Monadica* autem quae non habet homogeneam, v. probl. 7.

18. *Caput multiplicabile* dicitur, cujus partes possunt variari.

19. *Res repetita* est quae in eadem variatione saepius ponitur, v. probl. 6.

20. Signo + designamus additionem, − subtractionem, ∩ multiplicationem, ∪ divisionem, f. facit, seu summam, = aequalitatem. In prioribus duobus et ultimo convenimus cum Cartesio, Algebraistis, aliisque: Alia signa habet Isaacus Barrowius in sua editione Euclidis, Cantabrig. 8vo, anno 1655.

## Problemata.

Tria sunt quae spectari debent: *Problemata, Theoremata, usus*; in singulis problematis usum adjecimus; sicubi operae pretium videbatur, et theoremata. Problematum autem quibusdam rationem solutionis addidimus. Ex iis partem posteriorem primi, secundum et quartum aliis debemus, reliqua ipsi eruimus. Quis illa primus detexerit ignoramus. Schwenterus Delic. l. 1. sect. 1. prop. 32 apud Hieronymum Cardanum, Johannem Buteonem, et

14. A *Class of things* is a lesser Whole made up of things gathered together in some other thing, as if they are parts, and in such a way that the remaining classes contain contradistinct things. For example, in Problem 3 below, in which I shall deal with the classes of opinions concerning the Highest Good delivered by St Augustine.

15. A *Head of a Variation* is a fixing of certain parts that holds in several variations; the *Form of a Variation* is a fixing of all the parts that holds in several variations. See below in Problem 7.[18]

16. *Common Variations* are those in which several heads share. See below in Problems 8 and 9.

17. A *homogeneous thing* is something that can be equally put down in a given place, with the only exception of the head. On the other hand, a *Monadic thing* is something that is not made of homogeneous things. See Problem 7.

18. A *Head* will be said to be *multipliable* if its parts can be varied.

19. A *repeated Thing* is something that appears more than once in the same variation. See Problem 6.

20. I designate addition by the sign +, subtraction by −, multiplication by ∩, division by ∪, "makes", or the result of a sum by f, and equality by =.[19] In the first two and the last I agree with Descartes, the Algebraists, and others. The other signs come from Isaac Barrow in his edition of Euclid, Cambridge 8vo, 1655.[20]

## The Problems

There are three things that have to be considered: *Problems, Theorems,* and *Applications.* To some of the problems I have appended an application and, where it seemed worthwhile, theorems. Moreover, to some of the problems I have added an explanation of the solution. Of these, I owe to others the latter part of the first, the second, and the fourth; the rest I have worked out for myself. I do not know who first discovered them. Schwenter, in *Mathematical Recreations* 1.1 Sect. 1, Prop. 32, says that they are found in Girolamo Cardano,

---

[18] The *head* of a variation is analogous to an *invariant* of the variation itself. Cf. Iommi Amunátegui (2015: 103–7).

[19] Leibniz employs the first letter of the Latin term *facit* ('makes') to designate the result of a sum (in general, of any arithmetical operation).

[20] Cf. Barrow (1655). Through his *Geometrie* of 1637, Descartes contributed much to the development of modern ways of notation in mathematics. On the *Geometrie*, see also above, note 10.

Nicolaum Tartaleam extare dicit. In Cardani tamen Practica Arithmetica quae prodiit Mediolani anno 1539 nihil reperimus. Inprimis dilucide, quicquid dudum habetur, proposuit Christoph. Clavius in Com. supra Joh. de Sacro Bosc. Sphaer. edit. Romae forma 4ta anno 1585, pag. 33 seqq. [6, *174*]

## Probl. I. DATO NUMERO ET EXPONENTE COMPLEXIONES INVENIRE.

1. I. Solutionis duo sunt modi, unus de omnibus Complexionibus, alter de Com2nationibus solum: ille quidem est generalior, hic vero pauciora requirit data, nempe numerum solum et exponentem; cum ille etiam praesupponat inventas complexiones antecedentes.

2. Generaliorem modum nos deteximus, specialis est vulgatus. Solutio illius talis est: "addantur complexiones exponentis antecedentis et dati de numero antecedenti, productum erunt complexiones quaesitae"; v. g.

Johannes Buteo, and Niccolò Tartaglia.[21] However, I can find nothing in Cardano's *Practical Arithmetic*, published in Milan, 1539.[22] But as far as we know, the first clear mention of them is in Christopher Clavius, who proposed them in his *Commentary on the* De Sphaera *of Johannes de Sacrobosco*, published at Rome in 4to, 1585, p. 33 seqq.[23]

## Problem I GIVEN A NUMBER AND EXPONENT, TO FIND THE COMPLEXIONS

1. There are two modes of solution, one concerning all the Complexions, the other concerning Com2nations only. The former is more general, but the latter requires fewer data, in fact, only the number and the exponent; while the former also presupposes that preceding complexions have been found.
2. I have developed the more general mode, the special mode being well-known. The solution of the former is as follows: 'Let the number of complexions of the preceding exponent with the preceding number be added to the number of complexions of the given exponent with the preceding number; the result will be the required number of

---

[21] In his *Deliciae Physico-Mathematicae* (1636), Daniel Schwenter names "Hieronymus Cardanus, Johannes Buteo, Nicolaus Tartalius and others" both in the Introduction and in the passage referred to by Leibniz, twice referring to the example of twelve persons changing places around a table in such a way that they "do not once sit in the same way." Cf. Schwenter (1991: 15, 66). Schwenter refers to Georg Henisch, who calculated the total number of positions as 479,001,600, which, according to Schwenter, would take 130 years to realize with 10,000 changes made every day. Cf. Henisch (1609: 399) and Schwenter (1991: 67-8). The Nuremberg scholar and poet Daniel S. Schwenter (1585-1636) was a librarian and professor at Altdorf University, where he taught Hebrew, oriental languages, as well as mathematics. Cf. DBE IX, 243; *Jocher* IV, cols 415-416; and especially Moritz Cantor, 'Schwenter: Daniel S.', in: ADB XXXIII, 413-414. Schwenter's book *Deliciae Physico-Mathematicae* forms part of a range of seventeenth-century publications on the subject of recreational mathematics and science that goes back to ancient and medieval traditions on arithmetic, secrets, tricks, and wonders. See: Heeffer (2004). The *Deliciae* were partly based on Etten (1624), for which see below, note 241, and later extended to two and subsequently three volumes by Georg Philip Harsdörffer. See below, note 158.
[22] Cf. Cardanus (1539).
[23] The reference is to the third edition of Clavius's work, the first of which had been published as Clavius (1570). Cf. Clavius (1999: 17-20), where the number and order of the elements are discussed, and where the figure may be found that Leibniz reproduced on the frontispiece of his own edition of the present treatise, in order to "show the origin of the Elements from primary Qualities in a pictorial manner." John of Sacrobosco's *Sphaera*, or *On the Sphere* was an early thirteenth-century university textbook which, even though it dealt with "only elementary spherical astronomy and geography, and hardly anything on planetary theory" would nevertheless become "one of the most widely studied astronomical books of all time [ . . . ]." Cf. North (1994: 234-5).

esto numerus datus 4, exponens datus 3, addantur de numero antecedente 3 com2nationes 3 et con3natio 1 (3 + 1 f. 4), productum 4 erit quaesitum.

3. Sed cum praerequirantur complexiones numeri antecedentis, construenda est Tabula אַ in qua linea suprema a sinistra dextrorsum continet *Numeros*, a 0 usque ad 12 utrimque inclusive, satis enim esse duximus huc usque progredi, quam facile est continuare: linea extrema sinistra a summo deorsum continet *Exponentes* a 0 ad 12, linea infima a sinistra dextrorsum continet *Complexiones simpliciter*. Reliquae inter has lineae continent Complexiones dato *numero* qui sibi in vertice directe respondet, et *exponente* qui e regione sinistra. [*175*]

4. *Ratio solutionis*, et fundamentum Tabulae patebit, si demonstraverimus, *Complexiones dati numeri et exponentis oriri ex summa complexionum de numero praecedenti exponentis et praecedentis et dati*. Sit enim numerus datus 5, exponens datus 3. Erit numerus antecedens 4, is habet con3nationes 4, per Tabulam אַ com2nationes 6. Jam numerus 5 habet omnes con3nationes quas praecedens (in toto enim et pars continetur), nempe 4 et praeterea tot quot praecedens habet com2nationes (nova enim res, qua numerus 5 excedit 4, addita singulis com2nationibus hujus, facit totidem novas con3nationes[)], nempe 6 + 4 f. 10. E. *Complexiones dati numeri* etc. Q. E. D.

complexions.' For example, let the given number be 4, and the given exponent 3. Let the 3 com2nations of the preceding number 3 be added to the 1 con3nation of the preceding number (3 +1 f. 4). The result 4 is the required number.[24]

3. But since the complexions of the preceding number are a prerequisite, one must construct Table ℵ, in which the uppermost line running from left to right contains *Numbers* from 0 to 12 inclusive.[25] (I take it to be enough to go only this far, as there is no difficulty in continuing.) The column on the extreme left contains the *Exponents* from 0 to 12. The bottom line * running from left to right contains the *overall Complexions*. The remaining lines in between these contain the Complexions with respect to the given *number* standing at the head of the column, and the given *exponent* corresponding to it in the left-hand column.

4. The *reason for the solution*, and the basis of the Table, will become clear if I show that *the Complexions of a given number and exponent arise from the sum of the number of complexions of the preceding number with the preceding exponent, and the number of complexions of the preceding number with the given exponent*. For example, let the given number be 5, and the given exponent 3. The preceding number will be 4, and it has, according to Table ℵ, 4 con3nations and 6 com2nations. The number 5 then has all the con3nations that the preceding number has (the part being contained in the whole), namely 4, and in addition as many as the preceding number has com2nations (for the new element with which the number 5 exceeds the number 4 adjoined to each of its com2nations makes up new con3nations), namely 6 + 4 f. 10. Therefore, *the Complexions of the given number are as stated*. Q.E.D.[26]

[24] If we denote with $\binom{n}{k}$ the number of $k$ combinations from a given set of $n$ elements, we may extract from Leibniz's text the general formula: $C\binom{n}{k} = C\binom{n-1}{k-1} + C\binom{n-1}{k}$.

[25] The table is based on the so-called 'arithmetical triangle'. Pascal's essay on the triangle was published posthumously (Pascal 1665), but the triangle as such was well known before this date. Leibniz here includes '0' among the exponents, even though he explicitly excludes combinations of no elements "since he gives the number of all possible combinations as $2^n - 1$" (Knobloch 1973: 26, 1974: 413).

[26] Cf. Introduction, § 6.

EXPONENTES.

Tab. N.

| Exp. | n=0 | 1 | 2 | 3 | 4 | 5 | 6 | 7 | 8 | 9 | 10 | 11 | 12 |
|---|---|---|---|---|---|---|---|---|---|---|---|---|---|
| 0 | 1 | 1 | 1 | 1 | 1 | 1 | 1 | 1 | 1 | 1 | 1 | 1 | 1 |
| 1 | 0 | 1 | 2 | 3 | 4 | 5 | 6 | 7 | 8 | 9 | 10 | 11 | 12 |
| 2 | 0 | 0 | 1 | 3 | 6 | 10 | 15 | 21 | 28 | 36 | 45 | 55 | 66 |
| 3 | 0 | 0 | 0 | 1 | 4 | 10 | 20 | 35 | 56 | 84 | 120 | 165 | 220 |
| 4 | 0 | 0 | 0 | 0 | 1 | 5 | 15 | 35 | 70 | 126 | 210 | 330 | 495 |
| 5 | 0 | 0 | 0 | 0 | 0 | 1 | 6 | 21 | 56 | 126 | 252 | 462 | 792 |
| 6 | 0 | 0 | 0 | 0 | 0 | 0 | 1 | 7 | 28 | 84 | 210 | 462 | 924 |
| 7 | 0 | 0 | 0 | 0 | 0 | 0 | 0 | 1 | 8 | 36 | 120 | 330 | 792 |
| 8 | 0 | 0 | 0 | 0 | 0 | 0 | 0 | 0 | 1 | 9 | 45 | 165 | 495 |
| 9 | 0 | 0 | 0 | 0 | 0 | 0 | 0 | 0 | 0 | 1 | 10 | 55 | 220 |
| 10 | 0 | 0 | 0 | 0 | 0 | 0 | 0 | 0 | 0 | 0 | 1 | 11 | 66 |
| 11 | 0 | 0 | 0 | 0 | 0 | 0 | 0 | 0 | 0 | 0 | 0 | 1 | 12 |
| 12 | 0 | 0 | 0 | 0 | 0 | 0 | 0 | 0 | 0 | 0 | 0 | 0 | 1 |
| * | 0 | 1. | 3. | 7. | 15. | 31. | 63. | 127. | 255. | 511. | 1023. | 2047. | 4095. |
| † | 1. | 2. | 4. | 8. | 16. | 32. | 64. | 128. | 256. | 512. | 1024. | 2048. | 4096. |

Complexiones simpliciter * (seu summa Complexionum dato exponente) addita unitate, quae coincidunt cum terminis progressionis geometricae duplae.

[7]  5. Majoris lucis causa apposuimus Tabulam ⊐ ubi lineis transversis distinximus Con3nationem de 3, et de 4, et de 5. Sic tamen ut con3nationes priores sint sequenti communes, et per consequens tota tabula sit con3nationum numeri 5 utque manifestum esset, quae con3nationes numeri sequentis ex com2nationibus antecedentis addito singulis novo hospite orirentur, linea deorsum tendente combinatione a novo hospite distinximus.

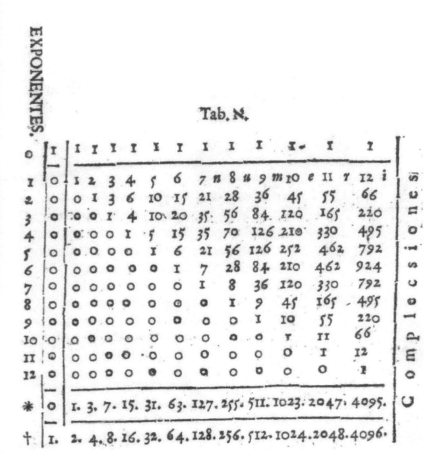

**Tab. N.**

EXPONENTES.

| | 0 | 1 | 2 | 3 | 4 | 5 | 6 | 7 | 8 | 9 | 10 | 11 | 12 |
|---|---|---|---|---|---|---|---|---|---|---|---|---|---|
| 0 | 1 | 1 | 1 | 1 | 1 | 1 | 1 | 1 | 1 | 1 | 1 | 1 | 1 |
| 1 | 0 | 1 | 2 | 3 | 4 | 5 | 6 | 7 | 8 | 9 | 10 | 11 | 12 |
| 2 | 0 | 0 | 1 | 3 | 6 | 10 | 15 | 21 | 28 | 36 | 45 | 55 | 66 |
| 3 | 0 | 0 | 0 | 1 | 4 | 10 | 20 | 35 | 56 | 84 | 120 | 165 | 220 |
| 4 | 0 | 0 | 0 | 0 | 1 | 5 | 15 | 35 | 70 | 126 | 210 | 330 | 495 |
| 5 | 0 | 0 | 0 | 0 | 0 | 1 | 6 | 21 | 56 | 126 | 252 | 462 | 792 |
| 6 | 0 | 0 | 0 | 0 | 0 | 0 | 1 | 7 | 28 | 84 | 210 | 462 | 924 |
| 7 | 0 | 0 | 0 | 0 | 0 | 0 | 0 | 1 | 8 | 36 | 120 | 330 | 792 |
| 8 | 0 | 0 | 0 | 0 | 0 | 0 | 0 | 0 | 1 | 9 | 45 | 165 | 495 |
| 9 | 0 | 0 | 0 | 0 | 0 | 0 | 0 | 0 | 0 | 1 | 10 | 55 | 220 |
| 10 | 0 | 0 | 0 | 0 | 0 | 0 | 0 | 0 | 0 | 0 | 1 | 11 | 66 |
| 11 | 0 | 0 | 0 | 0 | 0 | 0 | 0 | 0 | 0 | 0 | 0 | 1 | 12 |
| 12 | 0 | 0 | 0 | 0 | 0 | 0 | 0 | 0 | 0 | 0 | 0 | 0 | 1 |
| * | 0 | 1. | 3. | 7. | 15. | 31. | 63. | 127. | 255. | 511. | 1023. | 2047. | 4095. |
| † | 1. | 2. | 4. | 8. | 16. | 32. | 64. | 128. | 256. | 512. | 1024. | 2048. | 4096. |

(COMPLEXIONES)

'*' denotes the row with the numbers of overall complexions (i.e. the sums
of complexions with a given exponent). Increased by 1, these numbers
coincide with the terms of the geometric progression with
common ratio 2, listed in the row denoted by '†'.

5. For the sake of greater clarity, I append Table ב. In this, I mark off with
horizontal lines the number of *Con3nations* of 3, 4, and 5 things respectively. But in order that the earlier Con3nations may be shared with the
following number, and in consequence that the whole table may be of the
Con3nations of number 5, and in order that it may be clear which
Con3nations of the following number arise from which Com2nations
of the preceding number with the addition of a new element, I have
marked off the combinations from the new element with a vertical line.

[8]

Tab. ב

| N | 1 | ab | c | 3 |
|---|---|----|---|---|
| u | 2 | ab | d | 4 |
| m | 3 | ac | d | R |
| e | 4 | bc | d | e |
| r | 5 | ab | e | r |
| u | 6 | ac | e | um |
| s |   |    |   | N |
| Con | 7 | ad | e | um |
| 3 | 8 | bc | e | e |
| na | 9 | bd | e | r |
| ti | 10 | cd | e | us |
| onum |   |    |   | 5 |

6. Adjiciemus hic *Theoremata* quorum τὸ ὅτι ex ipsa Tabula א manifestum est, τὸ διότι ex Tabulae fundamento: 1. Si Exponens est major Numero, Complexio est 0. 2. Si aequalis, ea est 1. 3. Si Exponens est Numero unitate minor, complexio et Numerus sunt idem. 4. Generaliter: Exponentes duo, in quos numerus bisecari potest, seu qui sibi invicem complemento sunt ad numerum, easdem de illo numero habent complexiones. Nam cum in minimis exponentibus 1 et 2 in quos bisecatur numerus 3, id verum sit quasi casu, per Tab. א et vero caeteri ex eorum additione oriantur per solut. probl. I. si aequalibus (3 et 3) addas aequalia (superius 1 et inferius 1), producta erunt aequalia (3 + 1 f. 4 = 4) et idem eveniet in caeteris necessitate. 5. Si numerus est impar, dantur in medio duae complexiones sibi proximae aequales; sin par, id non evenit. Nam numerus impar bisecari potest in duos exponentes proximos unitate distantes; v. g. 1 + 2 f. 3. [P]ar vero non potest. Sed proximi in quos bisecari par potest sunt iidem. [Q]uia igitur in duos exponentes impar numerus bisecari potest, hinc duas habet Complexiones aequales per th. 4, quia illi unitate distant, *proximas*. 6. Complexiones crescunt usque ad exponentem numero ipsi dimidium aut duos dimidio proximos, inde iterum decrescunt. 7. Omnes numeri primi metiuntur suas complexiones *particulares* (seu dato exponente). 8. Omnes Complexiones simpliciter, sunt numeri impares.

Tab. ב

| | | | | |
|---|---|---|---|---|
| N | 1 | ab | c | 3 |
| u | 2 | ab | d | 4 |
| m | 3 | ac | d | Num |
| b | 4 | bc | d | b |
| e | 5 | ab | e | e |
| r | 6 | ac | e | r |
| of | | | | of |
| Con | 7 | ad | e | Th |
| 3 | 8 | bc | e | i |
| na | 9 | bd | e | ng |
| ti | 10 | cd | e | s |
| ons | | | | 5 |

6. I append here some *Theorems*, of which the *fact* is clear from Table א itself, and the *reason why* from the principles underlying it.[27] 1. If the Exponent is greater than the Number, the number of Complexions is 0. 2. If the Exponent is equal to the Number, the number of Complexions is 1. 3. If the Exponent is one less than the number, the number of Complexions and the Number itself are the same. 4. Generally: two Exponents that can partition the number, that is, which are complementary with respect to the number, have the same number of complexions with respect to the number.[28]

For since in the smallest exponents 1 and 2 into which the number 3 can be partitioned, it is true almost by chance, by Table א the rest will arise by addition of them as with the solution of Problem I. If to equals (3 and 3) you add equals (1 and 1), the result will be equal (3 + 1 f. 4 = 4), and the same will necessarily happen with the rest.[29] 5. If the number is odd, the two adjacent complexions in the middle are equal; but if the number is even it does not happen. For an odd number can be partitioned into two adjacent exponents differing by 1; for example, 1 + 2 f. 3; but an even number cannot. But the adjacent numbers into which an even number can be partitioned are the same, while an odd number can be partitioned into two exponents, and so has two *equal* Complexions according to Theorem 4, which are *adjacent*, since they differ by 1. 6. The Complexions increase up to the exponent halfway through the number itself, or to the two exponents nearest to the midpoint, and then decrease again. 7. Every prime number divides its *particular* complexions (that is, complexions with a given exponent). 8. The Number of overall Complexions is odd.[30]

---

[27] The Greek expressions here used mean, respectively, 'the that', 'how things are' (*to òti*), and 'the why', 'the reason', or 'the cause' (*to diòti*). Aristotle employs these two expressions at *Posterior Analytics* I 13, 78a22, to distinguish between an investigation oriented towards a mere description of how things are and an investigation focused on the reasons or causes of the phenomena under scrutiny. The distinction became a standard one in the Western philosophical tradition, with Aristotle himself giving various answers to the question of 'the why' with his theory of four causes in *Physics* II 3 and *Metaphysics* V 2.

[28] Theorems 1–4 can be expressed as follows: 1. If $k > n$, then $\binom{n}{k} = 0$; 2. If $k = n$, then $\binom{n}{k} = 1$; 3. If k = n − 1, then $\binom{n}{k} = n$; 4. If $k_1 + k_2 = n$, then $\binom{n}{k_1} = \binom{n}{k_2}$. As Knobloch (1973: 26) has shown, the same results are present in Pascal (1665: 557–64).

[29] Leibniz's proof lacks generality: it works only because the number of complexions of 3 with exponents 1 and 2 is 3, and the number of complexions of 4 with exponents 1 and 3 happens to be one more than this.

[30] Leibniz was probably the first to recognize the last two theorems; see Knobloch (1973: 26 7).

**Tab. 3.**

```
ab ac ad ae af
   bc bd be bf
      cd ce cf
         de df
            ef
```

7. Restat hujus Problematis altera pars quasi specialis: dato numero (A) com2nationes (B) invenire. Solutio: "ducatur numerus in proxime minorem, facti dimidium erit quaesitum, $A \cap A - 1 \cup 2 = B$." Esto v. g. Numerus $6 \cap 5$ f. $30 \cup 2$ f. 15. Ratio Solutionis: esto Tab. $\lambda$ [176] in qua enumeran[8] tur VI rerum: abcdef com2nationes possibiles, prima autem res a ducta per caeteras facit com2nationes V, nempe ipso numero unitate minores; secunda b per caeteras ducta tantum IV, non enim in antecedentem a duci potest, rediret enim prior com2natio ba vel ab (haec enim in negotio combinationis nihil differunt), ergo solum in sequentes quae sunt IV; similiter tertia c in sequentes ducta facit III. quarta d facit II, quinta e cum ultima f facit I. Sunt igitur com2nationes 5, 4, 3, 2, 1 + f. 15. Ita patet numerum com2nationum componi ex terminis progressionis arithmeticae, cujus differentia: 1, numeratis ab 1 ad numerum numero rerum proximum, inclusive; sive ex omnibus numeris Numero rerum minoribus simul additis. Sed quia uti vulgo docent Arithmetici, tales numeri hoc compendio adduntur, ut maximus numerus ducatur in proxime majorem, facti dimidius sit quaesitus; et vero proxime major h. l. est ipse Numerus rerum, igitur perinde est ac si dicas: Numerum rerum ducendum in proxime minorem, facti dimidium fore quaesitum.

Tab. λ.

ab ac ad ae af

bc bd be bf

cd ce cf

de df

ef

7. The other part of this Problem remains as a special case: 'Given a Number (A), to find the number of com2nations (B). The solution: let the number be multiplied by one less than the number; half of the product will be what is required. That is, $(A \cap (A - 1)) \cup 2 = B$. For example, let the Number be $6 \cap 5$, f. $30 \cup 2$, makes 15.' The Reason for the Solution: draw Table λ, in which the possible com2nations of 6 things *abcdef* are enumerated.

The first thing *a* set beside each of the others makes 5 com2nations, that is, one less than the number itself; the second thing *b* set beside each of the others makes only 4, as it cannot be set beside the preceding thing *a*, since that would repeat the earlier com2nation *ba* or *ab* (these being no different, so far as the process of combination is concerned), and therefore only beside the following things, which are 4 in number; similarly, the third thing *c* set beside the following things makes 3; the fourth thing *d* [set beside the following things] makes 2; and the fifth thing *e* with the last thing *f* makes 1. Therefore, the number of com2nations is $5 + 4 + 3 + 2 + 1$ f. 15. Thus, it is clear that the number of com2nations is composed of the terms of an arithmetical progression, whose difference is 1, and which are numbered inclusively from 1 to the number one less than the number of things; that is, from all numbers less than the number of things added together. But, as Arithmeticians usually teach us, such a series of numbers can be added together very concisely in this way, by multiplying the highest number in the series by the next greater number, the product of which divided by 2 gives the required total: and since this next number is none other than the Number of things, it is just as if one should say: Multiply the Number of things by the next smaller number, and this product divided by 2 will be what is required.[31]

---

[31] Leibniz here alludes to the formula $\binom{n}{2} = \frac{n(n-1)}{2}$, which was well known before him, as he rightly recognizes.

## Probl. II.
## DATO NUMERO COMPLEXIONES
## SIMPLICITER INVENIRE.

8. Datus Numerus quaeratur inter Exponentes progressionis Geometricae duplae, numerus seu terminus progressionis ei e regione respondens demta Unitate erit *quaesitum*. *Rationem*, seu τὸ διότι difficile est vel concipere, vel si conceperis explicare. Τὸ ὅτι ex Tabula א manifestum est. Semper enim complexiones particulares simul additae addita unitate terminum progressionis geometricae duplae constituent, cujus exponens sit numerus datus. Ratio tamen, si quis curiosius investiget, petenda erit ex discerptione in Practica Italica usitata, ᵥom Ƶerfallen. Quae talis esse debet, ut datus terminus pro[10]gressionis geometricae discerpatur in una plures partes, quam sunt unitates exponentis sui, id est numeri rerum; quarum semper aequalis sit prima ultimae, secunda penultimae, tertia antepenultimae, etc. donec vel, si in parem discerptus est numerum par-tium·exponente seu Numero rerum impari existente, in medio duae correspondeant partes per probl. I. th. 5 (v. g. 128 de 7 discerpantur in partes 8 juxtaTabulam א: 1, 7, 21, 35, [35], 21, 7, 1) vel si in imparem exponente pari existente, in medio relinquatur unus nulli correspondens (v. g. 256 de 8 discerpantur in partes 9 juxta Tab. א: 1, 8, 28, 56, 70, 56, 28, 8 1).

Putet igitur aliquis ex eo manifestum esse novum modum, eumque absolutum, solvendi probl. I. seu dato exponente inveniendi Numerum complexionum, si nimirum ope Algebrae inveniatur discerptio Complexionum simpliciter seu Termini Progr. Geom. duplae juxta modum

# Problem II
## GIVEN A NUMBER, TO FIND THE
## OVERALL COMPLEXIONS

8. Look up the given Number among the Exponents of the Geometric Progression of powers of 2; the number, or rather the term of the progression corresponding to it, with 1 subtracted, will be *what is required*.[32] The *Proof*, or *reason why*, is difficult to conceive of; or if you can conceive of it, difficult to explain. The *fact* is clear from Table ℵ. For the numbers of particular complexions added together plus 1 always constitute the term of the geometrical progression of powers of 2 whose exponent is the given number. But the reason, if anyone is interested enough to enquire into it, must be obtained *by partitioning*, vom 3erfallen customarily employed in the Italian Practice.[33]

[9]. Which must be such that the given term of the geometric progression be divided up into parts numbering one more than its exponent, which is the number of things; the first of which must be equal to the last, the second to the penultimate, the third to the antepenultimate, and so on until either, when the exponent, that is, the number of things, is odd, it is divided into an even number of parts and the two parts in the middle must correspond as in Problem I, Theorem 5 (for example, let 128, with exponent 7, be divided into 8 parts according to Table ℵ, 1, 7, 21, 35, 35, 21, 7, 1); or, when the Number of things is even, into an odd number of parts, and there must remain one number in the middle not corresponding to anything (for example, let 256, with exponent 8, be divided into 9 parts according to Table ℵ, 1, 8, 28, 56, 70, 56, 28, 8, 1).[34] One might conclude from this that there is clearly a new and independent

---

[32] What Leibniz says here may be summarized by the formula '$\sum_{k=1}^{n} \binom{n}{k} = 2^n - 1$', where 'n' is the given number.

[33] The German verb *zerfallen* that Leibniz uses here, means 'to break up into pieces', 'to partition'. In order to make easier multiplications, the decomposition of a number into its parts was systematically employed in Leibniz's time in commercial arithmetic. As David E. Smith explains: "When northern writers of the 16th century spoke of the Italian practice, they usually referred merely to Italian commercial arithmetic in general" (Smith 2007: 492). What Leibniz here calls 'Italian practice' was also known as 'the Welsh practice'. The *praxis Italica* was mentioned in Lantz (1616), a text well known to Leibniz since when he was a student at the Nikolai Schule in Leipzig. The fact that Leibniz mentions, here and at pages 217 and 219, the procedure of *partitioning* a natural number into summands of a given type, foreshadows his further interest in what in mathematics is known as 'additive number theory.' Cf. Knobloch (1974), where one finds a short account of Leibniz's studies on partitions after the *DAC* and his stay in Paris (1672–6).

[34] As Knobloch (1973: 28) remarks, we here have Leibniz's first steps into the additive theory of numbers, which, after Leonard Euler, became known as *partitio numerorum*.

datum. Verum non sunt data sufficientia, et idem numerus in alias atque alias partes eadem tamen forma discerpi potest. [*177*]

## USUS Probl. I. et II.

10. Cum omnia quae sunt aut cogitari possunt, fere componantur ex partibus aut realibus aut saltem conceptualibus, necesse est quae specie differunt aut eo differre, quod alias partes habent, et hic *Complexionum* Usus, vel quod alio situ hic *Dispositionum*; illic materiae, hic formae diversitate censentur. Imo Complexionum ope non solum species rerum, sed et attributa inveniuntur. Ut ita tota propemodum Logicae pars *inventiva* illic circa terminos simplices, hic circa complexos fundetur in Complexionibus; uno verbo et doctrina *divisionum* et doctrina *propositionum*. Ut taceam quantopere partem Logices Analyticam, seu Judicii diligenti de Modis syllogisticis scrutatione Exemplo 6 illustrare speremus.

11. In divisionibus triplex usus est Complexionum, 1. dato fundamento unius divisioni inveniendi species ejus, 2. datis pluribus divisionibus de eodem Genere, inveniendi species ex diversis divisionibus mixtas, quod tamen servabimus problemati 3. 3. datis speciebus inveniendi genera subalterna. Exempla per totam Philosophiam diffusa sunt, imo nec Jurisprudentiae deesse ostendemus, apud Medicos [11] vero omnis varietas medicamentorum compositorum et φαρμακοποιητική ex variorum Ingredientium mixtione oritur; at in eligendis mixtionibus utilibus summo opus Judicio est. Primum igitur exempla dabimus Specierum hac ratione inveniendarum:

12. 1. Apud JCtos *l. 2, D. Mandati, et pr. J. de Mandato* haec divisio proponitur: *Mandatum* contrahitur 5 modis: mandantis gratia, mandantis et mandatarii, tertii, mandantis et tertii, mandatarii et tertii. Sufficientiam divisionis hujus sic venabimur: Fundamentum ejus est

method for solving Problem I, i.e. given an exponent, of finding the Number of complexions; providing, of course, that one can find an algebraic division of the number of Complexions simply so-called, that is, of the Term of the Geometrical Progression of powers of 2 according to the method of Problem II. But the data are insufficient, and the same number can be divided into different sets of parts with the same form.

## The Applications of Problems I and II

10. Since all things that exist, or can be thought, are composed, in general, of parts, either real or at least conceptual, whatever differs in kind must necessarily differ either in parts, and here lie the Applications of *Complexions*, or by a different situs, hence the application of *Dispositions*. The former are characterised by diversity of matter, the latter by diversity of form. In fact, with the aid of Complexions one can find not only the species of things but also their attributes. Likewise, almost the whole of the *inventive* part of Logic (both that which concerns simple terms and that which concerns the complex ones) is founded on complexions; in short, both the theory of *divisions* and the theory of *propositions*; not to mention its use in the Analytic part of Logic, that is, judging the validity of syllogistic Moods, which I hope to illustrate in Example VI.

11. In divisions one can apply Complexions in three ways. 1. Given the basis of a division, to find its species. 2. Given several divisions of the same Genus, to find species compounded of several divisions. This, however, I shall reserve for Problem III. 3. Given species, to find the subaltern genera. Examples are diffused throughout the whole of Philosophy: in fact, as I shall show, they are not absent even from Jurisprudence, while among Physicians we find every variety of medical compounds, and the preparation of drugs by mixing various ingredients; but in the selection of useful mixtures the greatest care is required. Accordingly, I shall first give examples of how to find Species in the following way:

## I

12. Jurists (*Book 2* of the *Digest, Mandati*, and Book 1 of the *Institutions of Justinian, De Mandato*) propose this division: A Mandate may be contracted in 5 ways: on account of the mandator alone, on account

finis $\tilde{\varphi}$, seu persona cujus gratia contrahitur, ea est triplex: mandans, mandatarius et tertius. Rerum autem trium complexiones sunt 7: 1niones tres: cum solius 1 *mandantis,* 2 *mandatarii,* 3 *tertii* gratia contrahitur. Com2nationes totidem: 4 *Mandantis* et Mandatarii, 5 *Mandantis et Tertii,* 6 *Mandatarii et Tertii* gratia. Con3natio una, nempe 7 et *mandantis* et *mandatarii* et *tertii* simul gratia. Hic JCti 1nionem illam, in qua contrahitur gratia mandatarii solum, rejiciunt velut inutilem, quia sit consilium potius quam mandatum; remanent igitur species 6, sed cur 5 reliquerint, omissa con3natione, nescio.

13. II. Elementorum numerum, seu corporis simplicis mutabilis species Aristoteles libr. 2 de Gen. cum Ocello Lucano Pythagorico deducit ex numero Qualitatum primarum, quas 4 esse supponit, tanquam Fundamento, his tamen legibus, ut 1. quodlibet componatur ex duabus qualitatibus et neque pluribus neque paucioribus, hinc manifestum est 1niones, con3nationes et con4nationem esse abjiciendas, solas con2nationes retinendas, quae sunt 6. 2 ut nunquam in unam com2nationem veniant qualitates contrariae, hinc iterum duae con2nationes fiunt

of the mandator and the mandatary, on account of a third party alone, on account of the mandator and a third party, and on account of the mandatary and a third party.[35] I shall investigate the adequacy of this division as follows: Its basis is an end-for-whom,[36] that is, the person on whose account it is contracted, and there are three possibilities available: the mandator, the mandatary, and a third party. Now there are 7 complexions of these three things: three 1nions, since it may be contracted on account of (1) the *mandator* alone, (2) the *mandatary* alone, and (3) *a third party* alone. The Com2nations are three as well, on account of (4) the *Mandator* and the *Mandatary*, (5) the *Mandator* and *a Third Party*, and (6) the *Mandatary* and *a Third Party*. There is one Con3nation, namely (7) on account of the *mandator*, the *mandatary*, and *a third party* all together. Of these, Jurists reject as useless the 1nion in which the mandate is contracted on account of the mandatary alone, since this would be advice rather than a mandate. There remain, then, 6 species; but why they kept only 5 of them, while omitting the con3nation, I do not know.

## II

13. Aristotle, in *On Generation*, Book 2, together with the Pythagorean Ocello Lucano, infers the number of elements, or species of simple, mutable body, from the number of primary Qualities, which he assumes to be 4,[37] as its Principle, but with these laws: that (1) everything must be composed of two qualities, neither more nor fewer, so that it is obvious that 1nions, con3nations, and the one con4nation, must be ruled out, and the com2nations alone, to the number of 6, retained; and (2) contrary qualities must never enter into the same com2nation, so that again two of the com2nations are useless, since among the primary

---

[35] Justinianus, *Digesta* XVII 1, 'Mandati vel contra', in *Corpus Iuris Civilis*, vol. 1 (1894), esp. XVII 1, § 2, and *Institutiones* III 26, 'De Mandato'. Cf. *Institutes* (1975: 242). A *mandate* is a contract according to which someone (the *mandator*) entrusts the conduct of a given business to someone else (the *mandatary*), who should act on behalf of the first person.

[36] Leibniz here employs the Greek relative pronoun in the dative case: *hōi*.

[37] The number '4' is missing in the 1666 edition, but it is present in the *errata corrige*.

inutiles, quia inter primas has qualitates dantur duae contrarietates, igitur remanent com2nationes 4, qui est numerus Elementorum. Apposuimus Schema (vide *paginam titulo trac[178]tatus proximam),* quo origo Elementorum ex primis Qualitatibus luculenter demonstratur. Porro uti ex his illa Aristoteles, ita ex illis 4 temperamenta Galenus, horumque varias mixtiones medici posteriores elicuere: quibus omnibus jam superiori seculo [12] se opposuit Claud. Campensius Animadvers. natural. in Arist. et Galen. adject. ad Com. ej. in Aph. Hippocr. 5 ed. 8, Lugduni anno 1576.

qualities there are two pairs of opposites. Hence, there remain 4 com2nations, and this is the number of Elements.[38]

I have added this Scheme (*see the frontispiece of this Treatise*) to show the origin of the Elements from primary Qualities in a pictorial manner.[39] One should also note that just as Aristotle derived the former from the latter, so Galen derived from them his 4 temperaments, and the physicians who came after him their various mixtures: to all of whom opposed himself in the last century Claudius Champier, in his *Natural Remarks* on Aristotle and Galen appended to his *Commentary on the Aphorisms of Hippocrates*, 8th edition, published in Lyon in 1576.[40]

---

[38] Cf. Aristotle, *De generatione et corruptione*, Book II, esp. II 3, 330a30–331a6. Ocello Lucano, or Ocellus the Lucanian, has traditionally been regarded as one of the first followers of Pythagoras. The book *Peri tês tou Pantos Phuseôs*, however, though attributed to Ocello, is a mainly Aristotelian work of Hellenistic origin. Cf. Kahn (2001: esp. 78–9). English translations of the text may be found in Ocellus Lucanus (1831), (1976), and (1987: 203–11). Leibniz probably took the example of the four elements from Clavius (Cf. Knobloch 1973: 28). He may have known Ocellus Lucanus (1661), where the relevant passage may be found on pp. 157–8. Note, however, that Pierre Gassendi also linked Aristotle and Ocello in the context of his discussion of a fifth, 'Empiric', element for the heavenly spheres—as well as for our minds, as Cicero added. Cf. Gassendi (1658, vol 1: 237–8) (reprint (1964)). Gassendi's reference to Ocello in turn seems to derive from Sextus Empiricus, *Pros Phusikous* II, 316, who mentions Ocello next to Aristotle as having held that, besides the fourth element, there is also a fifth "body that revolves in a circle". Cf. Sextus Empiricus (1960, vol. 268: 364–5). See also Cicero (1979, vol. 268: 436–8). On Aristotle's theory of elements, see Aristotle, *De generatione et corruptione* II, 1–5, as well as *Meteorologica* I 1 and *De Caelo* III and IV. Gassendi also mentions Aristotle's allusion to the heavenly substance as the "first element" (*to prôton stoicheion*) in *Metaphysica* I 7, 988a 31–2, as well as Aristotle's comparison of "the warm substance" animating animal seed with "the Element of the Stars", at *De generatione animalium* II 3, 736b29–737a1. Note that the expression *to prôton stoicheion* also occurs in Aristotle, *Meteorologica* I 3.

[39] See footnote 23. A scheme similar to this can be found in Clavius (1585: 33–6).

[40] In *On the Elements* 7-9, Galen argues that Hippocrates (as well as Aristotle), rightly claimed that our bodies are affected by alterations in the ratio of the qualities of hot, cold, wet, and dry. Cf. Galen (1996, vol. 1: 119–39), (1821: 474–91). Right at the start of his *Peri Kraseôn*, Galenus refers to this earlier work as well as to the accepted opinion of physicians and philosophers of old who derived the four temperaments from the mixture of primary qualities. The number of temperaments is limited to four on account of the idea that neither hot and cold, nor dry and humid may form pairs. Cf. Galen (1996, vol. 1: 518). Note that, in criticizing the idea of a relation between the four elements and the four primary qualities, Champier focused on Aristotle in particular, though he did indeed consider Aristotle's theory the source for all other 'physiologists'. For Aristotle's theory of elements, see the references in footnote 38. We have not been able to trace a 1576 edition of Claude Champier's *Aphorisms of Hippocrates*. The 1579 Lyon edition, however, likewise contains both the *Animadversiones in Galenum* and the *Naturales animadversiones in Aristotelem*. Cf. Champier (1579), in which Champier's discussion of the elements may be found on pp. 81–5 (second series).

1[4]. III. *Numerus* communiter ab Arithmeticis distinguitur in *Numerum* stricte dictum ut 3, *Fractum* ut 2/3, *Surdum* ut Rad. 3, id est numerum qui in se ductus efficit 3, qualis in rerum natura non est, sed analogia intelligitur, et *denominatum*, quem alii vocant figuratum, v. g. quadratum, cubicum, pronicum. Ex horum commixtione efficit Hier. Cardanus Pract. Arith. c. 2 species mixtas 11. Sunt igitur in universum Complexiones 15, nempe: 1niones 4, quas diximus, com2nationes 6. *Numerus et Fractus*, v. g. 3/2, aut 1½, *Numerus et Surdus*, v. g. 7 ∩ R. 3, *Numerus et Denominatus*, v. g. 3 + cub. de A, *Fractus et Surdus*, 1/2 + R. 3, *Fractus et Denominatus*, v.g. ½ ∩ cub. de A, *Surdus et Denominatus*, v. g. cub. de 7. Con3nationes 4. *Numerus et Fractus et Surdus, Numerus et Fractus et Denominatus, Numerus et Surdus et Denominatus, Fractus et Surdus et Denominatus*; Con4natio 1: *Numerus et Fractus et Surdus et Denominatus*. Loco vocis: Numerus, commodius substituetur vox: *Integer*. Jam 4, 6, 4 + 1 f. 15.

1[5]. IV. Registrum Germanice ein 𝔷ug dicitur in Organis Pneumaticis ansula quaedam cujus apertura variatur sonus non quidem in se melodiae aut elevationis intuitu; sed ratione canalis, ut modo tremebundus modo sibilans, etc. efficiatur. Talia recentiorum industria detecta sunt ultra 30. Sunto igitur in organo aliquo tantum 12 simplicia, ajo fore in universum quasi 4095, tot enim sunt 12 rerum Complexiones simpliciter per Tab. ℵ grandis organistis, dum modo plura, modo pauciora; modo haec, modo illa, simul aperit, variandi materia.

# III

14. Arithmeticians generally distinguish *Numbers* in a *Number* in the strict sense, such as 3; a *Fraction*, such as $^2/_3$; a *Surd*, such as $\sqrt{3}$, that is, the number that when multiplied by itself makes 3, a kind of number that does not exist in the nature of things, but has to be understood by analogy; and a *named*, or as some say, figurate number, such as a square, a cube, or an oblong number.[41] By mixing these together, Girolamo Cardano, in Chapter 2 of his *Practical Arithmetic*, makes 11 hybrid species.[42] Hence, there are in total 15 complexions, namely: the 4 1nions that I have already mentioned; 6 com2nations, *Number* and *Fraction*, for example, $^3/_2$ or $1^1/_2$; *Number* and *Surd*, for example, $7 \cap \sqrt{3}$; *Number* and *Named Number*, for example, $3 + A^3$; *Fraction* and *Surd*, for example, $^1/_2 + \sqrt{3}$; *Fraction* and *Named Number*, for example, $^1/_2 \cap A^3$; *Surd* and *Named Number*, for example, $[\sqrt{3} + 7^3]$[43]; 4 Con3-nations, *Number* and *Fraction* and *Surd*; *Number* and *Fraction* and *Named Number*; *Number* and *Surd* and *Named Number*; *Fraction* and *Surd* and *Named Number*; and 1 Con4nation, *Number* and *Fraction* and *Surd* and *Named Number*. It is more convenient to substitute the word *Integer* for the word 'Number'. Then, $4 + 6 + 4 + 1$ f. 15.[44]

# IV

15. On Pipe-Organs a *Register* (in German, *ein Zug*) is the name of a certain little stop, by the opening of which the sound is varied, not actually in respect of the melody or pitch as such, but by making the pipe tremble, hiss, or suchlike. The industry of our times has devised more than 30 such registers. Suppose, then, that there are only 12 single registers on an organ. I say that there will in effect be about 4095, as that is the number of overall Complexions of 12 things according to Table א; providing matter for variation by a skilful organist, opening as many registers at once as he pleases.

---

[41] A 'pronic', or 'oblong number', is an integer that is the product of two consecutive integers (i.e. a number of the form 'n(n + 1)').

[42] Cf. Cardanus (1539: 3).

[43] The 1666 original and the Academy edition only mention 'cub. de 7', i.e. $7^3$.

[44] Note that Cardano himself, in his *Practica arithmetice* of 1539, used the term *numerus integer* and even simply *integer* (as in *De Numeratione integrorum*).

[16.] V. Th. Hobbes Element. de Corpore p. I. c. 5. Res quarum dantur Termini in propositionem ingredientes, seu suo stylo, Nominata, quorum dantur nomina, dividit in *Corpora* (id est substantias, ipsi enim omnis substantia corpus), *Accidentia, Phantasmata, et Nomina.* Et sic nomina esse vel *Corporum,* v. g. Homo, vel *Accidentium,* v. g. omnia abstracta, rationalitas, motus; vel *Phantasmatum,* quo refert Spatium, Tempus, omnes Qualitates sensibiles etc. vel *Nominum,* quo refert secundas intentiones. Haec [13] cum inter se sexies com2nentur, totidem oriuntur genera propositionum, et additis iis ubi termini homogenei com2nantur (corpusque attribuitur corpori, accidens accidenti, phantasma phantasmati, notio secunda notioni secundae), nempe 4, exurgunt 10. Ex iis solos terminos homogeneos utiliter combinari arbitratur Hobbes. Quod, si ita est, uti certe et communis philosophia profitetur, abstractum et concretum, accidens et substantiam, notionem primam et secundam [179] male invicem praedicari, erit hoc utile ad artem inventivam propositionum, seu electionem com2nationum utilium ex innumerabili rerum farragine, observare; de qua infra.

# V

[16]  Thomas Hobbes, *Elements of Philosophy*, Section 1: *On Body*, Part I, Chapter 5. He divides things that correspond to Terms entering into propositions (or in his usage, Nominates, things that are given names) into *Bodies* (that is, substances, since for him every substance is a body), *Accidents, Phantasms*, and *Names*. And Names are accordingly either of *Bodies*, for example, man; of *Accidents*, for example, all abstractions, such as rationality or motion; of *Phantasms*, by which he means space, time, every sensible Quality, and so on; or of *Names*, by which he means second intentions.[45] When these are com2ned with one another, they give rise to a total of 6 genera of propositions, and with the addition of those where homogeneous terms are com2ned (body being attributed to body, accident to accident, phantasm to phantasm, and second notion to second notion[46]), that is, 4, they come to a total of 10. Of these, Hobbes thinks that only homogeneous terms give rise to useful combinations. And if it is the case, as certainly seems to be the general opinion in philosophy, that abstract and concrete, accident and substance, first notion and second notion should not be predicated of one another, this will be useful for the art of inventing propositions, that is, to guide the choice of useful com2nations from an innumerable medley of things; of which more below.

---

[45] Discussing material errors in syllogistic reasoning in *De Corpore* I 5, 'De erratione, falsitate et captionibus' ('On Erring, Falsity, and Captions'), Hobbes offers four *genera* or 'kinds' of *res nominatae*, i.e. 'things to which we give names': bodies, accidents, phantasms, and names. Cf. Hobbes (1839a: 51–2); *English Works* (1839c: 57–8). Cf. Hobbes (1999: 51–2). The word *intentio* is synonymous with 'concept', 'notion', and Leibniz uses it in this way. In scholastic usage, a 'second intention' (*intentio secunda*) is a 'concept of concepts'. Thus, for instance, whereas the concepts corresponding to the words 'man', 'horse', and 'table' are *first intentions*, the concept associated with the word *species* is a second intention, since it refers to concepts like those of 'man', 'horse', 'table', etc. Likewise, in Hobbes (1839a: 53; 1839c: 59), (1999: 52–3), Hobbes argues that *genus, universale*, and *particular* "are names of names and not of things".

[46] For Leibniz's terminological change from *intention* to *notion*, here and below, see the former footnote.

[17.] VI. Venio ad exemplum complexionum haud paulo implicatius: determinationem numeri *Modorum Syllogismi Categorici.* Qua in re novas rationes iniit Joh. Hospinianus Steinanus Prof. Organi Basileensis vir contemplationum minime vulgarium libello paucis noto, edito in 8, Basileae, an. 1560 hoc titulo: *Non esse tantum 36 bonos malosque categorici syllogismi modos, ut Aristot. cum interpretibus docuisse videtur; sed 512 quorum quidem probentur 36, reliqui omnes rejiciantur.*

[18.] Incidi postea in controversias dialecticas ejusdem editas post obitum autoris Basileae 8 anno 1576. Ubi quae in Erotematis Dialecticis libelloque de Modis singularia statuerat, velut quadam Apologia, ex 23 problematibus constante, tuetur[.] Promittit ibi et libellum de inveniendi judicandique facultatibus, et Lectiones suas in universum Organon cum Latina versione, quas ineditas arbitror fortasse ab autore conceptas potius, quam perfectas[.] Etsi autem variationem ordinis adhiberi necesse est, quae spectat ad probl. 4, quia tamen potissimae partes complexionibus debentur, huc referemus. Cum libri hujus de Modis titulus primum se obtulit, antequam introspeximus, ex nostris traditis calculum subduximus hoc modo: *Modus* est dispositio seu forma syllogismi ratione quantitatis et qualitatis simul: Quantitate autem propositio est vel Universalis vel Particularis vel Indefinita vel Singularis; nos brevitatis causa utemur literis initialibus: U, P, J, S. Qualitate vel Affirmativa vel [14] Negativa, A, N. Sunt autem in Syllogismo tres propositiones, igitur ratione quantitatis, Syllogismus vel est aequalis, vel inaequalis. *Aequalis*, seu habens propositiones ejusdem quantitatis 4. modis: 1. Syllogismus talis est: U, U, U. 2. P, P, P. 3. J, J, J. 4. S, S, S ex quibus sunt utiles 2, *1mus* et *4tus*. Inaequalis vel ex parte vel in totum.

# VI

[17].[47] I now come to a somewhat more complicated example of complexions: the determination of the number of *Moods of Categorical Syllogism*. In this field, new pathways were initiated by Johannes Wirth of Stein-am-Rhein, Professor of Logic at Basel, a man of uncommon insight, in a little-known book, published in 8vo in Basel, 1560, under this title: *That there are not just 36 valid and invalid moods of Categorical Syllogism, as Aristotle and his interpreters appear to have taught; but 512, of which 36 can be proved, and all the others must be rejected.*[48]

[18].[49] I later happened upon his *Dialectical Disputes*, published posthumously in Basel, in 8vo, in 1576, where, in a kind of apologia, consisting of 23 problems, he reviews each of the things that he had stated in his *Dialectical Questions* and in his book on the Moods, and promises a book about the faculties of inventing and judging, and also his Readings in the entire body of the *Organon* together with a Latin version of it, which I think have not been published, but perhaps initiated by the author rather than finished.[50] And because the most important parts of it are due to complexions, if there is any need for variation of order to be applied, which is the subject of Problem IV, I shall refer you to this. When the title of his book on the Moods first came to my attention, before I looked into it I made a calculation according to my own method in this way: a *Mood* is a disposition or form of syllogism taking account of both quantity and quality. As to Quantity, a proposition may be Universal, Particular, Indefinite, or Singular. For the sake of brevity I shall use the initial letters U, P, J, S. As to Quality, it may be either Affirmative or Negative, A or N. And as there are three propositions in a Syllogism, with respect to quantity it may be either equal or unequal. It is *equal*, that is, it has propositions of the same quantity, in 4 moods. The Syllogisms of this kind are (1) UUU (2) PPP (3) JJJ (4) SSS, of which there are two useful moods, the *first* and the *fourth*.[51]

---

[47] The 1666 original mistakenly has '15.'

[48] Cf. Hospinianus (1560). Leibniz properly designates the Swiss philosopher Johannes Hospinianus (1515–75), or Johannes Wirth, as "Professor of [Aristotle's] Organon", i.e. as a teacher of Aristotle's logical doctrine. See also footnote 50.

[49] The 1666 original mistakenly has '16.'

[50] Cf. Hospinianus (1576). In the title of this work, Wirth is mentioned as professor in the *Aristotelian Organon* at Basel. The 'Readings' mentioned must have consisted of lecture notes Wirth made as a professor of logic. Wirth's *Dialectical Questions* were first published as Iohannes Hospinianus Steinanus, *Quaestionum Dialecticarum Libri Sex*, Basel: [s.n.], (1543). On Wirth, see also: Dürr (1955: 272–84), and Korcik (1955: 51–70), as well as Thomas (1957: 382).

[51] On this point cf. Introduction, § 8.

19. *Ex parte*, quando duae quaecunque propositiones sunt ejusdem quantitatis, tertia diversae. Et in tali casu duo genera Quantitatis sunt in eodem Syllogismo, etsi unum bis repetitur: id toties diversimode contingit, quoties res 4, id est genera haec quantitatum: U. P. J. S. diversimode sunt com2nabilia, nempe 6 𝔐𝔞𝔥𝔩, et in singulis 2 sunt casus, quia jam hoc bis repetitur, jam illud, altero simplici existente. Ergo 6 ∩ 2 f. 12. Atque ita rursus in singulis, ratione ordinis, sunt variationes 3, nam v. g. hoc U, U, P vel ponitur uti jam; vel sic: P, U, U vel sic: U, P, U. Ergo 12 ∩ 3 f. 36. Ex quibus utiles 18: 2 U(S), U(S), S (U). 2 U(S), S(U), U(S). 2 S(U), U(S), U(S). 4 U(S), U(S) P vel J. 4 U J(P), J(P) vel loco U, S. 4 J P(U), J(P) et S loco U.

20. *In totum inaequalis*, quando nulla cum altera est ejusdem magnitudinis, et ita quemlibet Syllogismum ingrediuntur genera 3, toties alia quoties 4 res possunt con3nari, nempe 4 𝔪𝔞[𝔥]𝔩. Tria autem ratione ordinis variantur 6 𝔪𝔞𝔥𝔩, v. g. U, P, [J]. U,[J], P. P, U, [J]. [*180*] P, [J], U. [J], U, P. [J], P, U. Ergo 4 ∩ 6 f. 24. Ex quibus utiles 12: 2 U, P(J), J(P). 2 J(P), U, P(J); totidem si pro U ponas S. 4 + 4 f. 8. 2 U(S), S(U), P; totidem si pro P ponas [J]. 2 + 2 f. 4. Addamus jam: 4 + 36 + 24 f. 64. Hae sunt variationes Quantitatis solius. Ex quibus sunt utiles: 2 + 18 + 12 f. 32. Caeteri cadunt per Reg. 1. ex puris particularibus, nihil sequitur, 2. Conclusio nullam ex praemissis quantitate vincit; etsi fortasse interdum ab utraque vincatur, uti in Barbari.

21. Porro cum Qualitatis duae solum sint diversitates A et N, Propositiones vero 3, [h]inc repetitione opus est, et vel Modus est *Similis*, id est ejusdem qualitatis, vel dissimilis: hujus nulla ulterius est variatio, quia nunquam ex toto, sed semper ex parte est dissimilis. Nunquam enim omnes propositiones sunt dissimiles quia solum 2 sunt diversitates.

19. It is unequal either in part or in whole; *in part*, when 2 of the propositions are of the same quantity and the third is of a different quantity.
And in such a case there are two kinds of Quantity in the same
syllogism, albeit that one of them is found twice. This happens in as
many different ways as 4 things, that is, the kinds of quantity U, P, J,
S are com2nable, namely, 6 ways; and in each there are two cases,
because one of the quantities is repeated in company with the other,
unrepeated quantity. The result is 6 ∩ 2 f. 12. And again, in each mood,
with respect to order, there are 3 variations, since, for example, UUP
may be ordered either as it is, or like this, PUU, or like this, UPU. The
result is 12 ∩ 3 f. 36. Of these, 18 are useful: 2 – U(S)U(S)S(U); 2 –U(S)S
(U)U(S); 2 – S(U)U(S)U(S); 4 – U(S)U(S)P or J; 4 – UJ(P)J(P), or
instead of U, S; 4 – J(P)UJ(P) and S instead of U.[52]

20. A Syllogism is *unequal in whole* when no Quantity is of the same
magnitude as another. Thus, there are three kinds of quantity making
up every syllogism, and as many different kinds [of mood] as 4 things
can be con3ned, namely, 4. Further, taking order into account, the
three can be varied in 6 ways. For example, UPJ, UJP, PUJ, PJU, JUP,
JPU. The result is 4 ∩ 6 f. 24, of which 12 are useful: 2 – UP(J)J(P); 2 – J
(P)UP(J); just as much as, putting S in place of U,[53] the result is 4 + 4 f. 8;
2 – U(S)S(U)P; just as much as, putting J in place of P,[54] the result
is 2 + 2 f. 4. Let us now add up: 4 + 36 + 24 f. 64. These are the
variations taking account of Quantity alone. Of these, the useful ones
are: 2 + 18 + 12 f. 32. The others are eliminated under Rule (1) from
pure particulars nothing follows, and Rule (2) the conclusion cannot
surpass in quantity any of the premises; even though, perhaps, it may
sometimes be surpassed by both of the premises, as in *Barbari*.[55]

21. Furthermore, while there are only two kinds of Quality, A and N, there
are three Propositions. Hence, there has to be repetition, and a mood is
either *similar*, that is, of the same quality, or *dissimilar*: there is no
further variation of the latter, because it is never dissimilar as a whole,
but only in part; for the propositions are never all dissimilar, as there are

---

[52] Leibniz here uses brackets to indicate alternatives. Thus, "U(S)U(S)S(U)" is equivalent to
'UUS and SSU', "U(S)S(U)U(S)" to 'USU and SUS', etc. The addition "or ....", moreover, refers
to the last term if not otherwise indicated. Thus, "U(S)U(S)P or J" has to be read as 'UUP, SSP,
UUJ, SSJ'; "UJ(P)J(P), or instead of U, S" as 'UJJ, UPP, SJJ, SPP', etc.
[53] Since we then get SPJ, SJP and JSP, PSJ, respectively.
[54] This yields: USJ, SUJ.        [55] Cf. Introduction § 8 and n. 93.

Similis species sunt [15] 2: A, A, A. N, N, N. Dissimilis 2: A, A, N. vel N, N, A. Dissimilis singulae variantur ratione ordinis 3 Mahl, v.g. A, A, N. N, A, A. A, N, A. Ergo 2 ∩ 3 f. 6 + 2 f. 8. Toties variatur Qualitas. Ex quibus utiles Variationes sunt 3: A, A, A. N, A, N. A, N, N. per reg. 1. ex puris negativis nihil sequitur. 2. Conclusio sequitur partem in qualitate deteriorem. Sed quia modus est variatio Qualitatis et Quantitatis simul, et ita singulae variationes Quantitatis recipiunt singulas Qualitatis; hinc 64 ∩ 8 f. 512. Numerum omnium Modorum utilium et inutilium.

22. Ex quibus utiles sic repereris: duc variationes utiles quantitatis in qualitatis, 32 ∩ 3 f. 96. De producto subtrahe omnes modos qui continentur in Frisesmo, id est qui ratione Qualitatis quidem sunt A N N, ratione quantitatis vero Major prop. est [J] vel P, Minor autem U vel S, et conclusio [J] vel P, quales sunt 8. Frisesmo enim etsi modus est, per se quodammodo subsistens, tamen est in nulla figura, v. infra. Jam, 96 – 8 f. 88. Numerum utilium Modorum. Hospiniano, cui nostra methodus ignota, aliter, sed per ambages procedendum erat. Primum igitur c. 2, 3, Aristotelicos modos 36 investigat ex complicatione U, P, J omisso S et conclusione. Ex quibus utiles sunt 8: UA, UA in Barbara vel Darapti, UA, PA in Darii et Datisi, PA, UA in Disamis, UA, UN in Camestres; UN, UA in Celarent, Cesare, Felapton; UA, JN in Baroco, UN, JA in Ferio, Festino, Ferison; JN, UA in Bocardo. Quibus addit cap. 4 Singulares similes aequales SA, SA. SN, SN, 2 inaequales 3ium generum singulis inversis, et quibuslibet vel A vel Neg. 3 ∩ 2 ∩ 2 f. 12 + 2 f. I4. Ex quibus Hospinianus solum admittit, UA, PA et ponit in Darii. Quia singulares ait particularibus aequipollere cum communì Logicorum schola, quod tamen mox falsum esse ostendemus. C. 5 addit singulares dissimiles totidem, nempe 14, ex quibus Hosp. solum admittit SN, UA in Bocardo; item UN, SA in Ferio. C. 6, addita Conclusione quasi denuo incipiens enumerat modos similes aequales 4 ∩ 2 f. 8, ex quibus utiles solum UA, UA, UA in Barbara. [181] [J]uxta Hospin. similes inaequales, sunt vel ex toto inaequales, de quibus infra; vel [16] ex parte, de quibus nunc. Ubi duae propositiones sunt ejusdem quantitatis, tertia quaecunque diversae; et tunc modo duae sunt universales una indefinita, quo casu sunt modi 6 (nam una vel initio vel medio vel fine ponitur 3; semperque aut omnes sunt A, aut N, 3 ∩ 2 fac. 6),

only two ways of varying them. Of similar moods there are two kinds, AAA or NNN. Of dissimilar moods, AAN or NNA. Each of the dissimilar moods can be varied with respect to order in 3 ways, for example, AAN, NAA, ANA. Therefore, Quality can be varied in (2 ∩ 3 f. 6) +2 f. 8 ways. Of these there are 3 useful variations, AAA, NAN, ANN, according to rule (1) from mere negative premisses nothing follows, and rule (2) the conclusion follows the part inferior in quality. But because a mood is a variation taking account of both Quality and Quantity at once, each variation of Quantity may be paired with each variation of Quality. Hence, the total number of Moods, useful and useless, is 64 ∩ 8 f. 512.

22. Of these, you may find the useful moods in this way: multiply the number of useful variations with respect to quantity by the number of useful variations with respect to quality, that is, 32 ∩ 3 f. 96. From this product subtract all the moods contained in *Frisesmo*, that is, which are ANN in respect of Quality, and in which in respect of Quantity the Major premise is J or P, the Minor premise U or S, and the conclusion J or P. There are 8 moods of this kind. For even though *Frisesmo* is a mood that in a certain manner holds good in itself, it is not, however, in any figure. See below. The result is 96 - 8 f. 88, the number of useful moods. Wirth, to whom my method was unknown, had to proceed otherwise, and in a roundabout manner. He therefore first investigates in Chapters 2 and 3 the 36 Aristotelian moods that arise from combining U, P, and J, with S and the conclusion being omitted, of which 8 are useful: UA, UA in *Barbara* or *Darapti*; UA, PA in *Darii* and *Datisi*; PA, UA in *Disamis*; UA, UN in *Camestres*; UN, UA in *Celarent, Cesare*, and *Felapton*; UA, JN in *Baroco*; UN, JA in *Ferio, Festino*, and *Ferison*; and JN, UA in *Bocardo*. To these he adds, in Chapter 4, the 2 Singular, similar, and equal moods SA, SA and SN, SN, and the unequal moods of three kinds with each inverted, and all with A or N, totalling 3 ∩ 2 ∩ 2 f. 12 +2 f. 14. Of these, Wirth admits only UA, PA, and places it in *Darii*. This is because he says, according to the common view of the Scholastic Logicians, that singulars are equivalent to particulars, which I shall soon nevertheless show to be untrue. In Chapter 5, Wirth adds the same number, namely 14, of dissimilar singulars, of which he admits only SN, UA in *Bocardo*, and UN, SA in *Ferio*. In Chapter 6, as if beginning afresh, and with the Conclusion added, he lists the similar equal moods numbering 4 ∩ 2 f. 8, of which, according to him, only UA, UA, UA are useful, in *Barbara*. The similar unequals, according to Wirth, are either wholly unequal, of which more below; or partly unequal (which I shall

vel contra etiam 6 per cap. 7, fac. 12. Ex solis prioribus 6 utilis est UA, JA, JA in Darii et Datisi, item JA, UA, JA in Disamis, item UA, UA, JA in Darapti, et, ut Hospinianus non inepte, in Barbari. Certe cum ex propositione UA sequantur duae PA, una conversa, hinc oritur modus indirectus Baralip; altera subalterna [J], v. g. Omne animal est substantia. Omnis Homo est animal. E. Quidam Homo est substantia, hinc oritur iste: *Barbari.* Totidem, nempe 12, sunt Modi per c. 8, si duae U et una P jungantur, vel contra; et iidem sunt modi utiles qui in proxima mixtione, si pro J substituas P. Totidem, nempe 12, sunt modi per c. 8, si jungantur duae U, et una S per c. 9 et quia Hospin. habet S pro P, putat solum modum utilem esse in Darii UA, SA, SA, v. infra. It. 12 J, J, P vel P, P, J omnes inutiles per c.10. Item 12 J, J, S vel S, S, J omnes, ut ille putatur, inutiles per c. 11. Item 12 P, P, S vel S, S, P omnes ut ille putatur inutiles per c. 12. Jam 6 ∩ 12 f. 72 + 8 fac. 80, Numerum modorum similium additis variationibus Conclusionis. Dissimiles modi sunt vel aequales vel inaequales. Aequales sunt ex meris vel U vel P vel J vel S. 4 genera quae singula variantur ratione qualitatis sic: N, N, A. A, N, N. etc. 6 𝔐𝔞𝔥𝔩 uti supra diximus n. 20. Jam 6 ∩ 4 f. 24 v. cap. 13. Utilis est: UA, UN, UN in Camestres.

23. Dissimiles inaequales sunt vel ex toto inaequales, ut nulla Propositio alteri sit aequalis, *de quibus infra*, vel ex parte, ut duae s[i]nt aequales una inaequalis, de quibus nunc. Et redeunt omnes variationes quantitatis, de quibus in similibus ex c. 7, 8, 9, 10, 11, 12 in singulis de binis contrariis diximus, modi autem hic fiunt plures quam illic, ob variationem qualitatis accedentem. Erat igitur in c. 7 U, U, J vel contra J, J, U. Ordo quantitatis variatur 3 𝔐𝔞𝔥𝔩, quia v. g. J modo initio, modo medio, modo fine ponitur. Qualitatis tum complexus variatur 2 𝔐𝔞𝔥𝔩,

discuss now), where two propositions are of the same quantity, and the third of a different quantity, and then only when two are universal and one indefinite, in which case there are 6 moods (as the indefinite one can be placed in 3 positions, at the beginning, in the middle, or at the end, and they are all either A or N, with the result 3 ∩ 2 f. 6); or, contrariwise, 6 more, according to Chapter 7, making 12. Of the 6 aforementioned moods alone, the useful ones are UA, JA, JA in *Darii* and *Datisi*; JA, UA, JA in *Disamis*; and UA, UA, JA in *Darapti* and (as Wirth says, not inappropriately) *Barbari*. To be sure, from the proposition UA there follow two propositions PA, with one converted, and from which the indirect mood *Baralip* arises, and the other sub-alternated J, as in *All animals are substances, All men are animals, Therefore some men are substances*, from which arises the mood *Barbari*. There are the same number of Moods, namely 12, according to Chapter 8, when two U's and one P are joined, or vice versa; and if for J you substitute P there are the same useful moods as in the previous mixture. If two U's and one S are joined as in Chapter 9, there are the same number of moods, namely 12, as in Chapter 8; and because Wirth has S instead of P, he reckons that the only useful mood is in *Darii*, UA, SA, SA. See below. And all 12 moods of JJP or PPJ are useless, so he reckons, according to Chapter 10. And again, all 12 moods of JJS or SSJ are useless, so he reckons, according to Chapter 11. And all 12 moods of PPS or SSP are useless, so he reckons, according to Chapter 12. The result is 6 ∩ 12 f. 72 + 8 f. 80 as the number of similar moods, with added variations of their Conclusion. Dissimilar moods are either equal or unequal. The equal moods are unmixed, either U, P, J, or S, 4 kinds that are each varied in respect of quality as NNA, ANN, etc., in 6 ways, as I stated above in Paragraph 20. The result is 6 ∩ 4 f. 24. See Chapter 13. The useful one is UA, UN, UN in *Camestres*.

23. The dissimilar unequals are either wholly unequal, so that none of the Propositions is equal to another (of which more below), or partly unequal, so that two of the Propositions are equal and one unequal, which I shall now discuss. And all the variations of similars in respect of quantity in each of the two contraries that I discussed as arising out of Chapters 7–12 reappear, only here more moods arise than in those, on account of the additional variation in respect of quality. Chapter 7, then, dealt with UUJ and its contrary, JJU. The order may be varied in 3 ways in respect of quantity, because, for example, J may be placed either at the beginning, or in the middle, or at the end. Next, the complex of

N N A vel A A N [17] tum ordo 3 𝔐ahl, uti supra dictum, ponendo A, vel N, initio aut medio aut fine, Ergo 3 ∩ 2 [∩] 3 f. 18 de U, U, J et contra etiam 18 de J, J, U f. 36 per c. 14. In prioribus 18 utiles sunt modi: UA, UN, JN; vel loco JN, PN aut SN et sunt in modo *Camestros*, uti supra Barbari. UN, UA, J(P, S)N similiter in modo Celaro et Cesaro et Felapton. UA, J(P, S)N, J(P, S)N in Baroco; UN, J(P, S)A, J(P, S)N in Ferio Festino et Ferison qui ultimus tamen in S locum non habet. J(P, S)N, UA, J(P, S)N in Bocardo. Similiter U U P. vel P P U 36 modos habent. Utiles designavimus proxime per P in ( ). Similiter U, U, S vel S, S, U faciunt simul modos 36 per c. 15. Modos utiles proxime signavimus per S. J, J, P vel P, P, J faciunt similiter 36 per c. 16, modi omnes sunt inutiles. J J S et S S J et P P S et S S P faciunt modos 2 ∩ 36 = 72 per c. 17 qui omnes sunt inutiles. [*182*] Huc usque distulimus Inaequales ex toto, ubi nulla propositio in eodem Syllogismo est ejusdem quantitatis, sunt autem vel similes, vel dissimiles. Inaequales ex toto similes sunt: U, J, P quae forma habet modos 12, nam 3 res variant ordinem 6 𝔐ahl. Qualitas autem variatur 2 𝔐ahl E. 6 ∩ 2 f. 12 per c. 18, ubi sunt [in]utiles: UA, J(P, S.)A, P(J. S.)A; UA, P(J. S.)A, J(P. S.)A in Darii et Datisi. J(P, S.)A, UA, P(J, S)A; P(J, S)A, UA, J(P, S)A in Disamis, nisi quod S non ingreditur Minorem in Figura Tertia. U P S et U J S quae habent modos 24 per c. 10. Utiles signavimus proxime per S. J, P, S quae habet modos 12 per c. 20, omnes autem sunt inutiles juxta Hosp.

[24.] Dissimiles omnino inaequales sunt eodem modo uti similes: U, J, P quae variant ordinem 6 𝔐ahl. Qualitas autem variatur 6 𝔐ahl. Ergo 6 ∩ 6 f. 36 per c. 21. Modi utiles sunt: UA, J(P, S)N, P(J, S)N in Baroco; UN, J(P, S)A, P(J, S)N in Ferio, Festino et Ferison. J(P. S.)N, UA, P(J, S)N in Bocardo. U, J, S. et U, P, S 36 ∩ 2 f. 72 per c. 22. Modos utiles signavimus proxime per S et P et J in ( ). J, P, S habet modos

quantities may be varied in respect of quality in 2 ways, NNA or AAN, and after that their order may be varied in 3 ways, as was stated above, by placing A or N at the beginning, the middle, or the end, so that there are 3 ∩ 2 ∩ 3 f. 18 moods of UUJ and 18 of its contrary, JJU, making 36, according to Chapter 14. Of the former 18 the useful moods are: UA, UN, JN; or in place of JN, PN or SN, and they are in the mood *Camestros*, just as those mentioned above were in *Barbari*. UN, UA, J(P,S)N are similarly in the moods *Celaro*, *Cesaro*, and *Felapton*. UA, J(P,S)N, J(P,S)N are in *Baroco*. UN, J(P,S)A, J(P,S)N in *Ferio*, *Festino*, and *Ferison*, the last of which, however, does not apply to S. J(P,S)N, UA, J(P,S)N are in *Bocardo*. Similarly, UUP, or PPU, have 36 moods. I have designated the useful moods alongside those with J by means of the letter P in brackets. Similarly, UUS, or SSU, together give 36 moods according to Chapter 15. I have designated the useful moods alongside those with J by means of the letter S in brackets. JJP, or PPJ, similarly give 36 moods according to Chapter 16, all moods of which are useless. JJS, SSJ, PPS, and SSP give 2 ∩ 36 = 72 moods according to Chapter 17, all of which are useless. I have left till now the wholly Unequals, in which no proposition in the same Syllogism is the same in respect of quantity, and these are either similar or dissimilar. Similar unequals in whole are: UJP, a form that has 12 moods, as 3 things permute in 6 ways, and quality is varied in 2 ways, with the result 6 ∩ 2 f. 12 according to Chapter 18, where the useless[56] ones are: UA, J(P,S)A, P(J,S)A and UA, P(J,S)A, J(P,S)A in *Darii* and *Datisi*, and J(P,S)A, UA, P(J,S)A and P(J,S)A, UA, J(P,S)A in *Disamis*, except that S does not appear in the Minor Term in the Third Figure. UPS and UJS have 24 moods according to Chapter [20].[57] I have signified the useful ones alongside one another by means of S. JPS has 12 moods according to Chapter 20. But they are all useless, in Wirth's view.

[24]. The completely dissimilar unequals are dealt with in entirely the same way as the similars. UJP permutes in 6 ways, and the quality may be varied in 6 ways, with the result 6 ∩ 6 f. 36, according to Chapter 21. Its useful moods are: UA, J(P,S)N, P(J,S)N in *Baroco*; UN, J(P,S)A, P(J,S)N in *Ferio*, *Festino*, and *Ferison*; J(P,S)N, UA, P(J,S)N in *Bocardo*. With UJS and UPS, the result is 36 ∩ 2 f. 72, according to Chapter 22.

---

[56] The 1666 edition here has *utiles* ('useful').
[57] 'Chapter 10' is a misprint for 'Chapter 20'.

36 per c. 23., omnes inutiles juxta hypothesin Hosp. Addamus jam omnes modos a cap. 6 incl. ad c. 23 computatos 15 (nam anteriores in his rediere) + 80, 24, 36, 36, 36, 36, 72, 12, 24, 12, 36, 72, 36 [18] seu 80 + 12 $\cap$ 36 f. 512. In his Hospiniani speculationibus quaedam laudamus, quaedam desideramus. Laudamus inventionem novorum modorum: Barbari, Camestros, Celaro, Cesaro; laudamus quod recte observavit, modos qui vulgo nomen invenere, v. g. Darii etc. habere se ad modos a se enumeratos velut genus ad speciem, sub Darii enim hi Novem continentur ex ejus hypothesi: UA, JA, JA; UA, SA, SA; UA, PA, PA; UA, JA, SA; UA, SA, JA; UA, JA, PA; UA, PA, JA; UA, SA, PA; UA, PA, SA. Sed non aeque probare possumus, quod Singulares aequavit particularibus, quae res omnes ejus rationes conturbavit, effecitque ei modos utiles justo pauciores, ut mox apparebit. Hinc ipse in controversiis dialect. c. 22 p. 430 errasse se fatetur, et admittit modos utiles 38, nempe 2 praeter priores 36. 1. in Darapti cum ex meris UA concluditur SA, quoniam Christus ita concluserit Luc. XXIII. v. 37, 38. 2. in Felapton cum ex UN et UA concluditur SN, quia ita concluserit Paulus Rom. IX, v. 13. Nos etsi scimus ita vulgo sentiri, arbitramur tamen alia omnia veriora. Nam haec: Socrates est Sophronisci filius, si resolvatur fere juxta modum Joh. Rauen, ita habebit: Quicunque est Socrates, est Sophronisci filius. Neque male dicetur: Omnis Socrates est Sophronisci filius; etsi unicus sit. (Neque enim de nomine sed de illo homine loquimur). Perinde ac si

I have signified the useful moods alongside one another by means of S, P, and J in brackets. JPS has 36 moods, according to Chapter 23, all useless, in Wirth's view. I shall now add all the moods from Chapter 6 to Chapter 23 inclusive, coming to a total (gathering together the earlier ones) of 80 + 24 + 36 + 36 + 36 + 36 + 72 + 12 + 24 + 12 + 36 + 72 + 36, or 80 + 12 ∩ 36 f. 512. In these speculations of Wirth there is something to be applauded, something to be desired. I applaud the discovery of new moods: *Barbari, Camestros, Celaro, Cesaro*;[58] I applaud what he correctly noted, that the moods which have attracted the traditional names, such as *Darii*, etc., relate to the moods subsumed under them in the same way as a genus relates to a species. For example, in his view *Darii* contains these Nine moods: UA, JA, JA; UA, SA, SA; UA, PA, PA; UA, JA, SA; UA, SA, JA; UA, JA, PA; UA, PA, JA; UA, SA, PA; and UA, PA, SA. But I cannot equally approve of the fact that he equated Singulars with particulars, a circumstance that distorted all his arguments, and resulted in fewer useful moods than should be the case, as will presently become clear. Because of this, he admits in his *Dialectical Controversies*, Chapter 22, p. 430, that he erred, and allows 38 useful moods, that is, 2 more than the earlier 36: one in *Darapti*, in which from UA alone one may conclude SA, as Christ did in *Luke* 23: 37–38;[59] and the second in *Felapton*, in which from UN and UA one may conclude SN, as Paul did in *Romans* 9:13.[60] This, I know, is the common sentiment, but I nevertheless hold that there are others at least as true. For if the proposition *Socrates is a son of Sophroniscus* is resolved in line with the method of Johannes Raue, we shall have this: *Whoever is Socrates, is a son of Sophroniscus.*[61] Nor would it be wrong to say: *Every Socrates is a son of Sophroniscus*, even though he is unique. For I am speaking not of the name but of the man, just as, if I were to say: *All the clothes*

---

[58] Contrary to what Leibniz writes, these moods were known before Wirth: cf. Introduction, § 11.

[59] In the text quoted by Leibniz (Hospinianus 1576: 430) Wirth constructs a syllogism that according to him has the form of *Darapti* (figure III) and is based on the assumption that "God is not the God of the dead but of the living"—a claim expressed in Luke 20:38 (not 23:36, as Leibniz suggests).

[60] Rom. 9:13: "As it is written, Jacob have I loved, but Esau have I hated". 'Therefore Jacob was elected, whereas Esau was damned' is the conclusion of a syllogism that, according to Wirth (Hospinianus 1576: 430), has the form of *Felapton* (figure III).

[61] Johannes Raue (1610–79) 'resolved' subject and predicate terms in such a way as to identify a 'third common' implicit in every statement of predication, allowing the sentence 'Man is an animal', for instance, to be analysed in logical terms as 'He who | is a man | is he, who | is an animal.' Cf. Raue (1638: 168). See on Raue, as well as on the relation of his logic to a Fregean 'subordination of concepts': Angelelli (1990: 184–90).

dicam: Titio omnes vestes quas habeo, do lego, quis dubitet etsi unicam habeam ei deberi? Imo secundum JCtos universitas quandoque in uno subsistit. l. municipium 7, D. quod cujuscunque univers. nom. Magnif. Carpzov. p. II. c. VI. def. 17. Vox enim: [O]mnis, non infert multitudinem, sed singulorum comprehensionem. Imo supposito quod Socrates non habuerit [*183*] fratrem, etiam ita recte loquor: Omnis Sophronisci filius est Socrates. Quid de hac propositione dicemus: Hic homo est doctus? Ex qua recte concludemus: Petrus est hic homo, E. Petrus est doctus. Vox autem: Hic, est *Signum Singulare*. Generaliter igitur pronunciare audemus: omnis Propositio singularis ratione modi in syllogismo habenda est pro Universali. Uti omnis indefinita pro particulari. Hinc etsi Modos utiles solum 36 numerat, sunt ta[**19**]men 88, de quo supra, omissa nihilominus variatione, quae oritur ex figuris. Nam modi diversarum figurarum *correspondentes*, id est quantitate et qualitate convenientes, sunt unus simplex v. g. Darii et Datisi. *Simplices* a[utem] modos voco, non computata figurarum varietate, *Figuratos* contra, tales sunt modi Figurarum quos vulgo recensent. Age igitur, ne quid mancum sit, et ad hoc descendamus dum fervet impetus. Ad figuram requiruntur termini tres: Major,

*I have, I give, I bequeath to Titius*, who would dispute it, even though I may have only a single garment to give him? Moreover, according to Jurists, a collective sometimes consists in its being One. See the entries "Municipality" and "What may be done in the name of a collective, or against it" in the excellent Carpzov, *Legal Definitions*, Part 11, Chapter VI, Definition 17.[62] For the word 'Every' implies not a multitude but a grouping of individuals. Indeed, on the supposition that Socrates did not have a brother, I am also justified in saying something like, *Every son of Sophroniscus is Socrates*. And what should I say of the proposition, *This man is learned*? From it I may rightly conclude: *Peter is this man, therefore Peter is learned*. The word 'this' is, after all, a *Singular Sign*. In general, therefore, I venture to claim that every singular Proposition should be taken in a Syllogism to stand, so far as mood is concerned, for a Universal. In the same way, every indefinite should be taken to stand for a particular. Hence, even though Wirth numbers the useful Moods at only 36, there are, nevertheless, the 88 of which I spoke earlier, omitting, however, the variation that arises from the figures. For the *corresponding* moods of the various figures, that is, agreeing in respect of quantity and quality, are one simple mood, such as *Darii* and *Datisi*. I call moods on the one hand *Simple*, when the figures are not taken into account, and on the other hand *Figurative*, as they are the moods of the traditional Figures. Let us then waste no more time, and get down to business while the fire still burns. Three terms are required for a figure: a Major term, which I shall signify by the Greek letter μ; a Minor term, which I shall signify by the Latin letter M; and a Middle term, which I shall signify by the German letter 𝔐 each appearing twice. From these arise 3 com2nations, which are called propositions, of which the last is the conclusion and the preceding ones the premises. The general Rules for com2ning them into each figure are: (1) The same two terms are never com2ned:

---

[62] In his 1638 *Iurisprudentia forensis Romano-Saxonica*, the German lawyer Benedikt Carpzov the Younger (1595–1666) discusses how the rights and corporate goods of deceased members of a corporate body befall to its other members and how those that are left behind by a corporate body as a whole befall to superior colleges or to the municipality. Cf. Carpzovius (1638: 425–6). Famous for his clarity and erudition and hugely influential especially for the standardisation of criminal processes throughout Germany, later generations would denounce Carpzov for his severity, especially with a view to the persecution of witches. Modern scholarship, however, tends to play down Carpzov's influence on the number of death penalties issued as a result of his work, and to put into perspective his role in the persecution of witches, thereby significantly modifying the negative image. Cf. Wolfgang Schild, 'Carpzov, Benedikt II', in DBE 5: 286–7 and *Zedler* 5, col. 1134.

quem signabimus graece: μ; minor quem latine: M; medius quem germanice: 𝔐; et singuli bis. Ex his fiunt com2nationes 3 quae hic dicuntur propositiones, quarum ultima conclusio est, priores praemissae. Regulae com2nandi generales cuique figurae sunt: 1. Nunquam com2nentur duo termini iidem, nulla enim propositio est: MM seu minor minor. 2. M et 𝔐 solum com2nentur in Conclusione, ita ut semper praeponatur M hoc modo: M𝔐. 3. In praemissarum 1ma com2nentur 𝔐 et M in secunda M et μ. Neque enim pro variatione figurae habeo, quando aliqui praemissas transponunt, et loco hujus: B est C, A est B, Ergo A est C, ponunt sic: A est B, B est C, Ergo A est C, uti collocant P. Ramus, P. Gassendus, nescio quis J. C. E. libello peculiari edito, et jam olim Alcinous lib. 1. Doct. Plat. Qui semper Majorem prop. postponunt, Minorem Prop. praeponunt. Sed id non variat figuram, alioqui tot essent figurae quot variationes numerant Rhetores, dum in vita communi Conclusionem nunc initio, nunc medio, nunc fine quam observant.

there is no such proposition as MM, or minor com2ned with minor. (2) M and [μ]⁶³ are com2ned only in the Conclusion, and in such a way that M is always placed first, like this, [Mμ].⁶⁴ (3) In the first of the premises 𝔐 and [μ] are com2ned, and in the second [𝔐 and M].⁶⁵ For I do not see it as a variation of figure when some authors transpose the premises, and in place of this, B is C, A is B, therefore A is C, arrange them like this, A is B, B is C, therefore A is C, as do Petrus Ramus, Pierre Gassendi, someone (I do not know who) called J.C.E., in a remarkable little book, and in ancient times Alcinous in *Platonic Doctrines*, Book 1.⁶⁶ These authors always place the Major proposition second, and the Minor proposition first. But that does not vary the figure, as otherwise there would be as many figures as Rhetoricians count variations when in everyday life they have the Conclusion sometimes coming at the beginning, sometimes in the middle, and sometimes at the end.

---

⁶³ The original erroneously has '𝔐'.

⁶⁴ The original erroneously has 'M 𝔐'.

⁶⁵ Here, the original equally erroneously gives 'M and 𝔐' for the first and 'M and μ' for the second premise.

⁶⁶ Though not counting the alternative as a variation of figure, Leibniz shows a clear preference for the traditional way of putting the major premise first. Ramus, Gassendi, 'J.C.E.' and Alcinous are offered as examples of authors who arrange syllogisms in the alternative way of putting the minor premise first. Alcinous is the otherwise unknown author of the second century AD *Didaskalikos tôn Platônos dogmatôn*, or *Handbook of Platonism*. For an English edition, see: Alcinous (1993: esp. 11), where it is mentioned that: "The first figure is that in which the common term is predicated of the first term, and is the subject of the other [etc.]." Such a way of putting 'S-M' in first and 'M-P' in second place is indeed contrary to common usage. Petrus Ramus, however, seems to have accepted the standard presentation, offering the *propositio* that relates the middle term to the predicate first, the *assumptio* that links the subject and middle term second, and the *complexio* or *conclusio* in third place, both in Ramus (1543: 20 ff), and in Ramus (1556, Book II: 180 ff). Despite his unrelenting criticism of Aristotle, moreover, Ramus discusses all three Aristotelian figures in the standard way. Cf. Ramus (1556: 186–211). Pierre Gassendi, on the other hand, in the third part of his *Institutiones Logicae*, offers an alternative way of presenting syllogisms, distinguishing "coherent" from "incoherent" figures. In both cases, the *propositio* relates subject to middle term and appears before the *assumptio* that links the middle term to the predicate or *attributum*. Cf. Gassendi (1658, vol. 1: 106–10). Gassendi next compares his own division with the three figures of Aristotle. Cf. *idem*, pp. 110–12 (Canones VI-VIII). The author "called J. C. E." is the otherwise virtually unknown Johann Chrysostomus Eggefeld, or Eggefelder, and the "remarkable little book" mentioned by Leibniz is Eggefeld (1661). Arguing that the minor premiss presents the subject from which the reasoning takes its leave, Eggefeld advocates his alternative arrangement of syllogisms in the very short fifth chapter of this work, 'Quod Aristotelicus Syllogisticarum propositionum Ordo praeposterus sit', pp. 131–2.

25. Manifestum igitur figurarum varietatem oriri ex ordine medii in praemissis, dum modo in majore praeponitur, in minore postponitur, quae
est Aristotelica I, modo in majore et minore postponitur, quae est Arist.
II, modo utrobique praeponitur, quae est III, modo in majore postponitur, in minore praeponitur, quae est IV Galeni (frustra ab Hospiniano
contr. Dial. Probl. 19 tributa Scoto, cum ejus meminerit Aben Rois)
quam approbat Th. Hobbes Elem. de Corp. P. I. c. 4, art. 11. Designa-
buntur sic: I. $\mathfrak{M}\mu$, M$\mathfrak{M}$, M$\mu$. II. $\mu\mathfrak{M}$, M$\mathfrak{M}$, M$\mu$. III. $\mathfrak{M}\mu$, $\mathfrak{M}$M, M$\mu$.
IV. $\mu\mathfrak{M}$, $\mathfrak{M}$M, M$\mu$. IVtae figurae hostibus unum [20] hoc interim
oppono: Quarta figura, aeque bona est ac ipsa prima; imo si modo,
non praedicationis, ut vulgo solent, sed subjectionis, ut Aristoteles, eam
enunciemus, ex IV. fiet I, et contra. Nam Arist. ita solet hanc v. g.
propositionem: omne $\alpha$ est $\beta$ enunciare: $\beta$ inest omni $\alpha$. IVtae igitur
figurae designatio orietur talis. $\mathfrak{M}$ inest $t\tilde{\varphi}$ $\mu$, M inest $t\tilde{\varphi}$ $\mathfrak{M}$, E. M est $\mu$.

25. It is therefore clear that the variety of the figures arises from the position of the middle term within the premises. When it comes first in the major premise and second in the minor premise, the figure is the Aristotelian First Figure. When it comes second in both the major and the minor premise, the figure is the Aristotelian Second Figure. When it comes first in both premises, the figure is the Aristotelian Third Figure; and when it comes second in the Major premise and first in the Minor premise, it is the Fourth Figure of Galen (notwithstanding Wirth's assertion to the contrary in his *Dialectical Problems*, 19, attributing it to Scotus, since the latter was only recalling it from Averroës),[67] as Thomas Hobbes agrees in his *Elements of Philosophy*, Part I, *On Body*, Chapter 4, Article 11.[68] They are represented thus: I. 𝔐μ, M𝔐, [Mμ].[69] II. μ𝔐, M𝔐, Mμ. III. 𝔐μ, 𝔐M, Mμ. IV. μ𝔐, 𝔐M, Mμ. To the enemies of the Fourth Figure I make this one objection for the time being, that the Fourth Figure is as

---

[67] Tradition may have attributed the fourth figure of simple categorical syllogism to Galen on account of the four figures of *compound*, i.e. 'four term' categorical syllogisms that Galen may have developed, and of which he found examples in two Platonic dialogues. See: Łukasiewicz (1951) (1957²: 38–42), and Morison (2008: 66–115, esp. 85–91). The traditional notion that Galen did invent the fourth figure of simple Aristotelian syllogism was again upheld by Rescher (1966). Discussing the question 'Whether there are only three figures of categorical syllogism, or four?' in Chapter 19 of his *De controversiis dialecticis*, Johannes Wirth argues that it seems improbable to him that, as some say, the fourth figure indeed originated with Galen. 'Scotus', according to Wirth, seems to be a more probable source. Cf. Hospinianus (1576: 351). The author of the fourteenth-century commentaries on logic formerly attributed to Duns Scotus is nowadays referred to as *Pseudo-Scotus*, whom some believe to have been the early fourteenth-century logician John of Cornwall. On Pseudo-Scotus, see: Lagerlund (2000: 165–83). His texts were still published as logical works by Duns Scotus in seventeenth-century editions of the latter's works, and even included in the late nineteenth-century re-edition of Duns Scotus's *Opera Omnia*. Considering Duns Scotus the originator of the fourth figure, Wirth does not mention Averroës. Pseudo-Scotus discusses the fourth figure in the 34th *Quaestio* of his commentary on the first book of Aristotle's *Prior Analytics*, without, however, discussing the question of its origin. Offering references to other authors, however, Duns Scotus (1639) already mentions Averroës as one of several, earlier and later, sources on the matter. Cf. Duns Scotus (1639, vol. 1: 325) and Duns Scotus (1891, vol. 2: 168). Leibniz may have mentioned Averroës on the basis of his own general knowledge of Galen's fourth figure and its purported history, rather than on any specific reference in (Pseudo-)Scotus, and may also have quoted Johannes Regius here. Cf. Regius (1615: 153) (second series): "Plurimi *Galeno* huius inventionem tribuunt. *Joan*. verò *Hospin. Scoto* tribui mavult." On Regius, see below, note 76.

[68] Hobbes agreed not so much to the naming or the attribution of the Fourth Figure, as to its acceptance. Cf. Hobbes (1839¹, vol. 1: 47); *The English Works* (1839², vol. 1: 53): "The figures, therefore, of syllogisms, if they be numbered by the diverse situation of the middle term only, are but three [...]. But if they be numbered according to the situation of the terms simply, they are four [...]. From whence it is evident, that the controversy among logicians concerning the fourth figure, is a mere *logómachia*, or contention about the name thereof: for, as for the thing itself, it is plain that the situation of the terms [...] makes four differences of syllogisms, which may be called figures, or have any other name at pleasure."

[69] The 1666 original mistakenly has '𝔐𝔐' for the conclusion in the first figure.

Vel ut conclusio etiam sic enuncietur, transponendae praemissae, et conclusio erit: Ergo μ inest τῷ M. Idem in aliis fieri figuris potest, quod reducendi artificium nemo observavit hactenus. [*184*]

26. Caeterum secunda oritur ex prima, transposita propositione majore; 3tia, transposita minore, 4ta, transposita conclusione, sed hic alius efficitur syllogismus, quia alia conclusio. Unde modi hujus 4tae sunt designandi modis indirectis primae figurae ut vulgo vocant, dummodo praeponas majorem propositionem minori, non contra, ut vulgo contra morem omnium figurarum hanc unicam ob causam, ut ·vitaretur quarta Galeni, factum est, v. g. sit Syllogismus in Baralip: Omne animal est substantia, omnis homo est animal, E. quaedam substantia est homo. Certe substantia est minor terminus, igitur praemissa in qua ponitur, est minor, et per consequens, propositio haec: O[mne] animal est substantia, non est ponenda primo sed secundo loco; tum prodibit ipsissima IVta figura.

27. Propter hanc transpositionem propositionum, quas vulgo Syllogismos in Celantes ponunt, sunt in *Fapesmo*, loco Frisesmo dicendum *Fresismo*, loco Dabitis *Ditabis*; Baralip manet. Hi sunt modi figurae IVtae quibus addo *Celanto* et *Colanto*. Erunt simul 6 Modi 1mae sunt 6: *Barbara, Celarent, Darii, Ferio; Barbari, Celaro*. Modi IIdae 6: *Cesare, Camestres, Festino,*

good as even the First; in fact, if I state it only as subjection, like Aristotle, rather than as predication (which is the usual practice), then Figure I may be derived from Figure IV, and the other way round. For Aristotle uses β *is in all* α to express the proposition *All* α *is* β. Such an interpretation of the Fourth Figure leads to: 𝔐 *is in*[70] μ, *M is in* 𝔐, *therefore M is* μ. Or, to express the conclusion also in the same way, we have to transpose the premises, and the conclusion will be *Therefore* μ *is in M*. The same can be done in the other figures, a technique of reduction that no-one has remarked up to now.

26. Next, the second figure can be derived from the first, with the major premise transposed; the third figure, with the minor premise transposed; and the fourth figure, with the conclusion transposed.[71] But in this last case, another syllogism is produced, because the conclusion is different. Hence, the moods of this figure should be regarded as what are usually called indirect moods of the first figure, provided that you place the major premise before the minor, and not the opposite, as is often done against the spirit of all the figures solely in order to avoid the fourth figure of Galen. For example, take a Syllogism in *Baralip*, *All animals are substances, all men are animals, therefore some substances are men*. Substance is undoubtably the minor term, so that the premise in which it is placed is the minor premise, and in consequence the proposition *All animals are substances* should be placed not first but second. Then the very fourth figure itself will appear.[72]

27. Because of this transposition of propositions, the Syllogisms that are usually placed in *Celantes* are in *Fapesmo*, in place of *Frisesmo* one must speak of *Fresismo*, in place of *Dabitis*, *Ditabis*; *Baralip* remains unaffected. These are the moods of the fourth figure, to which I add

[70] As usual, Leibniz here employs the Greek article. See also below, note 115.

[71] With 'transposition', Leibniz means the simple change of the terms in a categorical proposition. Thus, for instance, changing 'MP' to 'PM' in the first premise of the first figure, we have the second figure.

[72] If we define the first syllogistic figure as that in which the middle term plays the role of subject in one premise and of predicate in the other, two arrangements of the terms are possible:
  1) MP, SM/SP; 2) PM, MS/SP.
(1) corresponds to the traditional first figure and (2) to the fourth. Transposing the terms in the conclusion of (1), we obtain the 'anomalous' (indirect) first-figure: MP, SM/PS, with the major term subject and the minor predicate of the conclusion. In this case, as Leibniz remarks, the first figure is maintained but at the cost of inverting the premises. If we restore the traditional order, placing the premise with the major term (the predicate of the conclusion) first, and then the premise with the minor term (the subject of the conclusion), we suddenly obtain the fourth figure.

*Baroco; Cesaro, Camestros.* Modi IIItiae etiam 6: *Darapti, Felapton, Disamis, Datisi, Bocardo, Ferison.* Ita ignota hactenus figurarum harmonia detegitur, singulae enim modis sunt aequales. 1. 1mae autem et 2dae figurae semper Major Propositio est U. 2. 1mae et IIItiae semper Minor A. 3. In IIda semper Conclusio N. 4. In IIItia Conclusio semper est P[.] In IVta Conclusio nunquam est UA. Major nunquam PN. Et si Minor N. Major UA. Propter has regulas fit, ut non quilibet [21] 88 modorum utilium in qualibet figura habeat locum. Alioqui essent Modi utiles: 4 ∩ 96 f. 384. Modi autem figurati in universum utiles et inutiles 512 ∩ 4 f. 2048. Qui autem in qua figura sint utiles, praesens schema docebit:

*Celanto* and *Colanto*,[73] making 6 altogether. There are 6 moods of the first figure: *Barbara, Celarent, Darii, Ferio, Barbari,* and *Celaro.* The 6 moods of the second figure are *Cesare, Camestres, Festino, Baroco, Cesaro,* and *Camestros.* There are also 6 moods in the third figure: *Darapti, Felapton, Disamis, Datisi, Bocardo,* and *Ferison.* Thus, the secret harmony of the figures now stands revealed: for they all have the same number of moods. (1) The Major Premise of the first and second figure is always U. (2) The Minor Premise of the first and third figure is always A. (3) In the second figure the Conclusion is always N. (4) In the third figure the Conclusion is always P.[74] (5) In the fourth figure the Conclusion is never UA, the Major Premise is never PN, and if the minor is N, the major is UA. As a result of these rules, not all of the 88 useful moods find a place in each figure. Otherwise, there would be 4 ∩ 96 f. 384 useful moods. And the total number of figurative moods, useful and useless, would be 512 ∩ 4 f. 2048. But this table indicates which are the useful moods in each figure:[75]

[73] According to the so-called 'theory of distribution' (that Leibniz endorses), the predicate P in the first premise of *Colanto* (figure IV) is taken particularly, whereas in the conclusion it is taken universally. In other words, P is *distributed* in the conclusion and *undistributed* in one of the premise. Therefore *Colanto* cannot be a valid mood.
[74] P = particular.
[75] In his critical remarks in the 1691 issue of the *Acta Eruditorum* concerning the new edition of the *Dissertation*, Leibniz points out that this table contains some mistakes. He explains that in the fourth figure there is not an OAO mood, but a mood AEE and adds, in a quite cryptic way that "in the place of *Colanto* near *Bocardo* must be put *Calerent* near *Camestres* [Adeoque pro *Colanto* juxta *Bocardo* poni debet *Calerent* juxta *Camestres*]." This means that *Colanto*, which is near *Bocardo*, must be substituted by *Calerent*, and *Calerent* has to be put on the same line as *Camestres*. As Couturat observes (Couturat (1901: 6, n. 2), Gerhardt misunderstood Leibniz's words, putting *Calerent* above *Ditabis* and *Camestres* above *Disamis* (GP 4: 103–4). Leibniz also criticizes the idea that *Frisesmo* would belong to the fourth, instead of the 0 figure.

```
8 UA,UA,UA | SA,SA,SA | UA,UA,SA | UA,SA,UA | SA,UA,UA |
8 UN,UA,UN | SN,SA,SN | UN,UA,SN | UN,SA,UN | SN,UA,UN |
8 UA,UN,UN | SA,SN,SN | UA,UN,SN | UA,SN,UN | SA,UN,UN |
8 UA,UA,PA | UA,UA,JA | SA,SA,PA | SA,SA,JA | UA,SA,JA |
8 UN,UA,PN | UN,UA,JN | SN,SA,PN | SN,SA,JN | UN,SA,JN |
8 UA,UN,PN | UA,UN,JN | SA,SN,PN | SA,SN,JN | UA,SN,JN |
8 UA,JA,JA | UA,PA,PA | UA,PA,JA | UA,JA,PA | SA,JA,JA |    *
8 UN,JA,JN | UN,PA,PN | UN,PA,JN | UN,JA,PN | SN,JA,JN |
8 UA,JN,JN | UA,PN,PN | UA,PN,JN | UA,JN,PN | SA,JN,JN |
8 JA,UA,JA | PA,UA,PA | JA,UA,PA | PA,UA,JA | JA,SA,JA |
8 JN,UA,JN | PN,UA,PN | JN,UA,PN | PN,UA,JN | JN,SA,JN |
                   Restat:
8 JA,UN,JN | PA,UN,PN | JA,UN,PN | PA,UN,JN | JA,SN,JN |
```

| | | | | 0 | 4 | 3 | 2 | 1 |
|---|---|---|---|---|---|---|---|---|
| SA,SA,UA | SA,UA,SA | UA,SA,SA | 1 | ..... | -------- | ------- | -------- | Barbara |
| SN,SA,UN | SN,UA,SN | UN,SA,SN | 2 | ..... | -------- | ------- | Cesare | Celarent |
| SA,SN,UN | SA,UN,SN | UA,SN,SN | 3 | ..... | -------- | ------- | Camestres | -------- |
| UA,SA,PA | SA,UA,JA | SA,UA,PA | 4 | ..... | Baralip | Darapti | -------- | Barbari |
| UN,SA,PN | SN,UA,JN | SN,UA,PN | 5 | ..... | Celanto | Felapton | Cesaro | Celaro |
| UA,SN,PN | SA,UN,JN | SA,UN,PN | 6 | ..... | Fapesmo | ------- | Camestros | -------- |
| SA,PA,PA | SA,PA,JA | SA,JA,PA | 7 | ..... | -------- | Datisi | --------- | Darii |
| SN,PA,PN | SN,PA,JN | SN,JA,PN | 8 | ..... | Fresismo | Ferison | Festino | Ferio |
| SA,PN,PN | SA,PN,JN | SA,JN,PN | 9 | ..... | -------- | ------- | Baroco | -------- |
| PA,SA,PA | JA,SA,PA | PA,SA,JA | 10 | .... | Ditabis | Disamis | --------- | -------- |
| PN,SA,PN | JN,SA,PN | PN,SA,JN | 11 | .... | Colanto | Bocardo | --------- | -------- |
| | | | Restat: | | | | | |
| PA,SN,PN | JA,SN,PN | PA,SN,JN | 12 | Frisesmo | -------- | ------- | ------- | ------- |

[22]     In quo descripti sunt omnes modi utiles, ex quibus octo semper constituunt *modum figuratum generalem*, tales autem voco illos vulgo appellatos, in quibus U et S, item J et P habentur pro iisdem: Ipsae lineae modorum constant ex quatuor trigis, in qualibet lineae quantitate conveniunt, different pro tribus illis utilibus qualitatis differentiis. Ipsae autem trigae inter se differunt quantitate, positae eo ordine quo supra variationes ejus invenimus, in quarum quatuor reducuntur omnes supra inventae, quia hic U et S, item J et P reducuntur ad eandem. Cuilibet lineae ad marginem posuimus Modos figuratos generales, in quos quilibet ejus Modus simplex specialis cadit. In summo signavimus numeris figuram.

| 8 | UA,UA,UA | SA,SA,SA | UA,UA,SA | UA,SA,UA | SA,UA,UA |
|---|---|---|---|---|---|
| 8 | UN,UA,UN | SN,SA,SN | UN,UA,SN | UN,SA,UN | SN,UA,UN |
| 8 | UA,UN,UN | SA,SN,SN | UA,UN,SN | UA,SN,UN | SA,UN,UN |
| 8 | UA,UA,PA | UA,UA,JA | SA,SA,PA | SA,SA,JA | UA,SA,JA |
| 8 | UN,UA,PN | UN,UA,JN | SN,SA,PN | SN,SA,JN | UN,SA,JN |
| 8 | UA,UN,PN | UA,UN,JN | SA,SN,PN | SA,SN,JN | UA,SN,JN |
| 8 | UA,JA,JA | UA,PA,PA | UA,PA,JA | UA,JA,PA | SA,JA,JA |
| 8 | UN,JA,JN | UN,PA,PN | UN,PA,JN | UN,JA,PN | SN,JA,JN |
| 8 | UA,JN,JN | UA,PN,PN | UA,PN,JN | UA,JN,PN | SA,JN,JN |
| 8 | JA,UA,JA | PA,UA,PA | JA,UA,PA | PA,UA,JA | JA,SA,JA |
| 8 | JN,UA,JN | PN,UA,PN | JN,UA,PN | PN,UA,JN | JN,SA,JN |

There remains:

| 8 | JA,UN,JN | PA,UN,PN | JA,UN,PN | PA,UN,JN | JA,SN,JN |
|---|---|---|---|---|---|

| | | | # | 0 | 4 | 3 | 2 | 1 |
|---|---|---|---|---|---|---|---|---|
| SA,SA,UA | SA,UA,SA | UA,SA,SA | 1 | | -------- | ------- | --------- | Barbara |
| SN,SA,UN | SN,UA,SN | UN,SA,SN | 2 | | -------- | ------- | Cesare | Celarent |
| SA,SN,UN | SA,UN,SN | UA,SN,SN | 3 | | -------- | ------- | Camestres | -------- |
| UA,SA,PA | SA,UA,JA | SA,UA,PA | 4 | | Baralip | Darapti | --------- | Barbari |
| UN,SA,PN | SN,UA,JN | SN,UA,PN | 5 | | Celanto | Felapton | Cesaro | Celaro |
| UA,SN,PN | SA,UN,JN | SA,UN,PN | 6 | | Fapesmo | ------- | Camestros | -------- |
| SA,PA,PA | SA,PA,JA | SA,JA,PA | 7 | | -------- | Datisi | --------- | Darii |
| SN,PA,PN | SN,PA,JN | SN,JA,PN | 8 | | Fresismo | Ferison | Festino | Ferio |
| SA,PN,PN | SA,PN,JN | SA,JN,PN | 9 | | ------- | ------- | Baroco | -------- |
| PA,SA,PA | JA,SA,PA | PA,SA,JA | 10 | | Ditabis | Disamis | --------- | -------- |
| PN,SA,PN | JN,SA,PN | PN,SA,JN | 11 | | Colanto | Bocardo | --------- | -------- |

There remains:

| PA,SN,PN | JA,SN,PN | PA,SN,JN | 12 | Frisesmo | -------- | ------- | ------- | ------- |
|---|---|---|---|---|---|---|---|---|

Here are shown all the useful moods, 8 of which constitute each *general figurate mood,* as I call those moods with conventional names, in which U and S, and J and P, are taken to be the same. The lines of moods themselves make up four sets of three, the lines agree in each quantity, but differ in the three useful variations of quality. The sets of three also differ from one another in quantity, being placed in the order in which I discovered their variations above. In four of them, all those found above are reduced, inasmuch as U and S, and J and P, are here reduced to the same. Against each line I have placed in the margin the general figurate Moods in which any of the simple special Moods falls. Along the top, I have indicated the figure by means of numbers.

28. Ex eodem autem manifestum est, Modos figuratos generales esse vel Monadicos; vel correspondentes, et hos vel 2 vel 3 vel 4, prout plures paucioresve uni lineae sunt [a]ppositi. Singulae porro lineae habent unum modum simplicem generalem, quem explicare possumus sumtis vocalibus, uti vulgo, ut A sit UA (vel SA), E sit UN (vel SN), [J] sit P (vel [J])A, O sit P([J])N (ita omittendae sunt 4 praeterea vocales U pro [J]A; Y pro [J]N; OY, seu ου pro SA; ω, pro SN; quas ad declarandum Hospinianum posuit Joh. Regius, quem vid. Disp. Log. lib. 4, probl. 5.), et ita modus lineae 1. est A A A. 2. E A E. 3. A E E. 4. A A [J]. 5. E A O. 6. A E O. 7. A [J] [J]. 8. E J O. 9. A O O. 10. [J] A [J]. 11. O A O. 12. [J] E O abjectis nempe consonantibus ex vocibus vulgaribus, in quibus Scholastici per consonas figuram, per vocales modos simplices, designarunt. Ultimus vero modus: [J] E O, quem diximus Frisesmo, et collocavimus in figura nulla, propterea est inutilis, quia major est P, hinc locum non habet in 1 et 2 minor vero N, hinc locum non habet in 1 et 3. Etsi ex regulis modorum non sit [*186*] inutilis. Quod vero in 4 locum non habeat, exemplo ostendo: Quoddam Ens est homo. Nullus Homo est Brutum. E. quoddam brutum non est Ens.

29. Atque hic obiter consilium suppeditabo utile, quod vel ipso exemplo hoc comprobatur, in quo consistit Proba, ut sic dicam, seu ars examinandi modum propositum, et sicubi non formae, sed materiae vi concludit, celeriter instantiam [23] reperiendi, qualem apud Logicos hactenus legere me non memini. Breviter: Pro UA sumatur propositio quam materia non patitur converti simpliciter, v. g. sumatur haec potius: Omnis homo est animai, quam, omnis homo est animal rationale, et quo remotius genus sumitur, hoc habebis accuratius. Pro UN eligatur talis, qua negentur de se invicem species quam maxime invicem vicinae sub eodem genere proximo, v. g. homo et brutum: et quae non sit

28. From this same table it is also clear that general figurate Moods are either Monadic or correspondent, and the latter as either 2, 3, or 4, according to how many there are along a particular line. Moreover, each line has one simple general mood, which I can best explain with the aid of the usual symbols, such as A for UA (or SA), E for UN (or SN), J for PA (or JA), and O for PN (or JN). You will see that I have omitted 4 other symbols, U for JA, Y for JN, OY or ου for SA, and ω for SN, which Johannes Regius employed in his exposition of Wirth. (See his *Logical Enquiries*, Book 4, Problem 5).[76] Accordingly, the mood of line 1 is AAA, of line 2, EAE, of line 3, AEE, of line 4, AAJ, of line 5, EAO, of line 6, AEO, of line 7, AJJ, of line 8, EJO, of line 9, AOO, of line 10, JAJ, of line 11, OAO, and of line 12, JEO, that is, with the deletion of the consonants from the common names with which the Scholastics designated figures by clusters of consonants and moods by single vowels. But the final mood, JEO, which I have called *Frisesmo* and placed in no figure at all, is thereby useless, because the major is P, and so has no place in figure 1 and 2, and the minor is N and so has no place in figure 1 and 3, even though it is not useless according to the rules of the moods. And that it cannot have a place in figure 4 either, I show by means of this example: *Some beings are men, No man is a brute, therefore some brute is not a being.*

29. And here, in passing, I shall offer some useful advice (which is to some extent supported by this example) on what, so to speak, constitutes Proof, or the art of examining a proposed mood, and if by chance it does not follow by virtue of form but in virtue of matter, immediately finding an instance, something I do not recall reading that Logicians have ever done.[77] To summarise: for UA, choose a proposition whose matter does not permit of simple conversion, for example, this proposition, *All men are animals*, rather than *All men are rational animals*; and the more remote the genus you choose, the better it will serve. For UN, choose a proposition in which species are denied of each other as near neighbours under the same proximate genus, for example, man and brute;

---

[76] Regius (1601). Regius's bulky work knew many editions. In the Wittenberg edition by Berger and Schurer (1615), his use of symbols may be found on pp. 163–70 (2nd series). Johannes Regius Dantiscanus (1567–1605) was a Prussian philosopher and theologian who played a minor role in the biography of the young Johannes Kepler, whose wrath he aroused while being *Rektor* of the *Stiftsschule* in Graz. Cf. *Jöcher* III, col. 1964; Adelung and Rotermund (1819, cols 1564–5); and Schmidt (1970: 65, 223).

[77] What Leibniz says here is not entirely accurate: in the *Prior Analytics*, Aristotle employs instances as counterexamples to refute invalid syllogisms.

convertibilis per contrapositionem in UA, seu cujus neque subjectum neque praedicatum sit terminus infinitus. Pro P(J)A sumatur semper talis quae non sit subalterna alicujus UA, sed in qua de genere quam maxime generali dicatur species particulariter. Pro (J)PN sumatur quae non sit subalterna alicujus UN, et cujus neuter terminus sit infinitus, et in qua negetur de genere maxime remoto species.

30. Quod diximus de Terminis infinitis vitandis, ejus ratio nunc patebit: Prodiit cujusdam Joh. Christoph. Sturmii Compendium Universalium seu Metaphysicae Euclideae, ed. 8, Hagae anno 1660 apud Adrian. Vlacq. Cui annexuit novos quosdam modos syllogisticos a se demonstratos, qui omnes videntur juxta communem sententiam impingere in alteram vel utramque harum duarum regularum qualitatis: ex puris negativis nihil sequitur; et conclusio sequitur qualitatem debilioris ex praemissis. Ut tamen recte procedat argumentum, vel assum[a]t propositionem affirmativam infiniti subjecti, quae stet pro negativa finiti; aut contra, v.g. aequipollent: Quidam non lapis est homo: et quidam lapis non est homo. (Verum annoto, non procedere in universali,

and which is not convertible by contraposition into UA, that is, neither its subject nor its predicate is an infinite term.[78] For PA and JA, always choose a proposition that is not the subaltern of any UA, but in which the species is affirmed particularly of the widest possible genus. For JN and PN, choose a proposition that is not the subaltern of any UN, of which neither term is infinite, and in which a species is denied of the most remote genus.[79]

30. The reason why I spoke of the need to avoid infinite terms will now emerge.[80] A certain Johann Christoph Sturm attached to his *Compendium of Universals, or Euclidean Metaphysics* (8th edition, published in The Hague, 1660, by Adrian Vlacq) some new syllogistic moods proved by himself, which by common consent all impinge on one or other of the two rules of quality: from purely negative premises nothing follows; and the conclusion adheres to the quality of the weaker of the premises.[81] But in order to make good his argument he chooses either an affirmative proposition of infinite subject, which must do service for a negative proposition of finite subject, or the opposite. For example, *Some non-stones are men* and *Some stones are not men* are equivalent. (I take note of the fact that it does not hold of universals; on the contrary, for example, *All stones are not men, therefore all non-stones are men*). Or else he has to choose either a negative proposition of infinite predicate in place of an affirmative proposition of finite predicate, or the opposite. For example, *All philosophers are not non-men*, and *All philosophers are men* are equivalent. Or thirdly, he has to choose in place of a given proposition its converse by means of contraposition.[82] Now UA converts by

[78] An 'infinite term' is a term with the sign of negation prefixed, as, for example, 'non-man', 'non-animal', etc.

[79] On Leibniz's method for finding a counterexample, see above; Introduction, § 10.

[80] Here, the text of the Academy edition mistakenly has *partebit* instead of *patebit*.

[81] A full version of Sturm (1661) has been made available on the internet by the University of Göttingen's *Digitalisierungszentrum* (under the title *Universalia Euclidea*). The *Novi Modi Syllogizandi* appears on pp. 69–116 of the volume. On Sturm, see also: *Zedler* 40, cols 1417–23 and DBE 9: 617.

[82] According to traditional syllogistic, to *convert* a proposition of the general 'subject-predicate' form means to transform it into a proposition with the same predicate in the place of the subject and vice versa. There are, however, two kinds of *conversion*: simple and by accident. Universal negative and particular affirmative propositions admit of *simple conversion*, i.e. from 'No A are B' and 'Some A are B' we can infer, respectively, 'No B are A' and 'Some B are A'. Universal affirmative propositions, instead, are susceptible of a *conversion by accident* (*per accidens*): i.e. from 'All A are B' we can only infer 'Some B are A'. The particular negative proposition does not convert (even though in the *DAC* Leibniz says that it does). The conversion by contraposition is a direct inference according to which one may infer 'No B is not A' from 'All A are B'.

contra, v. g. Omnis lapis non est homo. E. omnis non lapis est homo.)
Vel assumat negativam infiniti praedicati pro affirmativa finiti; vel
contra, v. g. aequipollent: Omnis philosophus non est non homo; et:
est homo. Vel 3. assumat loco datae conversam ejus per contrapositionem. Jam UA convertitur per contrap. in UN; UN et PN in PA. Ita
facile illi est elicere ex puris neg. affirmantem, si negativae ejus tales
sunt ut stent pro affirmativis; item ex A et N elicere affirmantem, [24]
si ista stet pro negativa. Ita patet omnes illas 8 variationes Qualitatis
fore utiles, et per consequens modos utiles fore $32 \cap 8$ f. 256, juxta
nostrum calculum. Similis fere ratio est syllogismi ejus de quo Logici
disputant: Quicunque non credunt, damnantur. Judaei non credunt.
E. damnantur. Sed ejus expeditissima solutio est, minorem esse affirmantem; quia Medius terminus affirmatur de minore. Medius terminus autem non est: credere, sed: non credere, id enim praeextitit in
majore prop.

[31.] Non possum hic praeterire modum Daropti ex ingenioso invento Cl.
Thomasii nostri. Is observavit ex Ramo Schol. Dialect. lib. 7 c. 6, pag.
m. 214. Conversionem posse demon[*187*]strari per Syllogismum
adjiciendo propositionem identicam; v. g. UA in PA. Sic: Omne α est
γ; Omne α est α (si in 3tiae modo Darapti velis; vel omne γ est γ, si in
4tae modo Baralip); E. quoddam γ est α. Item PA in PA. Sic: Quoddam α
est γ; Omne α est α (si in 3tiae modo Disamis velis, vel omne γ est γ, si in
4tae modo Ditabis); E. quoddam γ est α. Item UN in UN (in Cesare
2dae) sic: Nullum α est γ; Omne γ est γ; Ergo nullum γ est α. Item PN in
PN vel in Baroco 3tiae sic: Omne α est *a*; Quoddam α non est γ;
E. quoddam γ non est α. (Vel in Colanto 4tae: Quoddam α non est γ;

contraposition into UN, and UN and PN into PA,[83] so it is easy in those cases to derive an affirmative proposition from purely negative propositions, if its negative premises are such that they can do service for affirmatives; and again, it is easy to derive an affirmative from premises in A and N, if it does service for a negative. Hence, it emerges that all the 8 variations of Quality will be useful, and that in consequence there will be, according to my calculations, 32 ∩ 8 f. 256 useful moods. The argument is broadly similar regarding a syllogism about which Logicians dispute: *Whoever does not believe is damned, The Jews do not believe, therefore they are damned.* But the most straightforward solution of it is that the minor premise is affirmative, since the Middle term is affirmative of the minor term, the Middle term being not *believe,* but *not believe*: which was there already in the major proposition.[84]

[31.] And here, I cannot pass by the mood Daropti, an ingenious invention of our own celebrated Thomasius.[85] He observed in Ramus's *Scholastic Dialectics,* Book 7, Chapter 6, p. 214, that the conversion of a proposition can be demonstrated through a Syllogism by adding to it an identical proposition.[86] For example, the conversion of UA into PA, thus: *All α are γ, All α are α,* (in the mood *Darapti* of the third figure, or *All γ are γ* in the mood Baralip of the fourth figure, whichever you prefer), therefore *Some γ are α.* Again, the conversion of PA into PA, thus: *Some α are γ, All α are α* (in the mood *Disamis* of the third figure, or if in the mood Ditabis of the fourth figure, *All γ are γ,* whichever you prefer), therefore *Some γ are α.* Again, the conversion of UN into UN (in Cesare of the second figure), thus: *No α are γ, All γ are γ,* therefore *No γ are α.* Again, the conversion of PN into PN either in *Baroco* of the third figure, thus: *All α are α, Some α are not γ,* therefore *Some γ are not α* (or in Colanto of the fourth figure: *Some α are not γ, All γ are γ,* therefore *Some γ are not α*). He then tried to do the same for Conversion through

---

[83] Here Leibniz seems to have in mind the so-called *partial contraposition,* according to which, from 'All S are P' one infers 'No not-P is S' and from 'No S is P' and 'Some S is not P' it is possible to infer 'Some not-P is S'.

[84] Leibniz's remark becomes clearer if we consider that, according to the traditional analysis of the proposition, any verb must be decomposed in two parts: the copula and the participle of the verb itself. Thus, *Quicunque non credunt* ('Whoever does not believe') becomes *Quicunque sunt non-credentes* ('Whoever is a non-believer'), the middle term being *non-credentes.*

[85] The philosopher Jacob Thomasius (1622–84) was Leibniz's teacher at the University of Leipzig, where he taught moral philosophy and rhetoric. Cf. DBE 10: 20–1. Thomasius's son, the philosopher and lawyer Christian Thomasius (1655–1728), would become known as 'father of the German Enlightenment'.

[86] Ramus (1581: 209–11).

Omne γ est γ; Ergo quoddam γ non est α). Idem igitur ipse in Conversione per Contrapositionem tentavit, v. g. hujus PN: Quidam Homo non est Doctus, in hanc PA infiniti subjecti[:] Quoddam non doctum est homo. Syllogismus in Daropti erit talis: Omnis homo est homo; Quidam homo non est doctus; E. quoddam quod non est doctum est homo.

[32.] Observari tamen hic duo debent, Minorem juxta Sturmianam doctrinam videri quasi pro alia positam: Quidam homo est non doctus; deinde [omnino] optime sìc dici: propositionis hujus: Quidam Homo non est doctus, conversam per contrapositionem proprie hanc esse etiam negativam: *Quoddam doctum non est non non homo*, et in conversione per contrapositionem identicam ipsam debere esse contrapositam; id ostendit Syllogismus jam non amplius in Daropti, sed Baroco: *Omnis homo est non non homo* (id est: omnis [25] homo est homo). *Quidam homo non est doctus.* Ergo *Quoddam doctum non est non non homo* (id est: quoddam non doctum est homo).

33. Caeterum Sturmianos illos modos arbitror non formae sed materiae ratione concludere, quia quod termini vel finiti vel infiniti sint non ad formam propositionis seu copulam aut signum pertinet, sed ad terminos. Desinemus tandem aliquando Modorum, nam etsi minime pervulgata attulisse speramus, habet tamen et novitas taedium in per se taediosis. Ab instituto autem abiisse nemo nos dicet, qui omnia ex intima Variationum doctrina erui viderit: quae sola prope per omne infinitum obsequentem sibi ducit animum; et harmoniam mundi, et intimas constructiones rerum, seriemque formarum una complectitur. Cujus incredibilis utilitas perfecta demum Philosophia, aut prope perfecta, recte aestimabitur.

Contraposition, for example, of this PN, Some men are not learned, into this PA of infinite subject, *Some non-learned things are men.* The Syllogism, in Darapti, will be this: *All men are men, Some men are not learned,* therefore *Some things that are not learned are men.*[87]

[32.] But here I must make a couple of observations. The Minor looks, according to Sturm's system, as if it has been chosen in place of another proposition, *Some men are non-learned.* Then, we say, in an excellent way: the converse through contraposition of this proposition, *Some men are not learned* should properly be this, also negative, proposition, *Some learned things are not non-non-men,*[88] and the fact that in conversion through contraposition the contrapositive should be one and the same shows that it is now no longer a Syllogism in Darapti, but in *Baroco*: *All men are non-non-men* (that is, *All men are men*), *Some men are not learned, therefore some learned things are not non-non-men (that is, Some non-learned things are men).* Secondly, I think that those Sturmian moods follow by virtue not of form but of matter, because whether terms are either finite or infinite depends on the terms, not on the form of the proposition, i.e. the copula or the sign.[89]

33. I shall at long last leave the Moods, for even though I hope I have brought something, however little, to such commonplace matters, in naturally tedious things even novelty becomes tedious. But no-one will say that I have deviated from my purpose who sees that all these things have been derived from the heart of my system of Variations, which alone leads the mind that yields to it almost through all infinity, and embraces at once the harmony of the world, the inner workings of things, and the series of forms.[90] The incredible benefit of this to the perfecting, or near-perfecting, of philosophy will in the end be appreciated at its proper worth.

---

[87] The argument, however, does not occur in Thomasius (2003), Chapter 43, 'De Reductione', pp. 101–8, where the rules of conversion are given.

[88] But the 'full contrapositive' of *Some men are not learned*, according to the prevailing scholastic tradition is *Some not learned things are not men*; whereas the 'partial contrapositive' is *Some not learned things are men*. Cf. Keynes (1906: 135).

[89] That is: the *form* of a proposition depends on the copula and the quantifiers ('the sign').

[90] The idea that the universe is organized according to a hierarchical order, the so-called Great Chain of Being, starting with God and progressing downwards to the angels, to man and other animals, plants, and minerals, was deeply rooted in the tradition of Western philosophy. It may be found in Plato's writings and was further developed by Plotinus and the Neo-Platonic school, to become very influential in the sixteenth and seventeenth centuries. See the classical study by Lovejoy (1936).

34. Nam Vllmus est in complicandis figuris geometricis usus, qua in re glaciem fregit Joh. Keplerus lib. 2. Harmonicωv. Istis complicationibus, non solum infinitis novis Theorematis locupletari Geometria potest, nova enim complicatio novam figuram compositam efficit, cujus jam contemplando proprietates, nova theoremata, novas demonstrationes fabricamus; sed et (si quidem verum est grandia ex parvis, sive haec atomos, sive moleculas voces, componi) unica ista via est in arcana naturae penetrandi. Quando eo quisque perfectius rem cognoscere dicitur, quo magis rei partes et partium partes, earumque figuras positusque percepit. Haec figurarum ratio primum abstracte in Geometria ac Stereometria pervestiganda: [*188*] inde ubi ad historiam naturalem existentiamque, seu id quod revera invenitur in corporibus, accesseris, patebit Physicae porta ingens; et elementorum facies, et qualitatum origo et mixtura, et mixturae origo, et mixtura mixturarum, et quicquid hactenus in natura stupebamus.

35. Caeterum brevem gustum dabimus quo magis intelligamur: Figura omnis simplex aut rectilinea aut curvilinea est. Rectilineae omnes symmetrae, commune enim omnium principium: Triangulus. Ex cujus variis *complicationibus congruis*, omnes *Figurae* rectilineae *coeuntes* (id est non hiantes) [**26**] oriuntur. Verum curvilinearum neque circulus in ovalem etc. neque contra reduci potest, neque ad aliquid commune. Neutra vero triangulo et triangulatis symmetros. Porro quilibet circulus cuicunque circulo est symmetros, nam quilibet cuilibet aut concentricus

# VII

34. The seventh application is to the complications of geometrical figures, a subject in which Johann Kepler broke the ice in Book 2 of his *Harmonics*.[91] With these complications, not only can geometry be enriched by an infinite number of new Theorems, as every complication brings into being a new, compound figure, by the contemplation of whose properties we may devise new theorems and new demonstrations; but (if it is indeed true that great things are made up of little things, whether you call them atoms or molecules) we also have a unique way of penetrating into the arcana of nature. This is because the more one has perceived of the parts of a thing, the parts of its parts, and their shapes and arrangements, the more perfectly one can be said to know the thing. The structure of these figures is first to be investigated abstractly, in plane and solid geometry: after which, when you enter upon natural history and the question of being, that is, upon the question of the real constitution of bodies, the vast portals of Physics will stand open, and the character of the elements, the origin and mixture of the qualities, the origin of mixtures, the mixing of those mixtures, and everything that formerly lay hidden in darkness will be revealed.[92]

35. I shall now give you a brief taste of this in order to make things clearer. Every simple figure is either rectilinear or curvilinear. All rectilinear figures are symmetrical, since they all have one principle in common: the Triangle, from whose *convex complications* all *convex* (that is, not concave) rectilinear *Figures* arise. But of curvilinear figures, a circle cannot be reduced to an oval, nor the opposite, nor can they be reduced to anything which is common to both. And neither is symmetrical with a triangle or anything composed of triangles. Indeed, every circle is symmetrical with every other circle, for each is concentric to the other either in fact or in thought. But an oval or Ellipse can be conceived as

[91] In Book II of his *Harmony of the World*, 'De Congruentia Figurarum Harmonicarum', also referred to as the *Architectonicus, seu ex Geometria Figurata, De Figurarum Regularum Congruentia in plano vel solido*, Johannes Kepler discusses the congruence of regular figures, offering twelve figures in total "which will form congruences", viz. eight "basic, or primary figures" (trigon, tetragon, pentagon, hexagon, decagon, dodecagon, and icosagon) and four "augmented or star figures" (star pentagon, star octagon, star decagon, and star dodecagon). Kepler (1997: 124). Cf. Kepler (1619: 65–6).

[92] Leibniz's Keplerian interest in the mathematics of compound figures is apparently fuelled by its potential relevance for corpuscularian physics.

est aut esse intelligitur. Ovalis vero vel Elliptica ea tantum symmetros quae concentrica esse intelligitur. Ita neque omnis ovalis ovali symmetros est etc. Haec de simplicibus, jam ad complicationes.

[36.] Complicatio est aut congrua aut hians. Congrua tum cum figurae compositae lineae extremae seu circumferentiales nunquam faciunt angulum extrorsum, sed semper introrsum. *Extrorsum* a. fit angulus, cum portio circuli inter lineas angulum facientes descripta ex puncto concursus tanquam centro, cadit extra figuram, ad cujus circumferentiam lineae angulum facientes pertinent: *introrsum*, cum intra. *Hians est complicatio*, cum aliquis angulus fit extrorsum. *Stella* autem est complicatio hians, cujus omnes *radii* (id est lineae stellae circumferentiales angulum extrorsum facientes) sunt aequales; ita ut si circulo inscribatur, ubique eum radiis tangat. Caeterum hiantes figurarum complicationes *texturas* voco, congruas proprie *figuras*. Sunt tamen et quaedam *Texturae figuratae*, quas et *figuras hiantes* ad opposition[e]m *coeuntium* voco.

37. Jam sunt theoremata: 1. Si duae figurae asymmetrae sunt contiguae (*complicatio* enim vel immediata est *contiguitas*; vel mediata, inter tertium et primum, quoties tertium contiguum est secundo, et secundum vel mediate vel immediate primo), complicatio fit hians. 2. Curvilinearum inter se omnis contiguitas est hians, nisi alteri circumdetur Zona alterius symmetri dato concentrici. 3. Curvilineae cum rectilinea omnis contiguitas est hians, nisi in medio Zonae ponatur rectilinea. *Zonam* autem voco residuum in figura curvilinea majori, exempta concentrica minori. In contiguitate Rectilinearum autem aut angulus angulo, aut angulus lineae, aut linea lineae imponitur. 4. Si angulus angulo imponitur aut lineae, contiguitas est in puncto. 5. Omnis curvilinearum inter se contiguitas hians est in puncto. 6. Omnis earum cum rectis contiguitas etiam non hians, [27] itidem. 7. Linea lineae nonnisi ejusdem generis imponi potest, v. g. recta rectae, curvilinea ejusdem generis et sectionis. 8. Si linea lineae aequali imponatur, contiguitas est congrua, si inaequali, hians. [*189*]

symmetrical only with one that is concentric. Thus, it is not the case that every oval is symmetrical with every other oval, etc. So much for simple figures, now for complications.

[36.] A complication is either convex or concave. It is convex when the outermost, or circumferential lines of a composite figure never make an obtuse angle, but always an acute angle. An *obtuse* angle is made when the portion of a circle described between the lines making the angle, with the point of intersection as centre, falls outside the figure to whose circumference the lines making the angle belong; an angle is *acute* when the portion of the circle falls inside the figure. A *complication* is *concave* when an obtuse angle is made. In particular, a *star* is a concave complication, all of whose *radii* (that is, circumferential lines of the star that make an obtuse angle) are equal; so that if it is inscribed within a circle all the radii must touch the circle. Finally, I call concave complications of figures *textures*, and convex complications *figures* in the strict sense. But there are some *figurate Textures* that I also call *concave figures*, as opposed to *convex figures*.

37. Now for some theorems. 1. If two asymmetrical figures are contiguous (for a complication is either an immediate *contiguity*, or it is mediate, between a third and a first, when the third is contiguous with the second, and the second is mediately or immediately contiguous with the first), the complication always becomes concave. 2. Every contiguity between curvilinear figures is hiant, unless one curve is surrounded by a Zone of another symmetrical curve concentric with the given curve. 3. Every contiguity of a curvilinear figure with a rectilinear figure is concave, unless a straight line is situated in the middle of a Zone: and by Zone I mean a straight section in a mostly curvilinear figure with a smaller concentric section removed. And as regards the contiguity of Rectilinear figures, either an angle is placed on an angle, an angle on a line, or a line on a line. 4. If an angle is placed on either an angle or a line, the contiguity is in a point. 5. Every concave contiguity of curvilinear figures is in a point. 6. Every contiguity of curvilinear figures with straight lines, even if it is non-concave, is similarly a point. 7. A line cannot be placed on a line except on one of the same kind, for example, a straight line on a straight line, a curve on one of the same kind and section. 8. If a line is placed on an equal line, the contiguity is convex; if it is placed on an unequal line, the contiguity is concave.

3[8.]  Observandum a. est plures figuras ad unum punctum suis angulis componi posse, quae est textura omnium maxime hians. Sed et hoc fieri potest, ut duae vel plures contiguae sint hiantes, accedat vero tertia vel plures, et efficiatur una figura, seu complicatio congrua. Unde nova contemplatio oritur, quae figura vel textura quibus addita faciat ex textura figuram. Quod nosse magni momenti est ad rerum hiatus explendos. Restat ut computationem ex nostris praeceptis instituamus, ad quam requiritur ut determinetur numerus figurarum ad conficiendam texturam; et determinentur figurae complicandae; utrumque enim alias infinitum est. Sed hoc facile cuilibet juxta enumeratos casus et theoremata praestare; nobis ad alia properantibus satis est prima lineamenta duxisse tractationis de Texturis hactenus fere neglectae. Decebat fortasse doctrinam hanc illustrare schematibus, sed intelligentes non indigebunt; imperiti, uti fieri solet, nec intelligere tanti aestimabunt.

39.  VIIIvus Usus est in casibus apud Jureconsultos formandis. Neque enim semper expectandum est praecipue legislatori, dum casus emergat; et majoris est prudentiae leges quam maxime initio sine vitiis ponere, quam restrictionem ac correctionem fortunae committere. Ut taceam, rem judicariam in qualibet republica hoc constitutam esse melius, quo minus est in arbitrio judicis. Plato lib. 9 de Leg. Arist. 1. Rhet. Menoch. Arbitr. Jud. lib. 1, prooem. n. 1.

40.  Porro Ars casuum formandorum fundatur in doctrina nostra de Complexionibus. Jurisprudentia enim cum in aliis Geometriae similis est, tum in hoc quod utraque habet Elementa, utraque casus. Elementa sunt simplicia, in Geometria figurae triangulus, circulus etc., in Jurisprudentia actus promissum, alienatio etc. Casus: complexiones horum, qui utrobique variabiles sunt infinities. Elementa Geometriae composuit Euclides, Elementa juris in ejus corpore continentur, utrobique tamen admiscentur Casus insigniores. Terminos autem in jure

38. Further, it must be observed that several figures can be brought together at a point by their angles, a texture that is the most concave of all. But it can also happen that two or more contiguous figures are concave, and that a third is added, or even more, resulting in a single figure that is a convex complication. In this way, we get a new complication which can make a figure out of a texture with a figure or texture added to it. The recognition of this is of great importance to filling in the gaps in things. All that remains is for me to carry out the calculation using my precepts, requiring the number of figures making up the texture and the complications of the figures, for both are sometimes endless. But it is easy for anyone to prove the cases and theorems just stated. As I am anxious to move on to other matters it is enough for me to have sketched the first outlines of the hitherto neglected subject of Textures. It would perhaps have been better to illustrate this subject with diagrams, but intelligent people will not need this; as for the ignorant, they will not place a high value on understanding it.

## VIII

39. The eighth Application is to the forming of cases among Jurists. For the legislator, in particular, cannot always wait for a case to emerge; and it is more prudent to make laws free of faults at the very outset than to trust to haphazard restrictions and amendments, not to mention that judicial affairs in any state are better constituted the less they are subject to the decisions of judges. See Plato's *Laws*, Book 9; Aristotle, *Rhetoric*, Book 1; Giacomo Menochio, *On the Decisions of Judges*, Book 1, Introduction, paragraph 1.[93]

40. Further, the Art of forming cases is founded on my doctrine of Complexions. For Jurisprudence is in some respects similar to geometry, especially in that both have Elements and both have cases. The Elements are the simple things, in geometry such figures as the triangle, the circle, etc., and in Jurisprudence acts such as contract, transfer, etc. The Cases are complexions of these things, which are in both instances infinitely variable. Euclid laid down the Elements of Geometry; while the Elements of law are contained in its Corpus. In both instances, however, the

---

[93] Cf. Plato, *Laws* IX, 875d–876a and Aristotle, *Rhetoric* I 1, 1354a31–1354b16. Leibniz here copies Menochio's references, who quoted both passages from Plato and Artistotle in the *Prooemium* to the first part of Menochio (1615: 1).

simplices, quorum [28] mixtione caeteri oriuntur, et quasi Locos communes, summaque genera colligere instituit Bernhardus Lavinheta Monachus ordinis Minorum Com. in Lullii Artem Magnam, quem vide. Nobis sic visum: Termini quorum complicatione oritur in Jure diversitas casuum, sunt: Personae, Res, Actus, Jura.

41. *Personarum* genera sunt tum naturalia, ut: Mas, Foemina, Hermaphroditus, Monstrum, Surdus, Mutus, Caecus, Aeger, Embryo, Puer, Juvenis, Adolescens, Vir, Senex, atque aliae differentiae, ex Physicis petendae quae in jure effectum habent specialem: Tum artificialia, nimirum genera vitae, corpora seu Collegia et similia. Nomina officiorum huc non pertinent, quia complicantur ex potestate et obligatione, sed ad jura.

42. RES sunt mobiles, immobiles, dividuae (homogeneae), individuae, corporales, incorporales; et speciatim: Homo, animal cicur, ferum, rabiosum, noxium; Equus, aqua, fundus, mare etc. Et omnes omnino res de quibus peculiare est jus. Hae differentiae petendae ex Physicis. [*190*]

43. ACTUS (a. non actus, s. status) considerandi qua naturales: ita dividui, individui, relinquunt ἀποτελέσμα vel sunt facti transeuntis; Detentio quae est materiale possessionis, traditio, effractio, vis, caedes, vulnus; noxa, huc temporis et loci circumstantia, hae differentiae itidem petendae ex physicis; qua morales: ita sunt actus spontanei, coacti, necessarii, mixti; significantes, non significantes; inter significantes verba, consilia, mandata, praecepta, pollicitationes, acceptationes, Conditiones. Hic omnis verborum varietas et interpretatio ex Grammaticis. Denique actus sunt vel juris effectum habentes, vel non habentes; et illi quidem pertinent ad catalogum jurium quae efficiunt, hi ex Politicis Ethicisque uberius enumerandi.

more prominent Cases are included as well. And the simple terms in law, out of whose mixing everything else arises, and which are, as it were, its Common Places, Bernardo de Lavinheta, a Monk of the Order of the Friars Minor, set out to collect in his *Explanation and Compendious Application of the Art of Ramon Lull*, to which I refer you.[94] It seems to me that the terms from whose complication the diversity of cases arises in Law are Persons, Things, Acts, and Laws.

41. The kinds of *Persons* are, first, the natural kinds, such as Male, Female, Hermaphrodite, Monster, Deaf, Mute, Blind, the Sick, the Unborn, Boy, Youth, Adolescent, Man, Old Man, and other differences to be ascertained from natural philosophers that have some special significance in law. Then there are, of course, the artificial kinds of life, Corporations or Colleges, and suchlike. The names of offices are not relevant here, because they are involved not through power or obligation, but through laws.

42. *THINGS* are movable, immovable, divisible (homogeneous), indivisible, corporeal, incorporeal, and in particular, man, domestic animal, wild animal, savage animal, dangerous animal, Horse, water, land, sea, etc. And in general, everything governed by a peculiar law. These differences have to be ascertained from natural philosophers.

43. *ACTS* (as well as non-acts, or states) are to be considered, firstly, as natural: divisible or indivisible, they either leave behind something completed[95] or they are of transient effect. For example, Detention (which is what is material to possession), surrender, burglary, force, killing, wounding, and injury, with the addition of circumstances of time and place. These differences have likewise to be ascertained from natural philosophers. Secondly, as moral, such as spontaneous acts, compulsory acts, necessary acts, mixed acts, significant acts, and insignificant acts; and among significant acts, words, counsels, mandates, precepts, promises, acceptances, and conditions. All this variety of words and their interpretations has to be ascertained from Grammarians. Finally, acts either have the force of law or they do not: the former belong to the corpus of the laws that are in force, while the latter may be collected in abundance from politicians and moralists.

---

[94] Bernard de Lavinheta (*c*. 1475–*c*. 1530) was a Franciscan monk of Basque origin who taught at Salamanca and Paris. He published an *Explanatio compendiosaque applicatio artis Raymundi Lulli ad omnes facultates* (Lavinheta, 1523). The discussion related to the juridical sciences occurs at pp. 417–31. Leibniz probably used Alsted's edition. Cf. Bernhardus de Lavinheta (1612, vol. 2: 520–39, esp. Chapter VI: 536–9),: 'De modo reducendi omnia jura positiva & particularia ad veram scientiam & artem.'

[95] The Greek word ἀποτέλεσμα means 'outcome'.

44. JURIUM itidem enumerandae vel species vel differentiae. Et hae quidem sunt v. g. realia, personalia; pura, dilata, suspensa; mobilia vel personae aut rei affixa etc. Species v. g. Dominium, directum, utile; Servitus, realis, personalis; Ususfructus, usus, proprietas, Jus possidendi, Usucapiendi conditio. Potestas, obligatio (active sumta). Potestas administratoria, rectoria, coercitoria. Tum actus judiciales sumti [**29**] pro jure id agendi; tales sunt: postulatio, seu jus exponendi desiderium in judicio, cujus species pro ratione ordinis: Actio, Exceptio, Replica etc. nempe in termino; tum in scriptis aut alias extra terminum; supplicatio pro impetranda citatione, pro Monitorio etc. Jurium a. catalogus ex sola Jurisprudentia sumitur.

45. Nos hic festini quicquid in mentem venit attulimus, saltem ut mens nostra perspiceretur; alii termini simplices privata cujusque industria suppleri possunt. Sed ita ut eos tantum ponat terminos, qui revera sunt simplices, id est quorum conceptus ex aliis homogeneis non componitur. Quanquam in locis communibus, quorum disponendorum artificium potissimum huc redit, licebit terminos complexos simplicibus valde vicinos etiam tanquam peculiarem titulum collocare, v. g. Compensationem, quae componitur ex *obligatione* Titii Cajo, et *ejusdem* Caji Titio *in rem dividuam, homogeneam* seu *commensurabilem* quae utraque dissolvitur in summam concurrentem.

46. Ex horum Terminorum simplicium, tum cum seipsis aliquoties repetitis, tum cum aliis, com2natione, con3natione etc. et in eadem complexione, variatione situs prodire casus prope infinitos quis non videt? Imo qui accuratius haec scrutabitur, inveniet regulas eruendi casus singulariores. Ac nos talia quaedam concepimus, sed adhuc impolitiora, quam ut afferre audeamus.

4[7]. Par in Theologia terminorum ratio est, quae est quasi Jurisprudentia quaedam specialis, sed eadem fundamentali ratione caeterarum. Est enim velut doctrina quaedam de Jure publico quod obtinet in Republica DEI in homines; ubi *Infideles* quasi rebelles sunt; *Ecclesia* velut subditi boni; *personae Ecclesiasticae*, imo et *Magistratus Politicus* velut Magistratus subordinati; *Excommunicatio* velut Bannus; Doctrina de *scriptura* [*191*] *sacra* et *verbo DEI* velut de Legibus et earum interpretatione; de *Canone*, quae leges authenticae; de *Erroribus fundamentalibus* quasi de Delictis capitalibus; de *Judicio extremo, et*

44. The species and differences of *LAWS* similarly have to be listed. Their differences are: real or personal; pure, expanded, or suspended; transferable, or affixed to persons or things, etc. Examples of species are: Absolute ownership, direct or practical; Servitude, real or personal; Usufruct, use, ownership, legal title to possession, the condition of Usucaption; Power, obligation (in the active sense), administrative, directive, or coercive Power. Next, judicial acts allowing a legal action, such as postulation, that is, the legal expression in judgement of a demand, the kinds of which are, in logical order, Action, Exception, Replication, that is, acts within term; then, acts in writing or otherwise out of term, supplication to plead for a summons, for Monition, and so on; to take the corpus of Jurisprudence alone.

45. I have jotted down whatever came to mind, so as at least to offer what I mean to the examination of others; other simple terms can be supplied by the private effort of others; but in such a way that it may employ only those terms that are really simple, that is, whose concept is not composed of other, homogeneous things. Although this is the most effective strategy for disposing things in common places, it will be permissible to locate complex terms right next to simple terms as a special entitlement: for example, Compensation, which is composed of the *obligation* of Titius to Caius, and of the *same* Caius to Titius *in the matter of something divisible and homogeneous* or *commensurable*, each of which is *dissolved* into a *concurrent discharge.*

46. Who cannot see that an almost infinite number of cases may be produced from the com2nation, con3nation, etc., of these simple terms, first with themselves repeated several times, then with others, and finally in the same complexion with variation of situs? In fact, anyone who examines these things with extra care will discover rules for deriving ever more singular cases. And I myself have some ideas on this topic, but as yet so rudimentary that I would not presume to publish them.

47. The principles governing Terms are the same in Theology, which is something like a species of Jurisprudence, though founded on the same principles as other species. For it is a sort of doctrine of the civil Law to which men are subject in the Republic of God; where *Infidels* are like rebels; the *Church* like good subjects; Ecclesiastical persons, as well as State Officials, like lower Magistrates; *Excommunication* like Outlawry; the doctrine of *Holy Scripture* and the *word* of GOD like laws and their interpretation; the Doctrine of the *Canon*, about the laws that are genuine; the Doctrine of *fundamental errors* like that of Capital offences;

*novissima die*, velut de Processu Judiciario, et Termino praestituto; de *Remissione Peccatorum* velut de jure aggratiandi; de *damnatione aeterna* velut de Poena capitali etc.

48. Hactenus de usu Complexionum in [30] Speciebus Divisionum inveniendis, sequitur IXnus usus: datis speciebus divisionis, praedivisiones seu genera et species subaltemas inveniendi. Ac siquidem divisio cujus species datae sunt, est διχοτομία, locum problema non habet, neque enim ea est ulterius reducibilis; sin πολυτομία, omnino.

49. Esto enim τριχοτομία inter πολυτομίασ minima, seu dati generis species 3, a. b. c. con3natio igitur earum tantum 1 est in dato genere summo. 1niones vero 3. Illic ipsum prodit genus summum, hic ipsae species infimae, inter con3nationem autem et 1nionem, sola restat com2natio. Trium autem rerum com2nationes sunt 3, hinc oriuntur 3 genera intermedia, nempe abstractum, seu genus proximum των a. b., item των b. c., item των a. c. Ad genus a[utem] requiritur, tum ut singulis competat, tum ut cum omnibus disjunctive sumtis sit convertibile.

50. Exemplo res fiet illustrior. Genus datum sit Respublica, species erunt 3 loco A *Monarchia*, loco B *oligarchia Polyarchica* seu optimatum, loco [C] *Panarchia*, his enim terminis utemur commodissime, ut apparebit, et voce *Panarchiae*, etsi alio sensu, usus est Fr. Patritius, Tomo inter sua opera peculiari ita inscripto, quo Hierarchias coelestes explicuit. *Polyarchiae* voce tanquam communi oligarchiae et panarchiae usus est Boxhomius lib. 2, c. 5. Inst. Polit. Igitur 1. Genus subaltemum

the Doctrine of the *Last Judgement and the last day* like Judicial Process, and prescribed Term; *Remission of Sins* like a law of pardon; *eternal damnation* like capital Punishment, and so on.

## IX

48. After the application of Complexions to finding Species of Divisions, there now follows the ninth application: given the species of a division, to find predivisions, that is, subaltern genera and species. If a division into given species is a dichotomy, the problem does not arise, because the division is not further reducible; but if it is a polytomy, it is entirely relevant.

49. For consider a trichotomy, which is the least division among polytomies, that is, 3 species of a given genus, *a, b, c*. There is, accordingly, only 1 con3nation of them in the given highest genus, but there are 3 1nions, the highest genus coming from the former, and the lowest species from the latter. This leaves only com2ations. As there are 3 com2nations of three things, 3 intermediate genera arise, namely, the genus abstracted from, or proximate to *a, b*, the genus abstracted from, or proximate to *b, c*, and the genus abstracted from, or proximate to *a, c*. And it is required of a genus, firstly, that it should apply to the singulars, and secondly, that it should be convertible with all of them taken disjunctively.

50. An example will make the matter clearer. Let the given genus be a State. The species will amount to 3: *Monarchy*, represented by *A*, *Polyarchic oligarchy*, that is, the rule of the aristocracy,[96] represented by *B*, and *Panarchy*, represented by *C*. As we shall see, I use these terms for the sake of maximum convenience; and as for the word *Panarchy*, its use is due to Francesco Patrizi, albeit in a different sense, in one particular volume of his works with that title, in which he unfolded the celestial Hierarchies.[97] As for *Polyarchy* as a common word for oligarchy and panarchy, it is used in Boxhorn, *Political Institutions*, Book 2,

---

[96] 'Aristocracy': note that the term *optimati*, 'the best', is used here—a term equally occurring in a technical sense in early modern political theory.

[97] The Croatian Renaissance philosopher Francesco Patrizi (1529–97) divided his *Nova de universis Philosophia* (first published in 1591) into four main parts: *Panaugia, Panarchia, Pampsychia*, and *Pancosmia*, the second of which—*Panarchia*, or *Source of all principles*—explains the emanation of metaphysical principles in Neo-Platonic terms. Cf. Patrizi (1593: 1–48), (second series). A summary of Patrizi's *Panarchia* may be found in: Brickman (1941: 30–9). Note that Leibniz, by contrast, uses the term 'Panarchy' as a synonym for 'Democracy' here.

των A. B. seu Monarchiae et regiminis Optimatum, erit Oligarchia. Imperant enim vel non omnes, *Oligarchia* (*sed* vel *unus*, Monarchia, vel *plures* Oligarchia Polyarchica) vel omnes, *Panarchia*.

51. Genus subalternum των B. C. erit Polyarchia. Imperat enim vel unus, *Monarchia*, vel plures, *Polyarchia* (in qua iterum vel non omnes, *Polyarchia Oligarchica*, vel omnes, *Panarchia*).

52. Genus subaltenum των A. C. est Respublica extrema. Nam species reipublicae alia *intermedia* est optimatum (hinc et nomen duplex: oligarchia polyarchica), alia *Extrema*. Extremae autem sunt, in quibus imperat *unus*, item in quibus *Omnes*. Ita in minima των πολυτομιων, τριχοτομία usum complexionum manifestum fecimus, quantae, amabo in divisione virtutum in 11 species, similibusque aliis erunt Varietates? Ubi non solum singulae com2nationes, sed et con3nationes etc. usque ad con10nationes, eruntque computato [31] genere summo et speciebus infimis in universum complicationes seu genera speciesque possibiles 2047. [*192*]

53. Nam profecto tam est in abstrahendo foecundus animus noster, ut datis quotcunque rebus Genus earum, id est conceptum singulis communem, et extra ipsas nulli, invenire possit. Imo etsi non inveniat, sciet Deus, invenient angeli, igitur praeexistet omnium ejusmodi abstractionum fundamentum.

54. Haec tanta varietas generum subalternorum facit, ut in praedivisionibus, seu tabellis construendis, invenienda etiam datae alicujus in species infimas divisionis, sufficientia, diversas vias ineant autores, et omnes nihilominus ad easdem infimas species perveniant. Deprehendet hoc, qui consulet Scholasticos numerum praedicamentorum, virtutum cardinalium, virtutum ab Aristotele enumeratarum, affectuum, etc. investigantes.

Chapter 5.[98] Therefore (1) the subaltern Genus of *A, B*, that is, of Monarchy and the rule of the Aristocracy, will be Oligarchy. For either not all rule, as in *Oligarchy*, or all rule, as in *Panarchy*.

51. (2) The Subaltern Genus of *B, C*, will be Polyarchy. For either one rules, as in *Monarchy*, or many rule, as in *Polyarchy* (in which, again, either not all rule, as in *Polyarchy-Oligarchy*, or all rule, as in *Panarchy*).

52. (3) The subaltern Genus of *A, C*, is an extreme State. For one species of State is *intermediate*, that of the aristocracy (hence also the double name oligarchy-polyarchy), while the others are *Extreme*. States are Extreme in which either *one* rules or *All* rule. Here I have made an obvious application of complexions to a trichotomy, the minimum division among polytomies: imagine the number of varieties in the division of virtues into 11 species, and in other, similar divisions, where there will be not only single com2nations, but con3nations, etc., up to con10nations, and counting the highest genus and the lowest species, there will be in all 2047 complications, that is, possible genera or species.

53. For assuredly, the human mind is so prolific in abstraction, that given any number of things, it can find a Genus for them, that is, the concept shared by each of them and by nothing outside them. In fact, even if the mind does not find it, God will know it, and the angels will find it, so that the basis of all the mind's abstractions is pre existent.

54. All this profusion of subaltern genera means that in devising predivisions or schemes, even when enquiring into the sufficiency of any given division into its lowest species authors may approach it from different directions and yet all arrive at the same lowest species This will be readily understood by anyone who observes how the Scholastics investigate the number of predicaments, the number of cardinal virtues, the number of virtues as enumerated by Aristotle, the number of passions, and so on.

---

[98] Marcus Zuerius van Boxhorn (1612–53) was a Leiden professor of rhetoric (1633) and history (1648) who first formulated the idea of a kinship between a vast group of European languages, as well as Persian. Book II, Chapter 5, of his *Political Institutions*, discusses Aristocracy and its derivatives Tyranny (which, according to Boxhorn, often involves more than one person) and Olicharchy, as the main examples of Polyarchy. Cf. Boxhornus (1668: 319–32). On Boxhorn, see also Brugmans (1974, cols 178–80).

55. X. A Divisionibus ad Propositiones tempus est ut veniamus, alteram partem Logicae inventionis. Propositio componitur ex subjecto et praedicato, omnes igitur propositiones sunt com2nationes. Logicae igitur inventivae propositionum est hoc problema solvere: *1. dato subjecto praedicata, 2. dato praedicato subjecta invenire, utraque tum affirmative, tum negative.*

56. Vidit hoc Raym. Lullius Kabbalae Tr. I. c. I, fig. I, p. 46. et ubi priora repetit pag. 239 Artis Magnae. Is, ut ostendat, quot propositiones ex novem illis suis terminis Universalissimis: *Bonitas, magnitudo, duratio,* etc. quas singulas de singulis praedicari posse dicit, oriantur, describit Circulum, ei inscribit ἐννεάγωνον figuram regularem, cuilibet angulo ascribit terminum, et a quolibet angulo ad quemlibet ducit lineam rectam. Tales lineae sunt 56, tot nempe quot com2nationes 11 rerum. Cumque variari situs in qualibet com2natione possit bis, seu propositio quaelibet converti simpliciter, prodibit 36 ∩ 2 f. 72, qui est numerus propositionum Lullianarum. Imo talibus complexionibus omne artificium Lullii absolvitur, v. ejusdem Operum Argentorati in 8, anno I598 editorum pag. 49, 53, 68, 135, quae repetuntur p. 240, 244, 245. Idem tabulam construxit ex 84 columnis constantem, quarum singulae continent 20 complexiones, quibus [32] enumerat con4ternationes suarum regularum literis alphabeticis denominatarum; ea tabula occupat pag. 260, 261, 262, 263, 264, 265, 266. Con3nationum vero tabulam habes

X

55. It is time to pass from Divisions to Propositions, another part of the
Logic of Invention. A proposition is composed of a subject and a
predicate, so that all propositions are com2nations. Hence, in the
inventive Logic of propositions the problem is to solve the following:
1. *Given a subject, to find its predicates.* 2. *Given a predicate, to find its
subjects; and in each case both affirmative and negative.*

56. Ramon Lull, in his *First Treatise on the Kabbalah*, Chapter 1, Figure 1,
p. 46, and where he repeats earlier matters on p. 239 of his *Ars Magna*,
saw this. In order to show how many propositions can arise out of his
nine most Universal terms, *Goodness, magnitude, duration,* and so on,
each of which he says can be predicated of any other, he draws a Circle,
inscribes in it a regular *nonagon,* assigns a term to each angle, and draws
a straight line from each angle to every other angle.[99] There are 36 of
these lines, that is, the number of com2nations of 9 things.[100] Since the
situs of each com2nation can be varied in two ways, that is, each
proposition can be simply converted, it will produce 36 ∩ 2 f. 72 as
the number of Lullian propositions. In fact, the whole art of Lull is
embodied in these complexions. See pp. 49, 53, 68, and 135, repeated on
pp. 240, 244, and 245, of his *Works,* published in 8 volumes in Strassburg, 1598.[101] In the same place, he constructed a table consisting of 84
columns, each of which contained 20 complexions, and with which he
enumerates the con4nations of his rules, denominated by letters of the
alphabet. This table occupies pp. 260-6.[102] But in Cornelius Agrippa's
*Commentary on the* Ars Brevis *of Lull* you have a table of Con3nations

---

[99] See Lullius (1598: 46), for the circle; the insertion of the Tabula occurring in between
pp. 44 and 45. In the 1617 Strassburg edition of the same work, the diagram in the First Treatise
on the Kabbalah containing the nine universal terms *Bonitas, Magnitudo, Æternitas seu
Duratio, Potestas, Sapientia, Voluntas, Virtus, Veritas, and Gloria* is inserted between pp. 44
and 45, with the nonagon illustration occurring on p. 45. Both elements are repeated in the
Proemium and first chapter of the *Ars Magna.* Cf. idem, 1598 ed., pp. 238–9; 1617 ed.,
pp. 219–20. On the development and various aspects of Lull's Art itself, see also: Bonner (2007).
[100] The 1666 original edition (and that of the Akademie, as well) mistakenly has '11'.
[101] These are references to further circles and stair-like tables containing declining series of
combinations occurring both in the *Treatise on the Kabbalah* and the *Ars Magna,* as well as a
figure in the *Treatise on the Kabbalah,* p. 68, representing a set of circles that may be used as
wheels to offer new subject and predicate combinations.
[102] Lullius (1598: 260-6).

apud Henr. Corn. Agrippam Com. in artem brevem Lullii quae occupat 9 paginas, a pag. 863 usque 871 inclusive. Eadem ex Lullio pleraque exequitur, sed brevius, Joh. Henr. Alstedius in Architectura artis Lullianae, inserta Thesauro ejus Artis Memorativae pag. 47 et seqq.

57. Sunt autem Termini Simplices hi: I. *Attributa absoluta*: Bonitas, Magnitudo, Duratio, Potestas, Sapientia, Voluntas, Virtus, Veritas, Gloria. II. *Relata*. Differentia, Concordantia, Contrarietas, Principium, Medium, Finis, Majoritas, Aequalitas, Minoritas. III. *Quaestiones*: Utrum, Quid, de Quo, Quare, Quantum, Quale, Quando, Ubi, Quomodo (cum Quo). IV. *Subjecta*: Deus, Angelus, Coelum, Homo, Imaginatio, Sensitiva, Vegetativa,[193]Elementativa, Instrumentativa. V. *Virtutes*: Justitia, Prudentia, Fortitudo, Temperantia, Fides, Spes, Charitas, Patientia, Pietas. VI. *Vitia*: Avaritia, Gula, Luxuria, Superbia, Acedia, Invidia, Ira, Mendacium, Inconstantia. Etsi Jan. Caecilius Frey via ad Scient. et art. part. XI. c. I. classem 3tiam et 6tam omittat.

58. Cum igitur in singulis classibus sint 9 res, et 9 rerum sint complexiones simpliciter, 511 totidem in singulis classibus complexiones erunt, porro ducendo classem in classem per prob. 3. 511, 511, 511, 511, 511 ∩ 511 f. 17804320388674561, Zensicub. de 511. Ut omittam omnes illas Variationes, quibus idem terminus repetitur, item quibus una classis repetitur, seu ex una classe termini ponuntur plures.

59. Et hae solum sunt complexiones, quid dicam de Variationibus Situs, si in complexiones ducantur. "Atque hic explicabo obiter Problema hoc: *variationes situs, seu dispositiones, ducere in complexiones.* Seu datis certis rebus omnes variationes tam complexionis seu materiae, quam

that occupies 9 pages, from p. 863 to p. 871 inclusive.[103] Johann
Heinrich Alsted follows Lull in most things, although not at such length,
in his *Construction of the Art of Lull*, included in his *Thesaurus of the Art
of Memory*, p. 47ff.[104]

57. Now the Simple Terms are these: I. *Absolute Attributes*. Goodness,
Magnitude, Duration, Power, Wisdom, Will, Virtue, Truth, and Glory.
II. *Relations*. Difference, Concordance, Contrariety, Beginning, Middle,
End, Majority, Equality, and Minority. III. *Questions*. Whether, What,
From which, Why, How much, Of which kind, When, Where, and How
(With which). IV. *Subjects*. God, Angel, Heaven, Man, Imagination,
Sensitive, Vegetative, Elemental, and Instrumentative. V. *Virtues*. Just-
ice, Prudence, Fortitude, Temperance, Faith, Hope, Charity, Patience,
Piety. VI. *Vices*. Avarice, Gluttony, Luxury, Haughtiness, Torpor, Mal-
ice, Anger, Mendacity, and Inconstancy. Although Janus Caecilius Frey,
in his *A New Way into Divine Learning and Art*, Part XI, Chapter 1,
omits classes III and VI.[105]

58. Accordingly, as there are 9 things in each class, and the number of
overall complexions of 9 things is 511, by multiplying class by class, as in
Problem III, there will be in all the classes taken together complexions
totalling $511 \cap 511 \cap 511 \cap 511 \cap 511 \cap 511$ f. 17804320388674561,
which is the 6th power of 511; omitting all those variations in which the
same term is repeated, in which one class is repeated, or in which several
terms are taken from one class.[106]

59. And these are only the Complexions: what can we say in respect of
Variations of Situs, if they are multiplied by complexions? "So I shall
here, in passing, explain this Problem: *to multiply variations of situs, or
permutations, by complexions*. That is, given certain things, to find all

[103] Agrippa (1533), sigg. f6-g2 (= pp. [85-93]). Leibniz, however, most likely based his
information on the inclusion of Agrippa's work in the 1598 Strassburg edition of Lull's
*Opera*. Cf. Lullius (1598: 863–71), (1617: 841–[9]).
[104] Alsted (1612). Alsted previously wrote a book entitled *Key to the Lullian Art*: Cf. Alsted
(1609) (reprinted 1983).
[105] Janus Caecilius Frey was born towards the end of the sixteenth century in Kaiserstuhl
(Canton Aargau, Switzerland) and died as a victim of the plague in Paris on August 1, 1631. Cf.
Adolf Schaumann, 'Frey, Janus Cäcilius' in ADB 7: 361–2. Frey (1628) offers only the remaining
four classes I, II, IV, and V as examples of Lullian Art, giving slight variations to these as well, in
comparison to Leibniz's enumerations. The book would be published in various German
editions. Cf., e.g., the later edition published at Jena: P. Brössel (1674: 78–9).
[106] All complexions of 9 things amount to a total of $2^9 - 1 = 511$ (the power set of a set of 9
elements, minus the null set).

situs seu formae reperire. Sumantur omnes Complexiones particulares dati Numeri (v. g. de Numero 4: 1niones 4, com2nationes 6, con3nationes 4, con4natio 1), quaeratur variatio dispositionis singulorum Exponentium, per probl. 4 infra (v. g. 1 dat. 1, 2. dat 2, 3. dat 6, 4 dat 24) ea multiplicetur per complexionem suam [33] particularem, seu de dato exponente (v. g. 1 ∩ 4 f. 4. 2 ∩ 6 f. 12. 4 ∩ 6 f. 24. 1 ∩ 24 f. 24). Aggregatum omnium factorum erit factus ex ductu Dispositionum in Complexiones, id est quaesitum (v. g. 4, 12, 24, 24 + f. 64)."

60. Verum in Terminis Lullianis multa desidero. Nam tota ejus methodus dirigitur ad artem potius ex tempore disserendi, quam plenam de re data scientiam consequendi, si non ex ipsius Lullii, certe Lullistarum intentione. Numerum Terminorum determinavit pro arbitrio, hinc in singulis classibus sunt novem. Cur praedicatis absolutis, quae abstractissima esse debent, commiscuit Voluntatem, Veritatem, Sapientiam, Virtutem, Gloriam, cur Pulchritudinem omisit, seu Figuram, cur Numerum? praedicatis relatis debebat accensere multo plura, v. g. Causam, totum, partem, Requisitum, etc. praeterea Majoritas, Aequalitas, Minoritas est nihil aliud quam concordantia et differentia magnitudinis. Quaestionum tota classis ad praedicata pertinet; utrum sit, est existentiae, quae durationem ad se trahit; Quid, Essentiae; Quare, Causae; de quo, objecti; Quantum, magnitudinis; Quale, Qualitas, quae est genus praedicatorum absolutorum; Quando, Temporis; ubi, loci; Quomodo, formae; cum quo, adjuncti: omnes terminorum sunt, qui aut relati sunt inter praedicata, aut referendi. Et cur Quamdiu omisit, an ne durationi coincideret?

the variations of complexion or matter, as well as those of situs or form. Take all the particular Complexions of a given Number (for example, Number 4: 4 1nions, 6 com2nations, 4 con3nations, and 1 con4nation). Then look up the variation by disposition of each Exponent in Problem IV below (in this example, 1 gives 1, 2 gives 2, 3 gives 6, and 4 gives 24). Multiply this by its particular complexions, that is, those arising out of its particular exponent (in this example, 1 ∩ 4 f. 4; 2 ∩ 6 f. 12; 4 ∩ 6 f. 24; 1 ∩ 24 f. 24). The sum of these will be the aggregate of all the sums obtained from multiplying permutations by complexions, and this is what is required (in this example, 4 + 12 + 24 +24 f. 64)."[107]

60. To tell the truth, Lull's choice of Terms leaves much to be desired. His whole method is directed at the art of extempore debate rather than at complete knowledge of how to draw conclusions out of a given matter. This may not have been Lull's own intention, but it was certainly the intention of his followers. He arbitrarily limited the number of Terms in each class to nine. Why did he mix Will, Truth, Wisdom, Virtue, and Glory with absolute predicates, which should be extremely abstract? Why did he omit Beauty or Figure, why did he omit Number? He ought to have counted many more things as relative predicates, such as Cause, whole, part, Requisite, and so on. Moreover, Majority, Equality, and Minority are nothing more than concordance and difference of magnitude. The whole class of Questions belongs with the predicates: *whether something is*, is existence, which attracts duration to itself; *What?* belongs with Essence; *Why?* with Cause; *From which?* with object; *How much?* with magnitude; *Of which kind?* with Quality, which is a genus of absolute predicates; *When?* with Time; *Where?* with place; *How?* with form; *With which?* with adjunct. All of these are Terms that are either related to predicates or must be referred to them. And why did he omit *For how long?* Should it not coincide with duration? But

---

[107] Leibniz considers the case of a set composed of 4 elements ("things") and calculates all the *complexions*, i.e. all the *combinations* that can be made out of them: $2^4 - 1 = 15$. All complexions are distributed in 4 classes and each class is individuated by the number of elements (the 'exponent') giving rise to each combination. Thus, for instance, class 1 contains 4 complexions, each of 1 element, and 1 is the exponent corresponding to this class. Class 2 contains 6 complexions of 2 elements and 2 is the exponent of the class, etc. Then Leibniz multiplies the number of complexions in each class with the number of *variations* that can be made with the exponent corresponding to the class. Thus, for instance, because the number of *variations* that can be made with exponent 3 is 6 (= 3!, the factorial of 3), he multiplies 6 with 4 (the number of combinations that can be made with 4 elements taking three at the time): the result is 24. Finally, he adds together all the partial results of each multiplication and gets 64 as the number "obtained from multiplying permutations by complexions" in the case of a set of 4 elements.

cur igitur alia aeque coincidentia admiscet: Denique Quomodo, et cum quo, male confunduntur.

61. Classes vero ultimae Vitiorum et Virtutum sunt prorsus ad Scientiam hanc tam generalem ἀπροσδιόνυσοι. Ipsa quoque earum recensio quam partim manca, partim superflua! [*194*] Virtutum recensuit priores 4 cardinales, mox 3 theologicas, cur igitur addita Patientia quae in fortitudine dicitur contineri; cur Pieta[s], id est amo[r] DEI, quae in Charitate? scilicet ut novenarii hiatus expleretur. Ipsa quoque vitia cur non virtutibus opposita recensuit? An ut intelligeremus in virtute vitia opposita, et in vitio virtutem? at ita vitia 27 prodibunt. Subjectorum census placet maxime. Sunt enim hi inprimis Entium gradus: DEUS, Angelus, Coelum (ex doctrina [**34**] peripatetica Ens incorruptibile), Homo, Brutum perfectius (s. habens imaginationem), imperfectius (seu sensum solum, qualia de ζωοφύτοις narrant), Planta, Forma communis corporum (qualis oritur ex commixtione Elementorum, quo pertinent omnia inanima), Artificialia (quae nominat: Instrumenta). Haec sunt quorum complexu Lullius utitur, de quo judicium, maturum utique, gravis viri Petri Gassendi Logicae suae Epicureae T. I. operum capite peculiari. Quare artem Lullii dudum Com2natoriam appellavit Jordan. Brunus Nolanus Scrutin. praefat. p. m. 684.

62. Atque hinc esse judico, quod immortalis Kircherus suam illam diu promissam artem magnam sciendi, seu novam portam scientiarum, qua de omnibus rebus infinitis rationibus disputari, cunctorumque summaria cognitio haberi possit (quo eodem fere modo suam Syntaxin

why, then, does he equally mix other coincidences? Finally, he is wrong to identify *How?* and *With which?*

61. The last two classes, those of Vices and Virtues, are simply out of place in a Science of such generality.[108] In addition, their enumeration is sometimes lame, sometimes redundant! He enumerated, first, 4 cardinal virtues, then 3 theological virtues. Why, then, add Patience, which is regarded as part of fortitude; why Piety, meaning the love of GOD, which is part of Charity? No doubt in order to fill a gap in the nine. Why did he enumerate vices that are not the opposites of the virtues? Is this because we are to take as understood the opposing vices in virtue, and the opposing virtue in vice? But in this way, we shall arrive at 27 vices. The enumeration of subjects is the most satisfactory, as these are the pre-eminent grades of Beings: GOD; Angel; heaven (an incorruptible Being, according to the Aristotelian doctrine); man; high-grade Brute (that is, one having imagination); low-grade Brute (that is, one having only sense, as they say of such things as zoöphytes); Plant; the common form of bodies (such as arises from a mixture of Elements, and to which all inanimate things belong); and Artifacts (which he calls Instruments). This is the variety of things of which Lull makes use, concerning which see the considered judgement of that great man, Pierre Gassendi, in his *Epicurean Logic*, Volume 1 of his Works, the relevant Chapter.[109] Did not Giordano Bruno of Nola, in his *Scrutiny*, Preface, paragraph 684, describe Lull's Art as Com2natory?[110]

62. And I daresay this is why the immortal Kircher is going to present to the world under the name of Com2natory his long-promised art of learning, or new gateway into learning, with which one could dispute about anything with an infinite number of arguments, and obtain encyclopaedic knowledge of all things (Pierre Grégoire of Toulouse described his

---

[108] The Greek word *aprosdiónusos* that Leibniz uses here literally means 'which does not concern the cult of Dionysus' and therefore, in a broad sense, designates something inappropriate or out of place. Cicero, for instance, uses the word in this sense in *Letters to Atticus* 16, 13a 1.

[109] In *Syntagma Philosophicum*, Part I (*Logica*), Book 1, 'De Logicae origine, et varietate', Chapter 8, 'Logica Lulij', Gassendi offers a fine, entirely descriptive, summary of Lullian art. Cf. Gassendi (1658, vol. 1: 56–9).

[110] Leibniz refers to Bruno's Address to William of St. Clemens which served as an Introduction to both Bruno's *De Lampade* and the *Scrutiny*, and was included in the 1598 Strassburg edition of Lull's *Opera*. Cf. Lullius (1598: 684), where Bruno refers to Lull's art as a "divine Combinatorial art". In fact, however, the title of Bruno's *De Lampade* itself contains the term *Combinatoria*. Cf. Lullius (1598: 698); Bruno (1890, vol. 2: 242). On Bruno's interest in Lull, see also: Rossi (2006: 81–96); Yates (1989; 1999, vol. 8).

artis mirabilis inscripsit Petr. Gregor. Tholosanus), Com2natoriae titulo
ostentaverit. Unum hoc opto, ut ingenio vir vastissimo, altius quam vel
Lullius vel Tholosanus penetret in intima rerum, ac quae nos praecon-
cepimus, quorum lineamenta duximus, quae inter desiderata ponimus,
expleat: quod de fatali ejus in illustrandis scientiis felicitate desperan-
dum non est. Ac nos profecto haec non tam Arithmeticae augendae,
et si et hoc fecimus, quam Logicae inventivae recludendis fontibus
destinavimus, fungentes praeconis munere, et quod in catalogo desider-
atorum suis augmentis Scientiarum Verulamius fecit, satis habituri, si

*Syntaxis of Wonderous Art* in much the same way).[111] It is my one hope, that this man of immense genius, loftier than either Lull or Grégoire, may penetrate the innermost recesses of things, and complete what I have preconceived, the outlines of which I have sketched, and which I have set down as desirable aims: which is not to be despaired of in one gifted with such divine genius in enlightening our understanding.[112] I actually intended these things not so much to enlarge Arithmetic (if I have indeed done so) as to open new sources of the Logic of invention, performing the office of herald of things to come, which Bacon did with the catalogue of desiderata in his *Advancement of Learning*;[113] and I shall hold myself to

---

[111] Athanasius Kircher's *Ars Magna Sciendi* was to appear in 1669 (Kircher 1669). Cf. Kircher (1646: 935v). Though the *Ars Magna Sciendi* was first sent to the Jesuit censors in 1660, the book would eventually see the light only nine years later. See: Findlen (2004). Pierre Grégoire de Toulouse (c. 1540–97) made it clear even in the title of his work that, as Leibniz argues, he had a notion of the broad application of his *Ars* similar to Kircher's (see Grégoire 1575a). The first five books of this work elaborate on the general idea of knowledge and present Grégoire's *Speculum*, or method for inquiring into any subject based on a list of basically Aristotelian questions and predicates. Books six and seven offer a first application of the method to questions concerning divine things, demonology, and magic. Thirty-three further books, containing an application of the method on such diverse subjects as astronomy, metaphysics, physics, the mathematical sciences, grammar, dialectics, poetics, history, economics, as well as on shipping, warfare and various kinds of crafts, psychological, biological, and chemical subjects, anatomy, and passion theory, were published ten years after the first edition in a volume ranging over 1055 pages and an extra 123 unnumbered pages of indexes: Grégoire (1585). Later editions, such as the early seventeenth-century editions published with Zetner in Cologne, would hence refer to "40" instead of the original seven books included in the work. Grégoire initially also published another text, Grégoire (1575b), of another 304 pages in the combined 1583–5 edition that we were able to consult. These relate his art to ancient philosophical and scholarly traditions. True to his own plea for a methodological and resolute treatment of all possible questions according to similar norms, Grégoire was an encyclopaedic author rather than simply an advocate of Lullian art. See also Rossi (2006: 41): "In addition to the usual theme of an art capable of discovering the axioms common to all sciences and of elaborating absolute criteria of certainty, we find many of the same problems here which had already been confronted by Agrippa and Lavinheta in the same period. But the text of Grégoire was not a simple 'commentary' on the Lullian art. Unlike the commentators, after referring to Lull and the principal theorists of universal syntax, he elaborates a true encyclopaedia of the sciences not unworthy of being compared (at least in terms of volume and extent) to Bacon's *De augmentis scientiarum*." Further: Bonner and Bonner (1985: 66–71).
[112] Note that Leibniz would later distance himself from Kircher. Cf. Findlen (2004: 6).
[113] As Bacon argues in the *The Advancement of Learning*, he is "but a trumpeter, not a combatant [...]. Nor is mine a trumpet which summons and excites men to cut each other to pieces [...]; but rather to make peace between themselves, and turning with united forces against the Nature of Things, to storm and occupy her castles and strongholds, and extend the bounds of human empire, as far as God Almighty in his goodness may permit." Bacon (1858: 372–3, 1857: 579–80). Bacon's *De augmentis scientiarum*, the 1623 enlargement of Bacon (1605), may itself be considered a catalogue, in that it consists of an enumeration of all the various sciences, their divisions and defects. Leibniz's use of the term "catalogue" may imply that he also has in mind Bacon's famous *Catalogue of Particular Histories*, but the reference is obviously to the 'Novus Orbis Scientiarum, sive Desiderata' or 'The New World of Sciences, or Desiderata' at the end of *De augmentis*, Book IX. Cf. Bacon (1857: 121–3).

suspicionem tantae artis hominibus faciamus, quam cum incredibili fructu generis humani alius producat.

63. Quare age tandem artis complicatoriae (sic enim malumus, neque enim omnis complexus com2natio est) uti nobis constituenda videatur, lineamenta prima ducemus. Profundissimus principiorum in omnibus rebus scrutator Th. Hobbes merito posuit omne opus mentis nostrae esse *computationem*, sed hac vel summam *addendo* vel *subtrahendo*, differentiam colligi. Elem. de Corp. p. I. c. I. art. 2. Quemadmodum igitur duo sunt Algebraistarum et Analyticorum primaria signa + et –. Ita duae quasi copulae *est* et *non- est*: illic componit mens, hic dividit. In [35] tali igitur sensu τὸ Est non est proprie copula, sed pars praedicati, duae a. sunt copulae, una nominata, *non*, altera innominata, sed includitur in τῷ est, quoties ipsi non additum: non. Quod ipsum fecit, ut τὸ Est habitum sit pro Copula. Possemus adhibere in subsidium vocem: *revera*, v. g. Homo *revera* est animal. Homo *non* est lapis. Sed haec obiter.

64. Porro ut constet ex quibus omnia conficiantur, ad constituenda hujus artis praedicamenta, et velut materiam, analysis adhibenda est. Analysis haec est: 1. Datus quicunque [*195*] Terminus resolvatur in partes formales, seu ponatur ejus definitio; partes autem hae iterum in partes, seu terminorum definitionis definitio, usque ad partes simplices, seu terminos indefinibiles. Nam οὐ δεῖ παντὸς ὅρον ζητεῖν; et ultimi illi termini non jam amplius definitione, sed analogia intelliguntur.

have done enough if I convey to men such a taste of so great an art as may blossom forth with the incredible fruit of a new mankind.

63. For this reason, the time has at last come for me to sketch the first outlines of the art of complication (which I prefer, as not every complexion is a com2nation), as it seems to me it should be constituted. That most profound searcher-out of the principles of all things, Thomas Hobbes, has rightly contended that every work of the human mind consists in computation, but on this understanding, that it is effected either by *adding up* a sum or *subtracting* a difference. (*The Elements of Philosophy*, Part 1, *On Body*, Chapter 1, Article 2).[114] Accordingly, just as there are two primitive signs used by Algebraists and Analysts, + and -, so there are, as it were, two copulae, *is* and *is not*: with the former the mind puts things together, with the latter it takes them apart. In this sense, *is* is not strictly a copula,[115] but part of a predicate, and the copulae are two: one that is named *not*, and another that is unnamed, but included in the *is* whenever *not* is not added to it. This is how *is* has come to stand for a copula. I could use for this purpose the word *really*. For example, Man is *really* an animal. Man is *not* a stone. But this is just by the way.[116]

64. To resume: in order to ascertain what everything is made of, analysis should be applied to establishing the predicaments, and as it were, the matter of this art. The analysis is this: (1) To resolve any given Term into its formal parts, that is, to lay down a definition of it; and to resolve those parts again into parts, that is, to lay down a definition of the definition of the terms, right down to simple parts, or undefinable terms. For,

*One does not always look for a definition,*[117]

and these last terms are now better understood not by means of a definition, but by means of an analogy.

---

[114] Hobbes (1839a, vol. 1; 1839c, vol. 1: 3–5), esp. Note that the whole of Part 1 of *De Corpore* bears the title "Computatio sive Logica".

[115] Note that the use of the Greek article 'τό' applied to *Est* serves by way of quotation marks in the original. The scholastic logicians used to employ the expression 'ly' (the definite article in French *Langue d'oïl*) to the same effect.

[116] During the first half of the seventeenth century, there was a lively discussion amongst philosophers in Europe, and in Germany in particular, concerning the place of the copula in the structure of an elementary sentence. The discussion involved issues concerning modal logic as well as metaphysics and theology. Cf. Roncaglia (1996).

[117] *Ou dei pantos horon zētein*. Quotation from Aristotle, *Metaph.* IX 6, 1048a36–7. Cf. Aristotle (1984, vol. 2: 1655): "and we must not seek a definition of everything".

65. 2. Inventi omnes Termini primi ponantur in una classe, et designentur notis quibusdam; commodissimum erit, numerari.

66. 3. Inter Terminos primos ponantur non solum res, sed et modi, sive respectus.

67. 4. Cum omnes Termini orti varient distantia a primis, prout ex pluribus Terminis primis componuntur, seu prout est exponens Complexionis, hinc tot classes faciendae, quot exponentes sunt. Et in eandem classem conjiciendi termini, qui ex eodem numero primorum componuntur.

68. 5. Termini orti per com2nationem scribi aliter non poterunt, quam scribendo terminos primos, ex quibus componuntur, et quia termini primi signati sunt numeris, scribantur duo numeri duos terminos signantes.

69. 6. At Termini orti per con3nationem aut alias majoris etiam exponentis Complexiones, seu Termini qui sunt in classe 3tia et sequentibus, singuli toties varie scribi possunt, quot habet complexiones simpliciter exponens ipsorum spectatus non jam amplius ut exponens, sed ut numerus rerum. Habet hoc suum fundamentum in Usu IX, v. g. sunto termini primi his numeris signati 3, 6, 7, 9. Sitque terminus ortus in classe tertia, seu per con3nationem compositus, nempe ex 3bus simplicibus 3, 6, 9. Et sint in classe 2da combinationes hae: (1) 3, 6; (2) 3, 7; (3) 3, 9; (4) 6, 7; (5) 6, 9; (6) 7, 9. Ajo terminum illum datum classis 3tiae scribi [**36**] posse vel sic: 3. 6. 9. exprimendo omnes simplices; vel exprimendo unum simplicem, et loco caeterorum duorum simplicium scribendo com2nationem, v. g. sic: ½.9 vel: 3/2.6 vel sic: 5/2.3. Hae quasi-fractiones

65. (2) As soon as one has found all the primitive Terms they should be placed in one class, and designated by certain symbols. It will be most convenient if they are numbers.

66. (3) Among the primitive Terms should be placed not only things, but also modes, or respects.

67. (4) Since all the Terms that arise vary in their distance from the primitive Terms to the extent that they are composed of several primitive Terms, that is, according to the exponent of the Complexion, one must create as many classes as there are exponents. And terms that are composed of the same number of primitive terms should be included in the same class.

68. (5) Terms that arise through com2nation cannot be written except by writing the primitive terms of which they are composed; and because the primitive terms are designated by numbers, two numbers must be written to designate two terms.

69. (6) As for Terms that arise through con3nation or other Complexions of even greater exponent, that is, terms that are in the third or higher classes, they can all be written in as many different ways as their exponent (no longer viewed as an exponent but as the number of things) has complexions in total. This was established in Application IX. For example, let the primitive terms be designated by the numbers 3, 6, 7, and 9, and consider a derivative term of the third class, that is, a term composed by con3nation of the three simple terms 3, 6, and 9. Also, let the com2nations in the second class be: (1) 3, 6; (2) 3, 7; (3) 3, 9; (4) 6, 7; (5) 6, 9; (6) 7, 9. I say that the given term in the third class can be written either as 3. 6. 9, an expression composed entirely of simple terms, or as an expression composed of one simple term, and in place of the other two simple terms writing a com2nation such as: 1|2. 9; 3|2. 6; or 5|2. 3. What these quasi-fractions signify will be stated presently.[118] The more

---

[118] As Leibniz explains in the next paragraph, the numerator of each fraction denotes the position that a given complexion (combination) has in the class denoted by the denominator. Thus, for instance, 3|2 denotes the third complexion of class '2' (the one made of 'con2nations'), i.e. '3.9'. Joining 6 to '3|2', we have the term '3.6.9' chosen by Leibniz in his example. Obviously, this presupposes that all complexions of each class are ordered in a unique way, according to the series of the natural numbers. As Leibniz remarks, once each primitive concept has been associated with a number and all complexions have been ordered in each class, it is possible to express all definitions of a concept whatsoever. '3|2. 6' and '5|2. 3', for example, are different ways of expressing (i.e. different definitions of) the same concept denoted by '3.6.9'. Moreover, Leibniz observes that the number of all possible definitions, or different ways of 'presenting' a given concept, is the same as the number of all complexions of the exponent ("no longer viewed as an exponent but as the number of things") of the class to which the concept itself belongs. Thus, for instance, the concept designated by '3.6.9', belonging to class 3 (on a universe of 5 things: Leibniz's example), may be written in $2^3 - 1 = 7$ different ways. Leibniz will get back to this point in paragraph 80.

quid significent mox dicetur. Quo autem classis a prima remotior, hoc variatio major. Semper enim termini classis antecedentis sunt quasi genera subalterna ad terminos quosdam variationis sequentis.

70. 7. Quoties terminus ortus citatur extra suam classem, scribatur per modum fractionis, ut numerus superior seu numerator, sit numerus loci in classe; inferior, seu nominator, numerus classis. 8. Commodius est in terminis ortis exponendis non omnes terminos primos, sed inter- medios scribere, ob multitudinem, et ex iis eos qui maxime cogitanti de re occurrunt. Verum omnes primos scribere est fundamentalius.

71. 9. His ita constitutis possunt omnia subjecta et praedicata inveniri, tam affirmativa quam negativa, tam universalia, quam particularia. Dati enim subjecti praedicata sunt omnes termini primi ejus: Item omnes orti primis propiores, quorum omnes termini primi sunt in dato. [196] Si igitur Terminus datus qui subjectum esse debet scriptus est terminis primis, facile est eos primos qui de ipso praedicantur invenire, ortos vero etiam invenire dabitur, si in complexionibus disponendis ordo servetur. Sin terminus datus scriptus est ortis, aut partim ortis partim simplicibus, quicquid praedicabitur de orto ejus, de dato praedicabitur. Et haec quidem omnia praedicata sunt latioris de angustiori, praedicatio vero aequalis de aequali est, quando definitio de Termino, id est vel omnes termini primi ejus simul, vel orti, aut orti et simplices, in quibus omnes illi primi continentur, praedicantur de dato. Eae sunt tot, quot modis nuperrime diximus, unum Terminum scribi posse.

72. Ex his jam facile erit numeris investigare omnia praedicata, quae de omni dato subjecto praedicari possunt, seu omnes U A. Propositiones de dato subjecto, nimirum singularum classium a prima usque ad classem dati inclusive, numeri ipsas denominantes, seu exponentes ponantur ordine, v. g. 1 (de classe 1ma), 2 (de 2da), 3, 4 etc. Unicuique [37] tanquam non jam amplius exponenti sed numero assignetur sua

remote a class is from the first class, the greater is the variation, as the terms of a preceding class are like genera subaltern to the terms of a subsequent variation.

70. (7) Whenever a Term is referred to outside its class, it should be written in the form of a fraction in such a way that its upper number, or numerator, is the number of its place in the class, and its lower number, or denominator, is the number of the class. (8) With terms arising from exponentiation it is more convenient to write not all the primitive terms, but some intermediate terms, on account of the great number of the former, and to choose from them those that occur most readily to someone thinking of the matter in question. But it is more fundamental to write all the primitive terms.

71. (9) On this basis, all subjects and predicates can be found, negative as well as affirmative, particular as well as universal. The predicates of a given subject are all its primitive terms, and all derivative terms nearer to the primitive terms, whose own primitive terms are all in the given subject. Accordingly, if the given Term, a subject in this case, is written with primitive terms, it is easy to find those primitive terms that are predicated of it, and we can also find the derivative terms that are predicated of it if we arrange the complexions in order. But if the given term is written with derivative terms, or partly with derivative terms and partly with simple terms, then whatever is predicable of one of its derivative terms is predicable of the given term. And all these are predicates belonging to a wider subject applied to a narrower subject; in contrast with predication belonging to an equal subject of an equal subject when defining a Term, that is, when either all its primitive terms at once, or its derivative terms, or a mixture of derivative and simple terms in which all the primitive terms are contained, are predicated of the given Term. There are as many kinds of these definitions as ways in which (as I said just now) one Term can be written.

72. With this analysis, it will now be easy, taking recourse to numbers, to investigate all the predicates that can be stated of a given subject, that is, all the UA.[119] To do this, propositions of a given subject, and of course from each of the classes, from the first up to and including the class of the given subject, should be denominated by numbers, and the numbers denominating them, that is, their exponents, should be placed in order.

---

[119] To avoid misunderstandings, it is important to emphasize here that Leibniz's solution concerns UA propositions insofar as they are *logical truths*.

complexio simpliciter, v. g. 1, 3, 7, 15. Quaerantur complexiones particulares numeri classis ultimae seu de qua est terminus datus, v. g. de 4 cujus complexio simpliciter 15, 1niones 4, com2nationes 6, con3nationes 4, con4natio 1. Singulae complexiones simpliciter classium multiplicentur per complexionem particularem classis ultimae, quae habeat exponentem eundem cum numero suae classis, v. g. 1 ∩ 4 f. 4. 3 ∩ 6 f. 18. 4 ∩ 7 f. 28, 15 ∩ 1 f. 15. Aggregatum omnium factorum erit numerus omnium praedicatorum de dato subjecto ita ut propositio sit UA, v. g. 4, 18, 28, 15 + f. 65.

73. Praedicata per propositionem PA seu numerus Propositionum Particularium affirmativarum ita investigabitur: inveniantur praedicata UA dati termini, uti nuper dictum est; et subjecta UA, uti mox dicetur. Addatur numerus uterque, quia ex UA propositione oritur PA tum per conversionem simpliciter, tum per subalternationem. Productum erit Quaesitum.

74. Subjecta in propositione UA dati termini, sunt tum omnes termini orti, in quibus terminus datus totus continetur, quales sunt solum in

For example, 1 (of the first class), 2 (of the second class), 3, 4, and so on. Then, to each of them should now be assigned its complexion, such as 1, 3, 7, 15, no longer as to a particular exponent but to a number. We require the particular complexions of the number of the last class, that is, the class containing the given term, for example, of 4, whose total number of complexions is 15, 4 1nions, 6 com2nations, 4 con3nations, and 1 con4nation. The total number of complexions of the classes must be multiplied by the particular complexion of the last class, which has the same exponent as the number of the class, in this example, 1 ∩ 4 f. 4, 3 ∩ 6 f. 18, 7 ∩ 4 f. 28, and 15 ∩ 1 f. 15. The aggregate of all these subtotals will be the number of all predicates of the given subject that make propositions UA, in this example, 4 + 18 + 28 + 15 f. 65.[120]

73. The predicates that make propositions PA, that is, the number of Particular affirmative Propositions, should be investigated as follows: find predicates UA of the given term in the manner just explained; and subjects UA, as I am shortly going to explain. Then add both numbers, because from a proposition UA proposition PA can be derived either by simple conversion or by subalternation.[121] The Result will be what is required.[122]

74. The subjects of a given term in a proposition UA are: first, all the derivative terms in which the given term is wholly contained, but only

---

[120] Leibniz's alleged solution, however, is wrong (the true one being $2^4 - 1 = 15$). Suppose indeed, as Leibniz does, that 4 is the number of a complex concept. All prime divisors of 4 correspond to the primitive concepts out of which the complex concept in question arises. Thus, to calculate all possible concepts, which can be predicated of the concept associated with 4, we have simply to consider all the possible combinations of the divisors of 4. In a more abstract manner: if $k$ is the number of the primitive (simple) terms that compose a given term $t$, then the number of all predicates that can be truly attributed to $t$ is equal to the number of all combinations of $k$ terms, i.e. $2^k - 1$. Leibniz, instead, first rightly calculates the number of all the complexions (combinations) that may be obtained out of 4 elements ($16 - 1 = 15$), then distributes these complexions into classes, each denoted by the number of elements that are combined in each class (thus '1' denotes the class made of all complexions of 1 element; '2' denotes the class of all the complexions of 2 elements, etc.). The numbers denoting the classes coincide with those of the exponents attributed to the collections of 4 elements. At this point, Leibniz observes that these numbers must be considered no longer as exponents, but simply as numbers as such, of which we have to determine the complexions. He then, in quite an odd way, proceeds to calculate the number of complexions that may be obtained out of each of these numbers (1, 3, 7, 15) and multiplies it with the number of complexions in each class. In a final step, Leibniz adds the partial products thus obtained, determining what he thinks to be the number of all the predicates of the given subject. On this point, see also: Couturat (1901: 41–2, and notes through to 43) (where a conjecture is proposed about the reason that determined Leibniz to make the mistake); Dürr (1949: 22–3).

[121] By conversion *per accidens* not by simple conversion.

[122] In other words, the number of the predicates of a PA proposition is equal to the sum of the number of predicates *plus* the number of the subjects of the corresponding UA proposition.

classibus sequentibus, et hinc oritur subjectum angustius, tum omnes termini orti, qui eosdem cum dato habent terminos simplices, uno verbo ejusdem termini definitiones, seu variationes eum scribendi, invicem, sunt sibi subjecta aequalia.

75. Numerum subjectorum sic computabimus: *inveniatur numerus omnium Classium*. Eae sunt tot, quot termini sunt primi in prima classe, v. g. sunt termini in prima classe tantum 5, erunt classes in universum 5, nempe in 1ma 1niones, in 2da com2nationes, in 3tia con3nationes, in 4ta con4ternationes, in 5ta con5nationes. *Ita erit inventus etiam numerus omnium classium sequentium,* subtrahendo numerum classis termini dati, v. g. 2 de numero classium in universum 5, remanebit 3. Numerum autem classium seu terminorum primorum supponamus pro Numero rerum, numerum classis pro exponente, erit [197] numerus terminorum in classe idem cum complexionibus particularibus dato numero et exponente, v. g. de 5 rebus 1niones sunt 5, com2(3)nationes 10, con4nationes 5, [38] con5natio 1. Tot igitur erunt in singulis classibus exponenti correspondentibus termini, supposito quod termini primi sint 5. Praeterea Terminus datus cujus subjecta quaeruntur respondebit capiti complexionum; Subjecta angustiora ipsis complexionibus quarum datum est caput. Igitur dati termini subjecta angustiora inveniemus, si problema hoc solvere poterimus:

those in subsequent classes, thus giving rise to a narrower subject;[123] and then, all the derivative terms having the same simple terms as the given term, in a word, the definitions of the term itself, that is, the variations in writing it, which are subjects equal to it.[124]

75. I shall calculate the number of subjects on this basis: first, *find the number of all Classes*. There are as many of them as there are primitive terms in the first class. For example, if there are only 5 terms in the first class, there will be 5 classes altogether, namely, in the first class 1nions, in the second class com2nations, in the third class con3nations, in the fourth class con4nations, and in the fifth class con5nations. Thus, *even the number of all the subsequent classes will be found* by subtracting the number of the class of the given term, for example 2, from the total number of classes, 5, the difference being 3. Then, take the number of classes, or primitive terms, as the Number of things, and the number of the class as exponent. The number of terms in each class will be the same as the number of particular complexions with the given number and exponent.[125] For example, with 5 things there are 5 1nions, 10 com2nations, 10 con3nations, 5 con4nations, and 1 con5nation. Hence, on the supposition that there are 5 primitive terms, this will be the number of terms in each class corresponding to the exponent. Further, the given Term whose subjects are required will correspond to the head of the complexions; narrower Subjects will correspond to the complexions whose head is the given one.[126]

---

[123] Suppose S is a subject belonging to class $k$. Each subject S' belonging to some class $m > k$ is *narrower* than S, because it is 'made' of more concepts (conceptual determinations) than S (*rational animal living in Europe* is a subject narrower than *rational animal*).

[124] Leibniz distinguishes two cases: (1) the set of all primitive concepts of (the concept of) the subject in a UA proposition is a proper subset of the set of concepts constituting the predicate; (2) the concept of the subject and the concept of the predicate are composed by the same set of primitive concepts. In case (1), the concept can be the subject of all UA propositions having a predicate that belongs to one of the 'subsequent classes'. In case (2), the predicate is only a different description of the subject.

[125] That is, given a set of 5 elements, these will give rise to 5 classes of complexions. If we consider, for instance, the third class, we see that it is composed of 10 complexions (each made of 3 elements), which, in Leibniz's terminology, are the *terms* belonging to the class.

[126] Remember that Leibniz calls 'head' a complexion which is common to a given set of complexions. The complexion '1, 2' for instance, is the 'head' of the following combinations: '1, 2, 3', '1, 2, 7', '1, 2, 5, 9'. As Knobloch (1974: 414) points out, because Leibniz defines a 'head' as "a definite subset of definitely given elements which have to be contained in the desired combinations, the problem, in modern terms, reads: How many combinations of a certain size or of all possible sizes contain a certain number of given elements?" Leibniz will tackle the same problem below in the case of variations (permutations) (Problem VII). He would solve the problem for the case of variations with repetitions only ten years later (1676: Cf. Knobloch (1973: 70).

76. "Dato capite complexiones invenire; partim simpliciter (ita inveniemus subjecta angustiora omnia), partim particulares, seu dato exponente (ita inveniemus ea tantum quae sunt in data classe). Problema hoc statim impraesentiarum solvemus, ubi manifestus ejus usus est, ne, ubi seorsim posuerimus, novis exemplis indigeamus. Solutio igitur haec est: Subtrahatur de Numero rerum, v. g. 5, a, b, c, d, e, exponens capitis dati, v.g. a, b: 2 – 5 f. 3 aut a: 1 – 5 f. 4. Sive supponamus datum caput 1nionem, sive com2nationem esse; complexio enim ut sit necesse est. Proposito item exponente subtrahatur de eo itidem exponens capitis dati. Igitur: si datus sit quicunque exponens, in cujus complexionibus quoties datum caput reperiatur invenire sit propositum; quaeratur complexio exponentis tanto minoris dato, quantus est exponens capitis dati, in numero rerum, qui sit itidem tanto minor dato, quantus est exponens capitis dati per Tab. א probl. 1, inventum erit quod quaerebatur. At si Complexiones simpliciter capitis dati in omnibus complexionibus dati numeri quocunque exponente, quaerere propositum sit; complexio Numeri rerum, numero dato tanto minoris, quantus est exponens capitis dati, erit quaesitum":

76. We shall therefore find the narrower subjects of the given term if we can solve this problem: "Given a head, to find the complexions: sometimes *in total* (finding all the narrower subjects), sometimes *particular*, or *with a given exponent* (finding only those in a given class). I shall presently offer an interim solution of the problem, in which the application is obvious, so that I shall not need new examples when I employ it elsewhere. And the solution is as follows: From the Number of things, say 5, *abcde*, subtract the exponent of the given head.[127] For example, with *ab*, 5 – 2 f. 3; or with *a*, 5 – 1 f. 4. That is, let us suppose that the given head is either a 1nion or a com2nation; for there must be some complexion. Next, in the same way, subtract the exponent of the given head from the proposed exponent.[128] Therefore, if an exponent whatsoever is given, and our purpose is to find how many times a given head can be found among the complexions of this very exponent, then we have to look in Table ℵ of Problem I for the complexion of the given exponent diminished by the exponent of the given head, and with the given number of things diminished by the exponent of the given head.[129] The result will be what is required. If, however, our purpose is to seek the overall Complexions of the given head in all the complexions of a given number with every possible exponent, it will be found in the total number of complexions of the Number of things less than the given number by the exponent of the given head."[130]

---

[127] Here Leibniz discusses the first case presented above: to find the complexions *in total* (finding all the narrower subjects).

[128] This rule corresponds to the second case introduced above: to find the complexions *in particular*, or *with a given exponent*.

[129] To understand Leibniz's words we need to look again at Table ℵ, above. Here, suppose (to continue Leibniz's example) that the 'number of things' is 5: we want to know how many times a given head, for instance the couple (*ab*), occurs among the complexions of a given exponent, say among the complexions of class 3. First of all, we subtract the number of the given head from the number of things: 5 – 2 = 3. Then we subtract "the exponent of the given head", i.e. 2, from the number of the exponent of the class, i.e. 3. The result is 1. We now dispose of two numbers: 3 and 1, and if we look in the table at the crossing of their rows, we find 3, which is the number of the occurrences of the couple (*ab*). Cf. Introduction and Knobloch (1973: 32).

[130] In this case the solution is analogous to that of the preceding, less general case. Instead of subtracting the exponent of the head from the exponent of the class, we have simply to subtract the exponent of the head from the total number of things. Thus, for example, to find the number of all occurrences of the couple (1, 4) in all classes of complexions out of 5 elements, we must first subtract 2, i.e. the exponent of the couple (1, 4), from 5, then look at the number of complexions that we may have out of 3 elements. This number is 8, as is confirmed by the Table ℵ. This rule, however, even though correctly stated, is in conflict with what Leibniz says in the numerical examples of section 78.

77. E. g. in 5 rerum: a, b, c, d, e, 1nionibus datum caput a reperitur 1 vice (quae est nullio, seu 0llio de 4) datum caput a, b 0lla vice (quae est super0llio, ut ita dicam, de 3), in com2nationibus earundem illud reperitur vicibus 4 (quae sunt 1niones de 4) hoc 1 (quae est 0llio de 3), in con3nationibus illud 6 (com2natio de 4.) hoc 3 (1nio de 3), in con4nationibus illud 4 (con3natio de 4) hoc 3 (com2natio de 3), in con5nationibus utrobique 1 vice (illic con4natio, hic con3natio de 3). Hae complexiones sunt dato exponente, ex quarum aggregatione oriuntur comple[39]xiones simpliciter sed et sic: in 5 rerum complexionibus simpliciter (quae sunt 31) a. reperitur vicibus 15 (complexio simpliciter de 4), a, b 7. (complexio simpliciter de 3) vici[b]us.

78. Hae complexiones sunt numerus subjectorum angustiorum dati termini. Subjecta aequalia, quando definitiones definitionibus subjiciuntur, eadem methodo inveniuntur qua supra praedicata aequalia. Termini enim aequales, sunt servata quantitate et qualitate convertibiles, igitur ex praedicatis fiunt subjecta et contra, praedicata a[utem] tot sunt, quot dati termini (cujus subjecta quaeruntur), termini primi habent complexiones simpliciter, v. g. + a 1; a, b 2. Additis jam subjectis aequalibus ad angustiora 1 + 15 f. I6. 2. + 7 f. 9, prodibit numerus subjectorum omnium dati termini. Quem erat propositum invenire. [198]

77. For example, in the 1nions of 5 things, *abcde*, the given head *a* is found
1 time (which is the null or 0-complexion relative to 4), the given head
*ab* is found no time (which is, so to speak, beyond the null complexion
relative to 3).[131] In the com2nations of the same things, the former is
found 4 times (which are the 1nions of 4), the latter 1 time (which is the
0-complexion of 3). In the con3nations, the former is found 6 times (the
com2nations of 4), the latter 3 times (the 1nions of 3). In the con4na-
tions, the former is found 4 times (the con3nations of 4), the latter
3 times (the com2nations of 3); and in the con5nations, each is found
1 time (the former a con4nation of 4, the latter a con3nation of 3). These
complexions are with a given exponent, out of whose aggregation arises
the total number of complexions. Moreover, in the total number of
complexions of 5 things (which is 31) *a* is found 15 times (the total
number of complexions of 4 things), *ab* 7 times (the total number
of complexions of 3 things).[132] These complexions are the number of
narrower subjects of the given term.

78. Equal subjects, where definitions are made the subject of definitions, are
found by the same method as equal predicates above. For equal terms
are convertible with preservation of quantity and quality, making subjects out of predicates, and the reverse, and there are as many predicates
as the first terms of the given term whose subjects are sought have
complexions in total, for example, with *a* 1, and with *ab* 2. With the
equal subjects now added to the narrower subjects, the number of all
subjects of the given term comes out as 1 + 15 f. 16 and 2 + 7 f. 9
respectively.[133] Which was what I proposed to find.

---

[131] The Latin expressions '0llio' (to be read as 'nullio'), and 'super0llio' (to be read as
'supernullio') are coined by Leibniz. Obviously, 'super0llio' corresponds to the case in which
$C\binom{m}{k} = 0$ with $k > m$. On the fact that Leibniz considers the two possibilities according to
which $k = 0$ and $k = -1$, see Knobloch (1973: 32–3).

[132] From these examples it emerges quite clearly that Leibniz considers the number of
combinations as determined by the formula $C\binom{n}{k} = 2^{n-k}-1$ whereas the right formula is
$C\binom{n}{k} = 2^{n-k}$. Therefore '15' must be changed to '16': "*a* is found 16 times (the total number
of complexions of 4 things)"; and '7' changed to '8': "*ab* 8 times (the total number of
complexions of 3 things)". Cf. Knobloch (1973: 31).

[133] Leibniz here wrongly adds 2, instead of 1, to 7; probably because, as Couturat suggests, he
considers "as distinct subjects the permutations of terms of the identical subject". Cf. Couturat
(1901: 44, n. 1). At any rate, the correct numbers are: 17 (1 + 16) and 9 (1 + 8).

79. Subjecta hactenus Universalia, restant Particularia, ea tot sunt quot praedicata particularia. Praedicata et Subjecta negativa sic invenientur: computentur ex datis certis Terminis primis tanquam Numero rerum, omnes termini tam primi quam orti, tanquam complexiones simpliciter, v. g. si termini primi sint 5, erunt 31. De producto detrahantur omnia praedicata affirmativa universalia, et subjecta angustiora affirmativa universalia: Residuum erunt omnia praedicata negativa. De subjectis contra. Particularia negativa ex universalibus computentur, uti supra PA ex UA computavimus. Omisimus vero propositiones identicas UA, quarum sunt tot quot complexiones simpliciter Terminorum primorum; seu quot sunt omnino termini et primi et orti. Quia quilibet terminus vel primus vel ortus de se dicitur. Caeterum inter complexiones illas omisimus, in quibus idem terminus repetitur, quae repetitio in nonnullis producit variationem in infinitum, ut in numeris, et figuris Geometriae.

80. Methodus porro argumenta inveniendi haec est: Esto datus quicunque terminus tanquam subjectum A, et alius quicunque tanquam praedicatum B, quaeratur Medium. Medium erit praedicatum subjecti et subjectum praedicati, id est terminus quicunque continens A, et contentus a B. Continere a. terminus terminum dicitur, si omnes ejus termini primi sunt [40] in illo. Fundamentalis a. demonstratio est si uterque terminus resolvatur in primos, manifestum erit alterum alterius aut partem esse, aut partium earundem. Mediorum a. numerum sic inveniemus. Subjectum et praedicatum vel sunt in eadem classe, vel diversa. Si in eadem, necesse est utrumque terminum esse ortum, et variationem scriptionis saltem seu definitionis ejusdem termini; poterunt igitur

79. So far, the subjects have been Universal; this leaves Particular subjects, of which there are as many as there are particular predicates.[134] Negative Predicates and Subjects can be found in this way: from certain given primitive Terms regarded as the Number of things, calculate all the terms, derivative as well as primitive, as the complexions in total. For example, if there are five primitive terms, the total number of complexions will be 31.[135] From this result subtract all the universal affirmative predicates and universal affirmative narrower subjects: the result will be all the negative predicates. With subjects the reverse holds.[136] The Particular negatives can be calculated from the universals, just as we calculated PA from UA above. But have we not left out identical UA propositions? There are as many of them as there are complexions in total of the primitive Terms; that is, as many as there are primitive and derivative terms altogether, since any term, primitive or derivative, may be said of itself. Moreover, we have left out those complexions in which the same term is repeated, repetition that produces infinite variation in certain things, such as numbers and geometrical figures.

80. And this is a method for finding arguments: let a given term be the subject A, and some other term the predicate B. We require a Middle term. A Middle term will be a predicate of the subject and a subject of the predicate, that is, a term containing A and contained by B.[137] And a term is said to contain a term if all the primitive terms of the latter are in the former. But the fundamental demonstration is that, once each term has been resolved into its primitives, it will be clear either that one is a part of the other or that they are made of the same parts. In this way, we shall arrive at a number of Middle terms. The Subject and Predicate are either in the same class or in different classes. If in the same class, both must be derivative terms, and only variations in writing, or at least in the definition of the same term.[138] But two definitions of the same

---

[134] This is easily explained by the fact that particular affirmative propositions are susceptible of simple conversion (i.e. they legitimate the inference from 'Some S are P' to 'Some P are S', and vice versa).

[135] This again corresponds to the general formula $\sum_{k=1}^{n} \binom{n}{k} = 2^n - 1$, thus for n = 5, $32 - 1 = 31$.

[136] That is: the number of UN propositions, in which a given concept C plays the role of predicate, is the same as the number of UN propositions, in which C plays the role of subject.

[137] Clearly, Leibniz here thinks of arguments as *arguments in syllogistic form*. Moreover, as the conditions specified for the containment of the middle term show, he supposes the conclusion of the syllogism to be a UA proposition; therefore the middle term he is looking for must belong to a syllogism having the form of *Barbara*.

[138] In paragraph 69, Leibniz explains how different combinations of numbers can represent the same concept. If subject and predicate belong to the same class, they can only be complex, not simple concepts, because simple concepts cannot be predicated the one of the other.

duae definitiones ejusdem termini non nisi per tertiam de se invicem probari. Igitur de numero definitionum ejusdem termini orti, quem investigavimus supra n. 69 subtrahatur 2: residuum erit numerus mediorum possibilium inter terminos aequales.

81. Sin non sunt in eadem classe, erit praedicatum in classe minoris exponentis, subjectum in classe majoris. Jam supponatur praedicatum velut caput complexionis, exponens classis subjecti supponatur pro numero rerum. Inveniantur omnes complexiones dati capitis particulares per singulas classes a classe praedicati ad classem subjecti inclusive; in singulis classibus complexiones dati capitis particulares ducantur in complexiones simpliciter, Exponentis ipsius classis pro numero rerum supposititi. Aggregatum omnium factorum subtracto 2 erit quaesitum.

term can be proved only through a third definition of each of them in turn.[139] Then, from the number of definitions of the same derivative term (which I investigated above in Paragraph 69) subtract 2, and the remainder will be the number of possible middle terms between equal terms.[140]

81. But if they are not in the same class, the predicate will be in the class of smaller exponent, and the subject will be in the class of greater exponent. Now suppose the Predicate to be the head of a complexion, and the exponent of the class of the subject to be the number of things. Find all the particular complexions of the given head in each class, from the class of the predicate to the class of the subject inclusive; then put into correspondence the particular complexions of the given head in each class with the overall complexions, considering all the complexions of the Exponent of that class taken as the number of things. The aggregate of all these factors, minus 2, will be what is required.[141]

---

[139] Given two propositions, to prove that they define the same term, a third proposition is needed, which analyses each of them in its first or primitive terms.

[140] That is, between terms belonging to the same class. In this case, we have the same term 'presented' or defined in different ways. Thus, to find the number of all middle terms, we must first determine the total number of definitions of this term, according to the procedure suggested in paragraph 69. Next, from this number we must subtract the identical proposition having this very term as subject and predicate.

[141] Suppose n = the number of simple terms constituting the subject S, and $k$ = the number of simple terms constituting the predicate P; suppose further, as Leibniz does, that $n > k$. The number of all complexions of the n terms of S containing the k terms of P (excluding S and P) is $2^{n-k} - 2$. Leibniz, however, suggests the following quite 'empirical' procedure to calculate the number of middle terms. First, consider $k$ as the number corresponding to a given head of complexions and n as the total 'number of things', i.e. as the number of elements in general. Once all complexions of n elements have been displayed and distributed into classes, find the number of occurrences of the head in each class from $k$ to $n$ included. As a final step, add together all these numbers and subtract 2 from the total. The resulting number is the solution of the problem. Thus, for example, suppose that the couple (1, 4) is the given head and 4 = n the number of 'things' (primitive terms). 15 is the number of complexions of 4 elements, and once distributed into classes, they give rise to classes from 1 to 4. Because the head (1, 4) belongs to class 2, consider only the classes from 2 to 4, and find in each of them all complexions containing the head (1, 4). In class 2, the given head is contained 1 time, 2 times in class 3, and 1 time in class 4 (which is a complexion of only four elements). Thus, we have: $1 + 2 + 1 = 4$. If from 4 we subtract 2, we have 2, which is the number of middle terms that can be subjects of P and predicates of S in our example. Remember that Leibniz wants to find the number of the middle terms, because he aims to determine the number of all arguments (syllogisms) having S as subject and P as predicate of their conclusion. Another point to be emphasized is that Leibniz's solution is correct only if the following three conditions are met: (1) Only concepts (middle terms) which contain all the first concepts belonging to the predicate P, are involved; (2) All first terms belonging to the middle term must belong to the subject S as well; (3) The calculus concerns logical truths only. Cf. Dürr (1949: 29).

82. Praedicatum autem de subjecto negari facile inveniemus, si utroque termino in primos resoluto manifestum est neutrum altero contineri. Probari tamen negativa sic poterit: inveniantur omnia praedicata subjecti, cum de omnibus negetur praedicatum, totidem erunt media probandi negativam. Inveniantur omnia subjecta praedicati, cum omnia negentur de subjecto, etiam erunt totidem media probandi negativam. Utrisque igitur computatis numerum mediorum probandi negativam habebimus. [*199*]

83. Admonendum denique est, totam hanc artem complicatoriam directam esse ad theoremata, seu propositiones quae sunt aeternae veritatis, seu non arbitrio DEI sed sua natura constant. Omnes vero propositiones singulares quasi *historicae*, v. g. Augustus fuit Romanorum imperator, aut observationes, id est propositiones universales, sed quarum veritas non in essentia, sed existentia fundata est; quaeque verae sunt quasi casu, [**41**] id est DEI arbitrio, v. g. omnes homines adulti in Europa habent cognitionem DEI. Talium non datur demonstratio sed inductio. Nisi quod interdum observatio per observationem interventu Theorematis demonstrari potest.

84. Ad tales observationes pertinent omnes propositiones particulares, quae non sunt conversae vel subalternae universalis. Hinc igitur manifestum est, quo sensu dicatur singularium non esse demonstrationem, et cur profundissimus Aristoteles locos argumentorum posuerit in Topicis,

82. Moreover, we shall easily find whether a predicate is denied of a subject if, when each term is resolved into its primitive terms, it is clear that neither contains the other. A negative can, however, also be proved like this: find all the predicates of the subject. Since the predicate must be denied of all of them, they will altogether be middle terms in proving the negative. Find all the subjects of the predicate: since all of them must be denied of the subject, they too will altogether be middle terms in proving the negative. With each of these calculated, we shall have the number of middle terms in proving the negative.[142]

83. I must finally draw attention to the fact that this whole art of complication is directed towards theorems, or propositions that are eternally true, that is, propositions that hold not because of the will of GOD, but by their very nature.[143] But as for all singular propositions, which might be called *historical* (for example, *Augustus was Emperor of the Romans*), and as for *observations*, that is, universal propositions (for example, *All adult men in Europe have cognisance of GOD*), whose truth is founded not on essence but on existence; and which are all true as if by chance, that is by the will of GOD—of these propositions there is no demonstration, but only induction; except that sometimes an observation can be demonstrated from another observation through the medium of a Theorem.[144]

84. All particular propositions that are not converted or subalternated from a universal proposition belong with such observations.[145] Accordingly, it is clear in what sense it may be said of singular propositions that no demonstration is possible, and why that most profound man Aristotle treated the places of the arguments in his *Topics*, where propositions are

---

[142] In case of negative propositions, Leibniz gives only a sketch of a solution.

[143] The thesis according to which 'eternal' truths (truths, for instance, of logic and mathematics), are independent of God's will, is a central tenet of Leibniz's philosophy and will remain unchanged from the *Dissertation on Combinatorial Art* (1666) to the *New Essays* (1702) and the *Essays of Theodicy* (1710).

[144] This paragraph foreshadows Leibniz's mature distinction between *truths of reason* and *truths of fact*. Cf. *Theodicy*, sec. 170; *Monadology* § 33, as well as Rescher (1991: 120–4), for further references and a detailed commentary.

[145] That is, all particular propositions that are not obtained by means of a conversion or a subalternation. For the operation of conversion, see footnote 82. The operation of subalternation permits the inference from a universal affirmative to the corresponding particular affirmative proposition (from *All men are animals* to *Some men are animals*) and from a universal negative to the corresponding particular negative (from *No man is a stone* to *Some men are not stones*). According to the so-called Aristotelian square of oppositions, particular affirmative and particular negative propositions are subalterns of the corresponding universals.

ubi et propositiones sunt contingentes, et argumenta probabilia, Demonstrationum autem unus locus est: definitio. Verum cum de re dicenda sunt ea quae non ex ipsius visceribus desumuntur, v. g. Christum natum esse Bethleemi, nemo huc definitionibus deveniet: sed historia materiam, loci reminiscentiam suppeditabunt. Haec jam locorum Topicorum origo, et in singulis maximarum, quibus omnibus qui sint fontes, ostenderemus itidem, nisi timeremus ne in progressu sermonis cupiditate declarandi omnia abriperemur.

85. Uno saltem verbo indigitabimus omnia ex doctrina metaphysica relationum Entis ad Ens repetenda esse, sic ut ex generibus quidem relationum Loci, ex theorematis autem singulorum maximae efformentur. Hoc vidisse arbitror, praeter morem compendiographorum solidissimum Joh. Henr. Bisterfeld in Phosphoro Catholico, seu Epitome artis meditandi ed. Lugd. Bat. anno 1657 quae tota fundatur in immeatione et περιχορήσει, ut vocat, universali omnium in omnibus, similitudine item et dissimilitudine omnium cum omnibus, quarum principia: Relationes. Eum libellum qui legerit, usum artis complicatoriae magis magisque perspiciet.

86. Ingeniosus ille, quem saepe nominavimus, Joh. Hospinianus, libellum promisit de inveniendi et judicandi facultatibus, in quo emendationem doctrinae Topicae paraverat, locosque recensuerat 180, maximas 2796, v. controvers. dial. p. 442. Hunc ego insignì rei logicae damno nunquam editum arbitror. Abibimus hinc, cum primum γεῦμα quoddam praxeos artis com2natoriae dederimus.

contingent and arguments probable.[146] Yet there is only one place for demonstrations, namely, in definition. But when it comes to what can be said of a thing that does not follow from its inner being, for example, that *Christ was born in Bethlehem*, nobody will arrive at it by means of definitions: rather, history will furnish the matter, places the remembrance.[147] This is the origin of the places of the *Topics* (and of the maxims set in each of them),[148] on which, as sources, I might similarly expatiate, if I did not fear that in the course of my delivery I might get carried away by the desire to make everything clear.

85. I shall finally and briefly summon everything that can be culled from the metaphysical doctrine of the relations of Being to Being in order to form Places out of the different kinds of relation, and appropriate maxims characteristic of each of them out of the theorems. I believe that I have seen this done, in more than just the summary fashion of encyclopaedists, by Johann Heinrich Bisterfeld in his *Universal Lamp, or Epitome of the Art of Meditating*, published in Leiden, 1657, which is entirely based on what he calls immeation and *perichòresis*, that is to say, the universal likeness of all things in all things, as well as the universal unlikeness of all things to all things, the principles of which are Relations. Anyone who reads that book will gain an increasing understanding of how to apply the art of complication.[149]

86. That ingenious man Wirth, whom I have cited several times, promised a book on the faculties of invention and judgement, in which he had planned to reform the theory of Topics, and had assembled 180 places and 2796 maxims. (See his *Dialectical Disputes*, p. 442.[150]) I do not think that he ever published it, to the great detriment of the cause of logic. I shall now leave this subject, having given you a first *taste* of the practice of the art of com2nation.

---

[146] Aristotle's *Topics* is concerned with the invention of arguments with respect to subjects that lend themselves only to a certain amount of probability (*Topica* I 2, 101a25–8). Leibniz here uses the term *loci argumentorum*, referring to the *loci* or commonplaces that are also mentioned further on in this paragraph, the Greek original of which (*topoi*) provided the title of Aristotle's book.

[147] In Aristotle's view, non-demonstrative types of reasoning belong to the dialectical sciences, the method of which is found in the *Topics* with its *loci*, 'places' for finding arguments. Mnemonic techniques were later developed on the basis of 'places' within the 'art of memory' tradition.

[148] The maxim associated to a place (*locus*) was a general rule that was supposed to warrant the correctness of an act of judgement. For the distinction between a *topic* in the proper sense and the *maxim* associated to it, see de Pater (1965).

[149] Bisterfeld (1657a). On Bisterfeld and Leibniz, see Loemker (1961), Mugnai (1973), Antognazza (1999). See also Introduction, § 4.

[150] Hospinianus (1576). See also above, note 50.

87. Commodissima Mathesis extempora[42]neo conatui visa est: hinc non a primis simpliciter terminis orsi sumus, sed a primis in Mathesi; neque omnes posuimus, sed quos ad producendos complicatione sua terminos ortos propositos sufficere judicabamus. Potuissemus eadem methodo omnes definitiones ex Elementis Euclidis exponere, si tempus superfuisset. Quoniam autem non a primis simpliciter terminis orsi sumus, hinc necessarium erat signa adhibere, quibus casus vocabulorum aliaque ad sermonem complendum necessaria intelligentur. Nam [*200*] siquidem a primis simpliciter terminis incepissemus, pro ipsis casuum variationibus, quorum ex relationibus et Metaphysica originem exposuit Jul. Caesar Scaliger lib. de Caus. L. L., terminos posuissemus. Adhibuimus autem articulos graecos. Numerum pluralem signavimus adscripto in ( ) 15, si quidem indefinitus: 2, 3, etc. si determinatus.

88. Esto igitur Classis I. in qua termini primi: 1. Punctum. 2. Spatium. 3. Intersitum. Adsitum seu Contiguum. 5. Dissitum, seu Distans. 6. Terminus, seu quae distant. 7. In·situm. 8. Inclusum (v. g. centrum est insitum circulo, inclusum peripheriae). 9. Pars. 10. Totum. 11. Idem. 12. Diversum. 13. Unum. 14. Numerus. 15. Plura, v. g. 1. 2. 3. 4. 5. etc. 16. Distantia. 17. Possibile. 18. Omne. 19. Datum. 20. Fit. 21. Regio. 22. Dimensio. 23. Longum. 24. Latum. 25. Profundum. 26. Commune. 27. Progressio, seu Continuatum.

   Classis II. 1. *Quantitas* est 14 τῶν 9(15). 2. *Includens* est 6. 10.
   
   III. 1. *Intervallum* est 2. 3. 10. 2. *Aequale*, A τῆς 11 ½. 3. *Continuum* est A ad B si τοῦ A ἡ 9 est 4 et 7 τῷ B.
   
   IV. 1. *Majus* est A habens τὴν 9 2/3 τῷ B. 2. *Minus* B 2/3 τῇ 9 τοῦ A. 3. *Linea*, 1/3 τῶν 1(2) 4. *Parallelum*, 2/3 ἐν τῇ 16. 5. *Figura*, 24. 8 ab 18. 21.
   
   V. 1. *Crescens*, quod 20. ¼. 2. *Decrescens*, 2/4. 3. *Implexum* est 5/3 in τῇ 11. 22. 4. *Secans*, 5/3 in τῇ I2. 22.
   
   VI. 1. *Convergens*, 2/5 ἐν τῇ I6. 2. *Divergens*, 1/5 ἐν τῇ 16.
   
   VII. 1. *Superficies*, 1/3 τῶν 3/4. 2. *Infinitum*, ¼ quam I8. I9. 17. 3. *Peripheria*, 3/4. I3. 2/2. A dicitur *Mensura*, seu metitur B, si 10 ex A (15)2/3 est 2/3 τῷ B.
   
   VIII. 1. *Maximum* est 1/4 non 2/4. 2. *Minimum*, 2/4 non ¼. 3. *Recta* ¾. 2/3. τῶν 6. (2). 4. quae non talis, *Curva*. 5. *Arcus*, 9. τῆς 3/7.
   
   IX. 1. *Ambitus* est 1/7 2/2.
   
   X. 1. *Commensurabilia* sunt, quorum 4/7. 26 est [43] et 1 et 2.
   
   XI. 1. *Angulus* est quem faciunt 3/4(2). 4. 2/6.
   
   XII. 1. *Planum* est 1/7. 2/3 τῇ 16. τῶν 6.

87. Mathematics has shown itself to be very useful to this present effort. For this reason, I have not started from primitive terms pure and simple, but from the primitive terms of Mathematics; not that I have chosen all of them, only those that I judged to be sufficient to produce the proposed derivative terms by their complications. In this way, I could have reproduced all the definitions in Euclid's *Elements* if I had had enough time. But since I did not start simply from primitive terms, it was necessary to introduce some signs in order to convey the cases of words, and other things required for complete expressions. If indeed I had begun simply from primitive terms, I would have needed terms to represent all those various cases whose origin in relations and Metaphysics Julius Caesar Scaliger expounded in his book *On the Foundations of the Latin Language*.[151] Accordingly, I have introduced Greek articles, and signified plurals by adding '15' in brackets, if indefinite, 2, 3, etc., if determinate.

88. Let us begin with Class I, in which the primitive terms are: 1. Point. 2. Space. 3. In between. 4. Next to, or neighbouring. 5. Scattered, or distant. 6. Boundary, or things that are distant. 7. Within. 8. Enclosed (for example, the centre is within the circle, but is enclosed by the periphery). 9. Part. 10. Whole. 11. Same. 12. Different. 13. One. 14. Number. 15. Plurality, for example, 1, 2, 3, 4, 5 etc. 16. Distance. 17. Possible. 18. All. 19. Given. 20. Becoming. 21. Region. 22. Dimension. 23. Length. 24. Width. 25. Depth. 26. Common. 27. Progression, or Continued. Class II: 1. *Quantity* is 14 of 9s (15). 2. *Including* is 6, 10. Class III: 1. *Interval* is 2, 3, 10. 2. *Equal*, A of 11, 1|2 [as B]. 3. A is *Continuous* with B if for A there is a 9, 4, and 7 with B. Class IV: 1. A is *Greater* [than B] when it has a 9, 2|3 more than B. 2. B is *Smaller* [than A] when it is 2|3 to 9 of A. 3. *Line*, 1|3 of 1s (2). 4. *Parallel*, 2|3 in 16. 5. *Figure*, 24, 8 by 18, 21. Class V: 1. *Increasing*, 20, 1|4. 2. *Decreasing*, 20, 2|4. 3. *Intertwined* is 3|3 in 11, 22. 4. *Intersecting*, 3|3 in 12, 22. Class VI: 1. *Convergent*, 2|5 in 16. 2. *Divergent*, 1|5 in 16. Class VII: 1. *Surface*, 1|3 of 3|4s. 2. *Infinite*, 1|4 than 18, 19, 17. 3. *Periphery*, 3|4, 13, 2|2. 4. A is called a *Measure*, or measures B, if 10 from A (15), 2|3 is 2|3 to B. Class VIII: 1. *Maximum* is 1|4 not 2|4. 2. *Minimum*, 2|4 not 1|4. 3. *Straight Line*, 3|4, 2|3 to 16 of 6s (2). 4. *Curve*, what is not such. 5. *Arc*, 9 of 3|7. Class IX: 1. *Ambit* is 1|7, 2|2. Class X: 1. *Commensurables* are things of which [there is] 4|7, 26 to

---

[151] In his book on Latin grammar, the question of cases prompted Julius Scaliger to offer a lengthy discussion of the various metaphysical distinctions involved in their use. Cf. Iulius Caesar Scaliger (1540: 146–218).

XIII. 1. *Gibbus*, 1/7. ¼. τῇ 16 τῶν 6.

XIV. 1. *Rectilineum* est 5/4 cujus 2/2 est τῶν 3/8(15). 2. quae dicuntur *Latera*. 3. Si 3/8(3) *Triangulum*. Si 3/8(4) *Quadrangulum* etc.

XV. 1. *Lunula* est 1/3 τῶν 5/8(2) non 2/3. 4 (2) (subintelligo a. tam lunulam gibbosam qua arcus arcui concavitatem obvertit, quam falcatam qua interior alterius concavitati suam convexitatem).

XVI. 1. *Angulus rectus* est 1/11. 2/3 in τῷ 18. 21. 2. *Segmentum* est 3 τῶν 2/2 et 3/8. 7 τῇ 5/4.

XVII. 1. *Aequilaterum* est 5/4 cujus 2/2 est 8 τῶν 3/8(15). 2. *Triangulum aequicrurum* est 5/4 cujus 2/2 est 8 τῶν 3/8(3)2/3 (2). 3. *Scalenum* est 5/4 cujus 2/2 est τῶν 3/8(3) non 2/3(3). [*201*]

XVIII. 1. *Angulus contactus* est, quem faciunt ¾(2) 4. 2/6 non 4/5. 27 modo 17.

XIX. 1. *Inscriptum* est 5/4. 7. cujus 1/11(15) sunt 4 τῷ 2/2. 2. *Circumscripta* vero est ea figura cui inscripta est.

XX. 1. *Angulus obtusus* est ¼ quam 1/16. 2. Acutus, 2/4 quam 1/16.

XXI. 1. *Diameter* est 3/8. 1/8. 7 τῇ 5/4.

XXII. 1. *Circulus* est 1/12. 8 ab 18. 21 habens τὴν 16. 2/3 τοῦ 19 alicuius 1 (quod dicitur *Centrum Circuli*) ab 18. 6. 2. *Triangulum rectangulum* est 5/4 cujus 1/11(3) sunt omnes 2/3 sed 13 est in τῷ 18. 21.

XXIII. 1. *Centrum Figurae* est 1. 26 τοῖσ 1/21(15).

XXIV. I. *Semifigura* data (v. g. semicirculus, etc.) est 3 τῶν 1/22 et (dimidium τοῦ) 3/2. Hinc facile erit definitiones conficere, si observetur, quod n. 70 diximus: in iis notis, quae per fractiones scriptae sunt, *nominatorem*, designare numerum classis; *numeratorem*, numerum termini in classe, v. g. *centrum* est 1 (punctum) 26 (commune) τοῖσ 1/21 (diametris) 15 pluribus. *Diameter* est 3/8 (recta) 1/8 (maxima) 7 (insita) τῇ 5/4 (figurae).

89. Ex his quae de Arte complicatoria Scientiarum, seu Logica inventiva disseruimus, cujus quasi praedicamenta ejusmodi Terminorum tabula absolverentur, fluit velut Porisma: seu usus XI. Scriptura Universalis, id est cuicunque legenti, cujuscunque linguae perito intelligibilis, qualem hodie complures viri eruditi tentarunt, quorum diligentissimus Caspar Schottus hos recenset lib. 7. Techn. Curios. Primo Hispanum quendam, cujus meminerit Kenelm. Digbaeus tr. de Nat. Corp. c. 28. n. 8, quique

both. Class XI: 1. *Angle* is what 3|4 (2), 4, 2|6 make. Class XII: 1. *Plane* is 1|7, 2|3 to 16 of 6s. Class XIII: 1. *Hump*, 1|7, 1|4 than 16 to 6s. Class XIV: 1. *Polygon* is 5|4 of which 2|2 is of 3|8s (15). 2. Which are called *Sides*. 3. If 3|8 (3), *Triangle*. 4. If 3|8 (4), *Quadrilateral*, and so on. Class XV: 1. *Lunule* is 1|3 of 5|8s (2) not 2|3, 4(2) (and I understand this not only of a sickle-shaped crescent, in which the inner arc subtends with its concavity the convexity of another arc, but also of a gibbous lunule, in which one arc subtends the concavity of another arc). Class XVI: 1. *Right Angle* is 1|11, 2|3 in 18, 21. 2. *Segment* is 3 to 2|2 and 3|8, 7 to 5|4. Class XVII: 1. *Equilateral Triangle* is 5|4 whose 2|2 is 8 of 3|8s (15). 2. *Isosceles Triangle* is 5|4 whose 2|2 is of 3|8s (3), 2|3s (2). 3. *Scalene Triangle* is 5|4 whose 2|2 is of 3|8s (3), not 2|3s (3). Class XVIII: 1. *Angle of contact* is what 3|4 (2), 4, 2|6, not 4|5 make, 27, only 17. Class XIX: 1: *Inscribed* is 5|4, 7 whose 1|11 (15) are 4 to 2|2. 2. *Circumscribed* [figure] is that in which a figure is inscribed. Class XX: 1. *Obtuse Angle* is 1|4 than 1|16. 2. *Acute Angle* is 2|4 than 1|16. Class XXI: 1. *Diameter* is 3|8, 1|8, 7 in 5|4. Class XXII: 1. *Circle* is 1|12, 8 by 18, 21 having 16, 2|3 from 19, some 1 (which is called 2. *Centre of Circle*) by 18, 6. 2. *Right-Angled Triangle* is 5|4 whose 1|11 (3) are all but 13 is 2|3 to 18, 21. Class XXIII: 1. *Centre of a Figure* is 1, 26 to 1|21s (15). Class XXIV: 1. A given *Semi-Figure* (for example, a semicircle, etc.) is 3, 1|22 and (half of) 2|2. In this way, it will be easy to make up definitions if it is observed, as I said in Paragraph 70, that in these signs, which are written as fractions, the *denominator* designates the number of the class, and the *numerator* designates the number of the term within the class. For example, a *centre* is 1 (point), 26 (common) to 1|21 (diameters) (15) (many). A *diameter* is 3|8 (straight line), 1|8 (maximum), 7 (within) a 5|4 (figure).

# XI

89. Out of what I have said regarding the Art of Complication of the Sciences, or Logic of Invention, whose quasi-predicaments are crowned by such a Table of terms, there flows like a Corollary, or application XI: a Universal Notation, that is, intelligible to any reader, no matter what his language. In our own time, a number of learned men have attempted this, and the following ones have been mentioned by Kaspar Schott, in Book 7 of his *Artificial Wonders*. First, a certain Spaniard (memorialised also by Sir Kenelm Digby in his *Treatise on the Nature of Body*,

fuerit Romae anno 1653. Ejus methodus haec ex ipsa na[44]tura rerum satis ingeniose petita: distribuebat res in varias classes, in qualibet classe erat certus numerus rerum. Ita meris numeris scribebat, citando numerum classis et rei in classe; adhibitis tamen notis quibusdam flexionum grammaticarum et orthographicarum. Idem fieret per classes a nobis praescriptas fundamentalius, quia in iis fundamentalior digestio est. Deinde Athanasium Kircherum, qui Polygraphiam suam novam et universalem dudum promisit. Denique Joh. Joachimum Becherum Archiatrum Moguntinum, opusculo primum Francofurti Latine edito, deinde germanice anno 1661. Is requirit, ut construatur Lexicon Latinum, tanquam fundamentum, et in eo disponantur voces ordine pure alphabetico et numerentur; fiant deinde Lexica, ubi voces in singulis linguis dispositae non alphabetice, sed quo ordine Latinae dispositae sunt ipsis respondentes. Scribantur igitur quae ab omnibus intelligi

Chapter 28, Paragraph 8), who is said to have been in Rome in the year 1653.[152] He had a method which was quite ingeniously derived from the nature of the things: distributing things into various classes, with a certain number of things in each class, he described them by means of numbers alone, by citing the number of the class, and the number of the thing within the class; but with the addition of certain signs indicating grammatical and orthographical inflections.[153] The same could be done in a more fundamental way by means of the classes that I laid down, because in them the disposition of things is more fundamental. Next, we have Athanasius Kircher, who has recently published his *New and Universal Polygraphy*;[154] and finally, Johann Joachim Becher, Professor

---

[152]  The "certain Spaniard" is probably Pedro Bermudo (or Vermudo), a Jesuit of whom very little is known. According to Juan Caramuel (Caramuel 1681: 498), Bermudo published in Rome, in the second half of the seventeenth century, a *Universal Grammar* printed on only one loose leaf (*uno volanti folio*). Leibniz's description of Bermudo's work completely depends on Kaspar Schott's account of it. For more references to Caramuel and Bermudo, see Velarde Lombraña (1987: 3–4 in particular). For the references to Bermudo in Schott and in Digby, see next footnote.

[153]  Cf. Schott (1664: 483) and Digby (1655: 248–9): 'VIII. De nobili quodam Hispano, qui sonum oculis percipiebat'. Leibniz apparently took the reference to Digby (as well as those to Kircher and Becher, for which see below) from Schott's book. For the reference to Digby, see Schott (1664: 483), where Schott mentions a speech-impaired Spanish Jesuit who published, in Rome, in 1653, an "Artificium" that invited "all nations of the World" to come to "an agreement in language and speech". The Spanish Jesuit apparently reminded Schott of the noble Spaniard mentioned by Digby, but Schott's memory may not have been altogether precise, since in Digby's example, mention was made of a deaf young Spanish nobleman who had learned to talk with the help of a patient priest, presumably by lip-reading. Be this as it may, contrary to what Leibniz seems to have thought on the basis of reading Schott, the passage in Digby has nothing whatsoever to do with the idea of a universal notation. Kaspar Schott (1608–66) was a disciple of Athanasius Kircher, with whom he worked in Rome from 1652 to 1655. Cf. Findlen (2004: 34). The title of Schott's *Technica curiosa* literally translates as 'Curious Technique', but may also be translated as 'Technical Curiosities' or 'Artificial Wonders'. The book is a counterpart to Schott's *Physica curiosa* of 1662; the two books together dividing curiosities into 'natural' (though often 'monstrous') and 'artificial' (i.e. mechanical) marvels, despite the fact that the full title of the first book read *Physica curiosa, sive Mirabilia naturae et artis*, In the *Technica curiosa*, Schott also mentions Kircher and Johann Joachim Becher, both of whom Leibniz likewise refers to. See the next two footnotes. Simplifying the graphic representations Johann Joachim Becher (1635–82) made for the Latin words of his lexicon, Kaspar Schott designed a box-type representation for these words and their inflections based on the abacus principle. See also: Slaughter (1982: 124).

[154]  Kircher (1663). The book contains a method for reducing all languages to one, using Latin, Italian, French, Spanish, and German words, names and phrases as examples, as well as a method for translating a language into any other language, even those not mastered, together with a secret script, both of which latter inventions Kircher claims to have construed from the *Polygraphia* of Trithemius (for which, see below), and to have adequately explained for the first time here. Schott mentions Kircher's work, adding a premonition to the reader, however, that it had so far only appeared in a few copies offered to a selected number of princes. He did, however, obtain a title page from Kircher's work, which he quotes in full. Cf. Schott (1664: 482). The German humanist abbot and occultist Johann Heidenberg, Johann von Trittenheim, or Johannes Trithemius (1462–1516) is the author of a book entitled *Polygraphia*, produced in Würzburg in 1508, and later published as Trithemius (1518), possibly the first printed work on cryptography. On Trithemius's *Polygraphia*, see: Glidden (1987: 183–95) and Brann (1999), as well as Brann (1981: 47, 91–4, esp.).

debent, numeris, et qui legere vult, is evolvat in Lexico suo vernaculo vocem dato numero signatam, et ita interpretabitur. Ita satis erit legentem vernaculam intelligere et ejus Lexicon evolvere, scribentem necesse est (nisi habeat unum adhuc Lexicon suae linguae alphabeticum ad numeros se referens) et vernaculam et latinam tenere, et utriusque Lexicon evolvere. Verum et Hispani illius et Becheri artificium et obvium et impracticabile est. Ob [202] synonyma, ob vocum ambiguitatem, ob evolvendi perpetuum taedium (quia numeros nemo unquam memoria mandabit), ob ἑτερογένειαν phrasium in linguis.

90. Verum constitutis Tabulis vel praedicamentis artis nostrae complicatoriae majora emergent. Nam Termini primi, ex quorum complexu omnes alii constituuntur, signentur notis, hae notae erunt quasi alphabetum. Commodum autem erit notas quam maxime fieri naturales, v. g. pro uno punctum, pro numeris puncta; pro relationibus Entis ad Ens lineas, pro variatione angulorum aut terminorum in lineis genera relationum. Ea si recte constituta fuerint et ingeniose, scriptura haec universalis aeque erit facilis quam communis, et quae possit sine omni lexico legi, simulque imbibetur omnium rerum fundamentalis cognitio. Fiet [45] igitur omnis talis scriptura quasi figuris geometricis; et velut picturis, uti olim Aegyptii hodie Sinenses, verum eorum picturae non reducuntur

of Medicine at Mainz, who in a work first published at Frankfurt in Latin, and afterwards in German, in the year 1661, calls for the construction of a Latin Dictionary as a basis, in which words would be listed in purely alphabetical order, and numbered.[155] Dictionaries would henceforth be constructed in which words in the various languages are listed not alphabetically, but in the order to which they correspond to the Latin. Something that needs to be universally understood would then be written in numbers, and anyone wishing to read it would then turn in his dictionary of the vernacular to the word designated by the given number, and so interpret it. Thus, while it would be sufficient for the reader to understand the vernacular, and to consult his Dictionary, the writer (unless he had in addition an alphabetical Dictionary of his language that referred him to the numbers) would need to know both the vernacular and Latin, and to consult a Dictionary of each. But the truth is that the systems of both Becher and that Spaniard are obviously impracticable, because of synonyms, the ambiguity of words, the infinite tedium of looking up the numbers (since nobody would ever be able to remember them), and the lack of correspondence between phrases in different languages.[156]

90. But once the Tables, or predicaments, of my art of complication have been constituted, new vistas will open. For the primitive terms from whose complication all the others are constituted are to be designated by symbols, and these symbols will be a kind of alphabet. And it will be desirable for the symbols to be as natural as possible: for example, a point for one, points for numbers in general; lines for the relations of Being to Being, and different kinds of relations for the variation of angles or limits in lines. If all this is correctly and skilfully accomplished, we shall have a universal script that will be as easy as it is general, readable without the aid of a dictionary, and allowing fundamental knowledge of everything to be absorbed at once. Accordingly, every such script should be made up of geometrical figures, and of pictures,

---

[155] Becher (1661). Besides assigning numbers to 9432 Latin terms, another 282 proper names and 569 names of places and regions, adding up to a total of 10283 numbered words in order to facilitate translation, the book also presents a new way of expressing numbers with the help of graphic signs that have variously placed dots and lines for unities, tens and hundreds, etc., thereby producing unique characters for every word. Note that Becher is also often quoted by Kaspar Schott, as Schott himself announces at the start of his discussion of 'Mirabilia Graphica', Schott (1664: 480). A discussion of Becher's 'Universal key' to the understanding of all languages occurs in Schott (1664: 505 ff.).

[156] Leibniz here uses the Greek word *heterogéneia*, 'heterogeneity'.

ad certum Alphabetum seu literas, quo fit ut incredibili memoriae afflictione opus sit, quod hic contra est. Hic igitur est Usus XI complexionum, in constituenda nempe polygraphia universali.

91. XIImo loco constituemus jucundas quasdam partim contemplationes, partim praxes ex Schwenteri Deliciis Mathematicis et supplementis G. P. Harsdörfferi, quem librum publice interest continuari, haustas. P. 1. sect. 1, prop. 32, reperitur numerus complexionum simpliciter, quem faciunt res 23 v. g. literae Alphabeti, nempe 8388607. P. 2 sect. 4, prop. 7 docet dato textu melodias invenire, de quo nos infra, probl. 6.

92. Harsdörfferus Parte ead. sect. 10, prop. 25 refert ingeniosum repertum Dni de Breissac, quo nihil potest arti scientiarum complicatoriae accommodatius reperiri. Is, quaecunque in re bellica attendere bonus imperator debet, ita complexus est: facit classes novem, in Ima quaestiones et circumstantias, in IIda status, in III personas, in IV actus, in V fines, in VI instrumenta exemtae actionis, seu quibus uti in nostra potestate est, facere autem ea, non est. VII instrumenta quae et facimus

such as the Egyptians used in ancient times and the Chinese use today, except that their pictures are not reduced to a fixed Alphabet, or to letters, and as a result impose an incredible strain on the memory, whereas exactly the contrary happens with my proposed script.[157] And this, then, is the XIth Application of Complexions, namely, the construction of a universal polygraphy.

## XII

91. In the XIIth place, I shall set down some partly leisurely reflections, partly exercises drawn from Schwenter's *Mathematical Recreations*, as supplemented by Georg Philipp Harsdörffer, whom it concerned that the book should be continued for the benefit of the public. In Part 1, Section 1, Proposition 32, he finds the total number of complexions made by 23 things (for example, the letters of the Alphabet), that is, 8,388,607.[158] In Part 2, Section 4, Proposition 7, he shows how to find melodies to fit a given text, a topic that I shall deal with below, in Problem VI.[159]

92. In the same Part, Section 10, Proposition 25, Harsdörffer adverts to the ingenious discovery of Seigneur de Breissac, which is excellently suited to the complicatory art of invention.[160] A good general must keep in mind everything that pertains to matters of warfare; and it is covered in this way: he creates nine classes, in the first, questions and circumstances, in the second, states, in the third, persons, in the fourth, acts, in the fifth, ends, in the sixth, instruments from which action is excluded, that is, such as it is in our power to make use of, but not in our power to bring about, in the seventh, instruments that we both bring about and

---

[157] Like many other authors of his time, Leibniz considered Chinese and Egyptian languages as 'pictorial' in nature.

[158] The example in fact occurs in Proposition 33 (Die XXXIII. Auffgab) of Schwenter's *Deliciae Physico-Mathematicae*. Cf. Schwenter (1991: 70). Schwenter, however, has the number '8,388,584'. As Knobloch (1973: 35) suggests, Leibniz has added the number of combinations of one letter to Schwenter's number. On Schwenter see also above, note 21. The poet, writer, and translator Georg Philipp Harsdörffer (1607–58) was a Nuremberg patrician, member of the Fruitbearing Society, and co-founder of the Nuremberg *Pegnesischer Blumenorden*, who extended Schwenter's *Deliciae* with two further volumes of problems (Harsdörffer 1990). Cf. DBE IV: 397; *Jöcher* II, cols 1377–8; and especially W. Creizenach, 'Harsdörfer: Georg Philipp' in ADB X: 644–6.

[159] This part contains additional problems by Harsdörffer himself: Harsdörffer (1990: 142–3). For Leibniz's own discussion of how to find melodies to fit a given text, see below, Problem VI, §§ 3–7.

[160] Harsdörffer (1990: 412–3).

et adhibemus. VIII instrumenta quorum usus consumtio est. IX actus finales seu proximos executioni. V. g.

| I. An. | Cum quo. | Ubi. | Quando. | Quomodo. | Quantum. |
|---|---|---|---|---|---|
| 2. Bellum. | Pax. | Induciae. | Colloquium. | Foedus. | Transactio. |
| 3. Patriotae. | Subditi. | Foederati. | Clientes. | Neutrales. | Hostes. |
| 4. Manere. | Cedere. | Pugnare. | Proficisci. | Expeditio. | Hyberna. |
| 5. Decus. | Lucrum. | Obedientia. | Honestas. | Necessitas. | Commoditas. |
| 6. Sol. | Aqua. | Ventus. | Itinera. | Angustiae. | Occasio. |
| 7. Currus. | Scalae. | Pontes. | Ligones. | Palae (Schauffeln). | Naves. |
| 8. Pecunia. | Commeatus. | Pulvis Torm. | Globi Torm. | Equi. | Medicamenta. |
| 9. Excubiae. | Ordo. | Impressio. | Securitas. | Aggressio. | Consilia. [203] |

[46] 93. Fiant novem rotae ex papyro, omnes concentricae, et se invicem circumdantes, ita ut quaelibet reliquis immotis rotari possit. Ita promota leviter quacunque rota nova quaestio, nova complexio prodibit. Verum cum hic inter res ejusdem classis non detur complexio, atque ita accurate loquendo non sit complexio terminorum cum terminis, sed classium cum classibus, pertinebit computatio variationis ad probl. 3. Quoniam tamen complexio etiam, quae hujus loci est, potest repraesentari rotis, ut mox dicemus, fecit cognatio, ut praeoccuparemus. Sic igitur inveniemus: multiplicetur 6. in se novies: 6, 6, 6, 6, 6, 6, 6, 6 ∩ 6 seu quaeratur progressio geometrica sextupla, cujus exponens 9, aut: Cubicubus de 6 f. 10077696. Tantum superest, ut sint solum 216 quaestiones, quod putat Harsdörfferus.

94. Caeterum quoties in Complexionibus singuli termini in singulos ducuntur, ibi necesse est tot fieri rotas, quot unitates continet numerus rerum: deinde necesse est singulis rotis inscribi omnes res. Ita variis rotarum conversionibus complexiones innumerabiles gignentur. Eruntque omnes complexiones quasi jam scriptae seorsim, quibus revera scribendis vix grandes libri sufficient.

95. Sic ipsemet doctissimus Harsdörff. P. 3. sect. 14, prop. 5, machinam 5, rotarum concentricarum construxit, quam vocat, Funffachen Denkring der teutschen Sprache Ubi in rota intima sunt 48 Vorsylben, in penintima 60 Anfangs-und Reim-Buchstaben, in media 12 Mittelbuchstaben, vocales nempe vel diphthongi; in penextima

apply, in the eighth, instruments whose application uses them up, and in the ninth, final acts, that is, acts about to be executed. For example:

| 1. Whether | With which | Where | When | How | How much |
|---|---|---|---|---|---|
| 2. War | Peace | Truce | Parley | Treaty | Agreement |
| 3. Natives | Vassals | Allies | Clients | Neutrals | Enemies |
| 4. Hold | Cede | Fight | Break up | Raid | Winter-quarters |
| 5. Glory | Booty | Obedience | Honour | Necessity | Convenience |
| 6. Sun | Water | Wind | Routes | Strait | Occasion |
| 7. Waggons | Ladders | Bridges | Mattocks | Shovels[161] | Ships |
| 8. Money | Communications | Gunpowder | Cannonballs | Horses | Medicine |
| 9. Watch | Order | Attack | Security | Advance | Counsels |

93. Let nine wheels be made out of paper, all concentric, and mounted on one another, so that each can be rotated without moving the others. Then, as each wheel is moved a little way, a new question, a new complexion will come into view. It is true that calculating the variations belongs to Problem III, since complexion is here not possible between things of the same class, that is, strictly speaking, we have complexion of classes with classes rather than complexion of terms with terms. But since the kind of complexion that we have been considering here can also be represented on wheels, as I shall presently describe, its kinship requires us to deal with it. I shall therefore find [the variations] in this way: multiply 6 by itself 9 times, $6.6.6.6.6.6.6.6 \cap 6$, that is, we need the [term of] the geometrical progression of 6 whose exponent is 9, or the ninth power of 6, which comes to 10077696; more than the mere 216 questions according to Harsdörffer's reckoning.[162]

94. Next, because in the Complexions each term is to be lined up with other terms, we need as many wheels as unities are contained in the number of things [in each class]. Finally, we have to write down all the things on the wheels. The various rotations of the wheels will thus generate innumerable complexions. All complexions will now be written separately, as it were, and huge books would hardly suffice for actually writing them down.

95. Likewise the learned Harsdörffer himself (Part 3, Section 14, Proposition 5) constructed a machine of five concentric wheels, *Fivefold Ring of the German Language*, as he calls it. This has, on the innermost wheel, 48 *Prefixes*; on the one next to innermost, 60 *Initial Letters*; on the

---

[161] Leibniz here uses the German word for 'shovels': *Schauffeln*.
[162] Harsdörffer (1990: 413).

120 End–Buchstaben, in extima Nachsylben. In has omnes voces germanicas resolvi contendit. Cum hic similiter classes sint in classes ducendae, multiplicemus: 48, 60, 12, 120, 24 factus ex prioribus per sequentem, f. 97209600. Qui est numerus vocum germanicarum hinc orientium, utilium seu significantium, et inutilium.

96. Construxit et rotas Raym. Lullius; et in Thesauro artis memorativae Joh. Henr. Alstedius, cujus rotis, in quibus res et quaestiones, adjecta est norma mobilis, in qua loci Topici, secundum quos de rebus disseratur, quaestiones probentur; et fraternitas Roseae Crucis in fama sua promittit gran[47]dem librum titulo Rotae Mundi in quo omne scibile contineatur. Orbitam quandam pietatis, ut vocat, adjecit suo Veridico Christiano Joh. Davidius Soc. J. Ex eodem principio Complicationum est Rhabdologia Neperi, et pensiles illae Serae, die Vorlegschlosser quae sine clave mirabili arte aperiuntur, vocant Mahl Schlosser, nempe superficies serae armillis tecta est, quasi annulis gyrabilibus, singulis annulis literae Alphabeti inscriptae sunt. Porro serae certum nomen impositum est, v. g. Ursula, Catharina, ad quod nisi casu qui nomen ignorat, annulorum gyrator pervenire non potest. At qui novit nomen, ita gyrat annulos invicem, ut tandem nomen prodeat, seu literae

middle one, 12 *Middle Letters*, that is, vowels or diphthongs; on the one next to outermost, 120 *Final letters*; and on the outermost one, [24] *Suffixes*.[163] He maintained that all German words can be analysed on these [wheels]. Since here also classes are to be multiplied by classes, I must form the product 48.60.12.120.24. The result of multiplying this sequence is 97,209,600, which is the number of German words arising from it, both useful, i.e. significant, and useless.[164]

96. Ramon Lull also constructed wheels;[165] and so did Johann Heinrich Alsted in his *Thesaurus of the Art of Memory*, to whose wheels, on which are things and questions, there is attached a moving pointer with Topical places, according to which the things may be examined, and questions resolved;[166] and the Fraternity of the Rosy Cross, in its *Fama*, promised a notable book entitled *The Wheels of the World*, on which everything knowable may be held;[167] and to his *Plain-speaking Christian* Jan David S.J. appended what he calls a *Volvelle of Probity*.[168] Out of the same principle of Complications comes the *Rhabdology* of Napier,[169] and those padlocks, *die Vorleg-Schlösser*,[170] which are opened in a wonderful way without the aid of a key, and which they call *Mahl-Schlösser*.[171] On these, the surface of a bolt is fitted with bracelets

---

[163] Leibniz cites in German the elements he takes over from Harsdörffer (1990).

[164] The problem actually occurs in Part 2 of the *Delitiae*. Cf. Harsdörffer (1990: 516–9).

[165] E.g. the construction of the 'fourth figure', in which two movable circles were placed on top of a bigger circle that is fixed. Lullius (1598: 55).

[166] Alsted (1612: 158–60).

[167] The oldest version of the *Fama* known today is *Fama* (1614). Along with the *Axiomata* and the *Proteus*, the *Rotae Mundi* is one of the three books of a 'Philosophical Library' the Rosicrucian manifesto mentions as extant sources of the Brotherhood, the Axiomata being mentioned as the "loftiest", the Proteus as the "most useful" and the Rotae as the "most artful" of the three. Cf. Gilly and Van der Kooij in *Fama* (1998: 86). Today, it is generally thought that the theologian Johann Valentin Andreae (1586–1654) was the central figure behind all the early documents associated with the Rosicrucian Brotherhood. He may have been inspired by the Tübingen Paracelsist Thobias Hess (1568–1614) and his spiritualist circle. Andreae was later to regret some of his earlier writings. On Leibniz and his possible relationships with the Rosicrucian Society, see: Schuchard (1998: 84–106, esp. 84–6) and the sources referred to there. On Leibniz's occupations with occultism, see: Knecht (1981), esp. Chapter 1, § 2, 'Le milieu intellectuel', and Coudert (1995).

[168] Cf. David (1601: 350–78). Jan David's *Orbita Probitatis* is a small *rotula* added at the end of the book intended to help the reader to choose between the maxims and emblems of the work. Such random reading may benefit the way in which God will secretly strike the reader's heart.

[169] Neperus (1617). Besides as an inventor of logarithms, the landowner, theologian, and mathematician John Napier, eighth laird of Merchiston (1550–1617), is known for his mechanical method of multiplication with the use of rods that came to be known as 'Napier's bones'. See: Margaret E. Baron, 'Napier, John', in DSB IX: 609–13. Extracts from John Napier's *Rabdology* were soon to be published in German in a short pamphlet as Ursinus (1623).

[170] German for 'padlocks'.    [171] German for 'combination locks'.

Alphabeti datum nomen conficientes sint ex diversis annulis in eadem linea, justa serie. Tum demum ubi in tali statu annuli erunt, poterit facillime sera aperiri. Vide de his Seris armillaribus Weckerum in Secretis, Illustrissi[204]mum Gustavum Selenum in Cryptographia fol. 489. Schwenterum in Deliciis Sect. 15, prop. 25. Desinemus Usus Problematis 1 et 2 enumerare, cum Coronidis loco de Coloribus disseruerimus.

97. Harsdörfferus P. 3. Sect. 3, prop. 16 ponit colores primos hos 5: Albus, flavus, rubeus, caeruleus, niger. Eos complicat, ita tamen ut extremi: albus et niger, nunquam simul coeant. Oritur igitur ex AF subalbus, AR carneus, AC cinereus; FR aureus, FC viridis, FN fuscus; RC purpureus; RN subrubeus; CN subcaeruleus. Sunt igitur 9 quot nempe sunt com2-nationes 5 rerum, demta Una, extremorum. Quid vero si tertii ordinis colores addantur, seu con3nationes primorum, et com2nationes secundorum, et ita porro, quanta multitudo exurget? Hoc tamen admoneo ipsos tanquam primos suppositos non esse primos; sed omnes ex albi et nigri, seu lucis et umbrae mixtione oriri.

resembling rotatable rings, and letters of the Alphabet are inscribed on each ring. Next, a certain name is chosen for the bolt, such as 'Ursula' or 'Catharina', which someone rotating the rings who does not know the name cannot get at, except by accident.[172] But if he knows the name, he rotates the rings one after the other until eventually the name appears, that is, the letters of the alphabet making up the given name are lined up in the correct order on the various rings. Then, at last, when the rings are set in this way, the bolt can be very easily withdrawn. Concerning these armillary Bolts, see Wecker's *Secrets*,[173] the *Cryptography* of the most illustrious Gustavus Selenus, folio 489,[174] and Schwenter's *Mathematical Recreations*, Section 15, Proposition 25.[175] I shall crown my survey of the Applications of Problems I and II with a discussion of Colours.

97. Harsdörffer (Part 3, Section 3, Proposition 16) posits these five primary colours: White, yellow, red, blue, and black, which he combines, though in such a way that the extremes, white and black, never come together.[176] From AF arises off-white, from AR, pink, from AC, grey; from FR, orange, from FC, green, from FN, brown; from RC, purple, from RN, burgundy; and from CN, indigo.[177] This comes to 9, that is, the number of com2nations of 5 things, less One, that of the extremes. But what if we add colours of the third order, that is, con3nations of the primary colours, and also com2nations of the secondary colours, and so on, then how great a multitude shall we obtain? However, I would caution that those so-called primary colours are not primary, but are all derived from mixing black and white, or light and dark.

---

[172] A remarkable case of using a password on a device before the computer era.
[173] Wecker (1582: 868–71).
[174] Selenus (1624: 489–93). *Gustavus Selenus* is a pseudonym for Herzog August II von Braunschweig-Lüneberg (1579–1666), the learned Duke and writer of books on chess after whom the famous library at Wolfenbüttel was named Herzog August Bibliothek. The book also carries the title of *Systema integrum Cryptographiae*. As has been documented by Rescher, Leibniz all of his life maintained a strong interest in cryptography. On two occasions (1679 and 1688) he proposed, to Duke Johann Friedrich of Hanover and Emperor Leopold I respectively, the construction of a machine to encode and decode messages. Leibniz never built the machine himself, but it was reconstructed by Nicholas Rescher, Richard Kotler, Klaus Badur, and Wolfgang Rottstett on the basis of Leibniz's notes and presented at the University of Pittsburgh's Hillman Library in 2012. On all these matters, see: Rescher (2012), as well as Rescher (2014: 103–15). See also: Beeley (2014: 111–22).
[175] Schwenter (1991: 548–50).     [176] Harsdörffer (1990: 233–5).
[177] Capital letters here refer to the initials of Latin names for colours, such as *Albus* (white), *Flavus* (yellow), *Rubeus* (red), *Caeruleus* (blue), and *Niger* (black).

98. Ac recordor legere me, etsi non succurrit autor, nobilem acupictorem nescio quem 80 colores contexuisse, vicinosque semper vicinis junxisse, ex filis tamen non nisi nigerrimis ac non nisi albissimis; porro varias alternationes alborum nigrorumque filorum; et immediationes modo plurium alborum, modo plurium nigrorum, varietatem colorum pro [48]genuisse; fila vero singula per se inermi oculo invisibilia pene fuisse. Si ita est, fuisset hoc solum experimentum satis ad colorum naturam ab ipsis incunabulis repetendam.

## Probl. III.
### DATO NUMERO CLASSIUM ET RERUM IN CLASSIBUS COMPLEXIONES CLASSIUM INVENIRE.

1. "*Complexiones* autem *classium* sunt, quarum exponens cum numero classium idem est; et qualibet complexione ex qualibet classe res una. Ducatur numerus rerum unius classis in numerum rerum alterius; et, si plures sunt, numerus tertiae in factum ex his: seu semper numerus sequentis in factum ex antecedentibus: factus ex omnibus continue, erit quaesitum."

2. Usus hujus problematis fuit tum in usu 6 probl. 1 et 2 ubi modos syllogisticos investigabamus, tum in usu 12 ubi et exempla prostant. Hic aliis utemur. Diximus supra Complexionum doctrinam versari in divisionum generibus subalternis inveniendis, inveniendis item speciebus unius divisionis; et denique plurium in se invicem ductarum. Idque postremum huic loco servavimus.

3. *Divisionem* a. *in divisionem ducere* est unius divisionis membra alterius membris subdividere, quod interdum procedit viceversa, interdum non.

[98.] And I recall reading (though I cannot bring the author to mind) of how a certain famous tapestry-maker used to interweave as many as 80 colours, which he always made by joining threads closely together, though only pure black and pure white; and he made various alternations of black and white threads, and textures that were sometimes mostly of white threads, sometimes of black threads; produced a variety of colours; each thread being in itself almost invisible to the unaided eye. If so, this experiment alone would be sufficient to trace the nature of colours to its true source.

## Problem III GIVEN A NUMBER OF CLASSES, AND OF THINGS IN THE CLASSES, TO FIND THE COMPLEXIONS OF THE CLASSES

1. "Now the *complexions* of *classes* are those of which the exponent and the number of classes are the same; and in each complexion there is only one thing from each class. Let the number of things in one class be multiplied by the number of things in the other class; and if there are several classes the number of things in the third by the product of the former, that is, always the number in the following class by the numbers of things in the preceding classes: then the product of them all together will be what is required."[178]

2. One of the applications of this problem was in application VI of Problems I and II, where we investigated syllogistic modes, and later in application XII, where examples were also featured. I shall now apply it to some more cases. I said above that the theory of Complexions can be used to find the subaltern genera of divisions, the species of a single division, and finally, the species of several divisions multiplied by one another. And it is this last-mentioned use that I shall reserve for treatment here.

3. *To multiply a division by a division* is to subdivide the members of one division by the members of the other division, which is sometimes reversible, sometimes not. Sometimes, all the members of one division can be

---

[178] Given $k$ classes, to calculate the number of all possible *complexions* of $k$ elements that can be made out of them, we have simply to multiply the number of elements belonging to the first class with the number of elements of the second, and so on, $k$ times. Thus, for instance, given two classes A and B, each, respectively, of 3 and 2 elements, the number of all complexions of 2 elements (i.e. of all complexions having the same exponent as the number of the classes) is $3 \times 2 = 6$.

Interdum omnia membra unius divisionis omnibus alterius subdividi possunt; interdum quaedam tantum, aut quibus[205]dam tantum. Si

vice versa, ita signabimus $A \begin{Bmatrix} a \\ b \end{Bmatrix} \begin{cases} c \\ d \\ e \end{cases}$ si quaedam tantum ita: $A \begin{cases} a \begin{cases} c \\ d \\ e \end{cases} \\ b \end{cases}$

si quaedam quibusdam tantum, ita: $A \begin{cases} a \begin{cases} c \\ d \end{cases} \\ b \dots e \end{cases}$

Ad nostram vero computationem primus saltem modus pertinet. In quo exemplum suppetit ex Politicis egregium. A esto Respublica, [49] a recta, b aberrans, quae est divisio moralis; c Monarchia, d Aristocratia, e Democratia, quae est divisio numerica: Ducta divisione numerica in moralem, orientur species mixtae 2 ∩ 3 f. 6: ac, ad, ae, bc, bd, be.

4. Hinc origo formulae hujus: divisionem in divisionem ducere, manifesta est, ducendus enim numerus specierum unius in numerum specierum alterius. Numerum autem in numerum ducere est numerum numero multiplicare, et toties ponere datum, quot alter habet unitates. Origo est ex Geometria, ubi si linea aliam extremitate contingens ab initio ad finem ipsius movetur, sic ut eam radat, spatium omne, quod occupabit linea mota, constituet figuram quadrangularem, si ad angulos rectos alteram contingit, ἑτερόμηκες aut quadratum; sin aliter, rhombum aut rhomboeides; si alteri aequalis, quadratum aut rhombum; sin aliter, ἑτερόμηκες aut rhomboeides. Hinc et spatium ipsum quadrangulare facto ex multiplicatione lineae per lineam aequale est.

subdivided by all the members of the other division; sometimes, only some can be subdivided, or they can be subdivided by only some. If all,

I shall signify it like this: $A \left\{ \begin{array}{l} a \\ b \end{array} \right. \left\{ \begin{array}{l} c \\ d \\ e \end{array} \right.$  If only some, like this: $A \left\{ \begin{array}{l} a \\ b \end{array} \right. \left\{ \begin{array}{l} c \\ d \\ e \end{array} \right.$

And if some by only some, like: $A \left\{ \begin{array}{l} a \\ b \end{array} \right. \left\{ \begin{array}{l} c \\ d \\ \dots e \end{array} \right.$

And the first mode, at least, pertains to our calculations, an obvious example of it being provided by Politics. Let $A$ be a State, $a$ just, $b$ deviating, which is a moral division; $c$ Monarchy, $d$ Aristocracy, and $e$ Democracy, which is a numerical division. When the numerical division is multiplied by the moral division, $2 \cap 3$ f. 6 hybrid species arise, namely, *ac, ad, ae, bc, bd,* and *be.*

4. This formula is derived as follows: it is clear that to multiply a division by a division, the number of species in one division must be multiplied by the number of species in the other division. And to multiply a number by a number in this sense is to multiply arithmetically a number by a number, that is, to apply the given number as many times as the other number has unities. It is derived from geometry, where if a line touching another line at its extremity is moved from the beginning to the end of the other line so that, as it passes the other line, the whole space that the moved line will sweep out will constitute a rectangular figure: an oblong[179] or a square if it touches the other line at right-angles; otherwise, a rhombus or a rhomboid; if the lines are equal, a square or a rhombus; but if otherwise, an oblong or a rhomboid. And here, the space itself is equal to the quadrangular figure formed by the multiplication of a line by a line.

---

[179] The Greek word *heteromekes* that Leibniz uses here, literally means 'of different length', i.e. oblong.

5. Caeterum ejusmodi divisionibus complicabilibus pleni sunt libri tabularum; oriunturque nonnunquam confusiones ex commixtione diversarum divisionum in unum, quod dividentibus conscientiam in rectam erroneam probabilem scrupulosam dubiam, factum videtur. Nam ratione veritatis in rectam et erroneam dispescitur; ratione firmitatis in apprehendendo in certam, probabilem, Dubiam; quid autem aliud dubia, quam scrupulosa?

6. Hujus problematis est etiam propria investigatio Varronis apud B. Augustinum lib.19 de Civ. DEI. cap. 1, numeri sectarum circa summum bonum possibilium. Primum igitur calculum ejus sequemur, deinde ad exactius judicium revocabimus.

7. Divisiones sunt 6. 1ma quadrimembris, 2da et 6ta trimembris; reliquae bimembres. I. *Summum Bonum* esse potest vel *Voluptas*, vel *Indoloria*, vel *utraque*, vel *prima naturae*. 4. II. Horum quodlibet vel *propter virtutem* expetitur, vel virtus *propter ipsum*, vel et *ipsum et virtus* propter se. 4 ∩ 3 f. 12. III. S. B. aliquis vel *in se* quaerit, vel *in societate*. 12 ∩ 2 f. 24. IV. Opinio autem de S. B. constat vel *apprehensione certa*, vel *probabili[tate] Academica*, 24 ∩ 2 f. 48. V. Vitae item genus *cynicum* vel [50] *cultum*. 48 ∩ 2 f. 96. VI. *Otiosum, negotiosum* vel *temperatum*. 96 ∩ 3 f. 288. Haec apud B. Augustinum Varro cap. 1.

[8.] At c. 2 accuratiorem retro censum instituit. Divisionem ait 3, 5 et 6 facere ad modum prosequendi, 4 ad modum apprehendendi S. B.

5. Furthermore, there are plenty of books of tables with divisions that can be combined; but confusions sometimes arise from the merging of different divisions into one, which seems to have happened to those who divide conscience into right, erroneous, probable, scrupulous, and doubtful.[180] For in respect of truth it is distinguished into right and erroneous, while in respect of firmness of apprehension it is distinguished into certain, probable, and Doubtful; for what is doubtful other than scrupulous?[181]

6. And now for a proper investigation of this problem by Varro (as it appears in St Augustine's *The City of God*, Book 19, Chapter 1) of the number of possible schools of thought regarding the Highest Good.[182] I shall first run through his calculations, then readdress myself to a more precise evaluation of them.

7. There are 6 divisions, the first with four members, the second and sixth with three members apiece; and the rest with two members apiece. I. The *Highest Good* can be either *Pleasure, Freedom from pain, both of these*, or *the primary wants of nature*. Total, 4. II. Each of these is sought *for the sake of virtue*, or virtue is sought *for its* [sc. the Highest Good's] *sake*, or *both the Highest Good and virtue* are sought *for their own sake*. 4 ∩ f.12. III. One seeks the *Highest Good* either *in oneself* or *in society*. 12 ∩ 2 f.24. IV. Opinion concerning the *Highest Good* consists either in *certain apprehension* or in *Academic probability*. 24 ∩ 2 f.48. V. One's mode of life is either *cynical* or *cultured*. 48 ∩ 2 f.96. VI. *Leisured, busy*, or *moderate*. 96 ∩ 3 f.288. Thus Varro, in St Augustine, Chapter 1.[183]

[8.] But in Chapter 2 he sets out a more accurate account than the foregoing. He says that divisions III, V, and VI belong to the means of pursuing the

---

[180] Distinctions between 'right', 'erroneous', 'probable', 'scrupulous', 'doubtful', etc. forms of conscience were a common feature of discussions in the context of moral theology. Whether such distinctions confusingly merge a variety of divisions into one, may be an even more complicated question than Leibniz here suggests, since the distinction between 'right' and 'erroneous' forms of conscience could itself be related to the ways in which a certain moral prescription was apprehended, instead of being made only "in respect of truth." Étienne Chauvin, accordingly, would develop the distinction between *conscientia recta* and *erronea* as a distinction between a well-informed (*conscientia rectè informata*) and a mistaken conscience (*conscientia errans*). Cf. *Chauvin*: 133–4.

[181] Leibniz must have meant it the other way round: 'for what is scrupulous other than doubtful?' Chauvin explains that conscience is called *scrupulosa* "when the judgement of the intellect is accompanied by a fearful apprehension that what someone deems good is by chance evil, or the other way around". Cf. *Chauvin*: 134.

[182] Aurelius Augustinus, *De Civitate Dei* XIX 1, in: Augustinus (1955: 657 ff.). The original work *De philosophia* by the Roman scholar and encyclopaedist Marcus Terentius Varro (116–27 BC) has been lost. On Varro and on the place of *De philosophia* within his oeuvre, see: Tarver (1997: 130–64).

[183] Augustinus, *De Civitate Dei* XIX 1, in: Augustinus (1955: 657–9).

corruunt igitur divisiones ultimae, et varietates 276 remanent 12. Porro capite 3, Voluptatem, indoloriam [206] et utramque ait contineri in Primis naturae. Remanent igitur 3 (corruunt 9): Prima naturae propter se, virtus propter se, utraque propter se. Postremam autem sententiam et quasi cribratione facta in fundo remanentem amplectitur Varro.

9. Ego in his noto, Varronem non tam possibiles sententias colligere voluisse, quam celebratas, hinc axioma ejus: qui circa summum bonum differant, secta differre; et contra. Interim dum divisionem instituit, non potuit, quin quasdam ἀδεσπότους admisceret. Alioqui cur divisiones attulit, quas postea summi boni varietatem non facere agnoscit; an ut numero imperitis admirationem incuteret? Praeterea si genera vitae admiscere voluit, cur non plura? nonne alii scientias sectantur alii minime; alii professionem faciunt ex sapientia, creduntque hac inprimis summum bonum obtineri? Etiam hoc ad S. B. magni momenti est in qua quis republica vivat: alii vitam rusticam urbanae praetulere: suntque genera variationum infinita fere, in quibus singulis aliqui fuere, qui hac sola via crederent ad S. B. iri posse.

10. Porro quando prima divisio ducitur in 1mum membrum secundae, facit 4. species: voluptas, 2. indoloria, 3. utraque, 4. prima naturae, propter virtutem, cum tamen in omnibus sit unum summum Bonum Virtus; qui prima naturae, is et caetera; qui voluptatem, is et indoloriam ad virtutem referet. Adde quod erat in potestate Varronis, non solum 2dam et 6tam, sed et 3. et 4. et 5. trimembrem facere, addendo 3tiam speciem, semper mixtam ex duabus, v. g. in se, vel in societate, vel utraque; apprehensione certa, probabili, dubia; cynicum, cultum, temperatum.

Highest Good, and that division IV belongs to the means of apprehending it, so that the final divisions (of which there are 276 varieties) fall, and there remain only 12.[184] Furthermore, in Chapter 3 he says that Pleasure, freedom from pain, and the two of them together, are included in the Primary Wants of nature. There remain, then, only 3 (9 more having fallen), the Primary Wants of nature for their own sake, virtue for its own sake, and both of these for their own sake.[185] In the end, Varro embraces this last opinion as if as a result of a kind of filtering process.

9.  I note in all this that his desire was to collect not so much all possible opinions, as the best known opinions, and so his maxim is: those who differ about the highest good, differ as a school of thought; and the other way round. Accordingly, when he made a division, he could not help but mingle some random divisions with it.[186] Why else did he offer divisions that he subsequently did not recognise as constituting a variety of the highest good, if not just to arouse the admiration of the uninformed at their number? Again, if he wanted to add the variety of modes of life, why stop there? Do not some study the sciences, while others have little time for them; do not others make a profession out of wisdom, and believe that the highest good is obtained chiefly from this? Also of great importance to the Highest Good is in what kind of State one lives; some have preferred the rural to the urban life: and there are almost infinitely many modes of life, in each of which there have been people who believe that in this way alone the Highest Good can be attained.

10.  When the first division is multiplied by the first member of the second division, it makes 4 species: 1. pleasure, 2. freedom from pain, 3. both, 4. the primary wants of nature, for the sake of virtue, where the Virtue in all of them is nonetheless the one Highest Good. Anyone who refers the primary wants of nature to virtue will also refer the others to virtue; anyone who refers pleasure to virtue will also refer freedom from pain to virtue. It was in addition open to Varro to make not only the second and the sixth divisions tri-membered, but also the third, fourth, and fifth, by adding a third species, in each case a mixture of the two; such as, in himself, in society, or in both; with certain apprehension, probable apprehension, or doubtful apprehension; with mode of life cynical, cultured, or moderate.

---

[184]  Augustinus, *De Civitate Dei* xix 2, in: Augustinus (1955: 661, esp.). Note that the new division, too, is Varro's, reported by Augustine.

[185]  The argument in fact occurs at the end of *De Civitate Dei* xix 2. Cf. *idem*, pp. 661–2.

[186]  The Greek word *adéspotos* that Leibniz uses here, properly means 'without master', 'independent'.

11. Fuit et sententia, quae negaret dari S. B. constans, sed faciendum quod cuique veniret in mentem, ad quod ferretur motu puro animi et irrefracto. Huc fere Academia nova, et hodiernus Anabaptistarum spiritus inclinabat. Ubi [51] vero illi qui negant in hac vita culmen hoc ascendi posse? quod Solon propter incertitudinem pronunciandi dixit, Christiani philosophi ipsa rei natura moti. Valentinus vero Weigelius nimis Enthusiastice, beatitudinem hominis esse Deificationem.

11. And there has even been a school of thought that would deny the existence of an unchanging Highest Good, in favour of doing whatever happens to come into one's head, to which one would be impelled by a pure and uninterrupted motion of the mind. To this the New Academy was mostly inclined, and in our own time the Anabaptist movement.[187] But where are those who deny that in the present life we can reach such an exalted state? This was the opinion of Solon,[188] on account of the uncertainty of what it is, and Christian philosophers are guided to the same conclusion by the nature of the thing. Valentin Weigel, however, perhaps too carried away by Enthusiasm, said that man's blessedness consists in DEIfication.[189]

---

[187] Marcus Varro saw the New Academy's scepticism with regard to the possibility of acquiring an indubitable knowledge of the 'ends' of good and evil as the major difference between their position and his own view, which he considered to be the established view of the Old Academy. Cf. Augustinus, De Civitate Dei xix 3, in: Augustinus (1955: 663). Leibniz brackets together the New Academics and the Anabaptists presumably on account of supposed antinomian tendencies within the latter movement.

[188] The Athenian statesman and lawgiver Solon (c. 640–558 BC), whose wisdom would become proverbial, is known, amongst other things, to have argued against calling anyone happy before he or she has died. The quotation derives from Solon's discussion with Croesus (c. 560–546 BC), the extremely rich and last king of Lydia in Herodotus, Histories I, 30–3. Cf. idem, § 32: "Until [a lucky man] is dead, you had better refrain from calling him happy, and just call him fortunate." Translation by Robin Waterfield, in: Herodotus (1998: 16). At the start of Nicomachean Ethics I 11, Aristotle famously quoted this line in the context of addressing the question whether one may call someone happy while still alive. See also: Pritzl (1983: 101–11), Gooch (1983: 112–6). Though Solon and Croesus never met in real life, the story of their acquaintance has been seen as an important factor in the development of the image of the Athenian as a man of wisdom. Cf. Masaracchia (1958: 4). On Solon, see also: Blok and Lardinois (2006).

[189] Valentin, or Valentine, Weigel (1533–88) was a Lutheran pastor from Saxony with mystical and spiritualist ideas, who taught that blessedness consist in the "unification of spirits", since "in spiritual things, one can be in the other, so that God is in me and I am in God, and I and God are one in will". See Weigel (2003: 138–9). Note that the goal of becoming 'godlike' is a common feature of moral theories in the Platonic tradition, after the idea presented by Socrates in the Theaetetus, that "a man becomes like God when he becomes just and pure". Cf. Plato, Theaetetus, 176 B. Translation from Plato (1992: 46). See also: Plato, Republic, 613 A–B. Weigel's position, however, is not only inspired by Neo-Platonic, mystical and spiritualist sources, but also by Biblical texts such as Ps. 82:6, John 10:34, and Gal. 3:36. Weigel himself uses the term Deificatio as well as vergottete Mensch ('deified man') and "wir sind Götter" and its variants at various passages in his works. Cf. Weigel (1996–2015, vol. 2: 62, 75, 76, 81; vol. 3: 36; vol 7: 14). On the seventeenth-century reception of Weigel's ideas, see Weeks (2000: 175 ff.). The editors are grateful to Andrew Weeks and Horst Pfefferl for their helpful comments. Leibniz is obviously aware of the fact that, to orthodox eyes at least, spiritualist thinkers may have taken the idea of a godlikeness in humans one step too far. On Leibniz's defence of the integrity of individual souls as against the idea of mystics like Weigel who interpreted the state of blessedness "as a 'spiritual union' with God", see also: Rutherford (1998: 30–3).

12. Apud illos quoque, quibus collocatur beatitudo in aeterna vita; alii asserunt, alii negant Visionem substantiae DEI beatificam. Hoc reformatos recordor facere, et extat de hoc argumento dissertatio inter Gisb. Voetii selectas; illud nostros, ac pro hac sententia scripsit Matth. Hoe ab Hoënegg peculiarem libellum contra Dnum Budowiz a Budowa.

13. In hac quoque vita omnes illos omisit Varro, qui bonum aliquod extemum, eorum quae fortunae esse dicunt, summum esse supponunt, quales fuisse, ipsa Aristotelis recensio indicio est. Corporis bona sane pertinent ad prima naturae, sed fieri potest ut aliquis hoc potissimum genus voluptatis sequatur, alius aliud. Et bonum animi jam aut habitus aut actio est, illud Stoicis hoc Aristoteli visum. Stoicis hodie se applicuit

12. As for those who place blessedness in everlasting life, some claim for it, others deny, the beatific Vision of God in His essence. I think that Calvinists hold the latter opinion, on which there is a dissertation in Gisbertus Voetius's *Selected Disputations*;[190] and that my own co-religionists hold the former opinion, on which Matthias Hoë von Hoënegg has written a little book of its own, attacking Baron Budowez von Budowa.[191]

13. Varro also omitted all those who, in the earthly life, take as the highest good some external good among those that they call their possessions; and that there were such people, Aristotle's own account is evidence.[192] Corporeal goods plainly belong with the primary wants of nature, but it can happen that someone may pursue one particular kind of pleasure, while another pursues another kind. Spiritual good is seen as either habit or action, the former by the Stoics, the latter by Aristotle.[193] In our own time, an outstandingly learned man, Eckard Leichner, a Physician

---

[190] Voetius is rather cautious on the whole matter, indicating that there are important differences of opinion among Church Fathers as well as theologians generally, despite an apparent neo-Scholastic consensus. Reformed sources generally tended to regard the question as vain curiosity, whilst Voetius's own position was additionally inspired by a wish to stay away as far as possible from overly confident Catholic and Socinian ideas on the matter and to stick to the views of his own Leiden tutors Franciscus Gomarus (1563–1641) and Lucas Trelcatius jr. (1573–1607). Cf. Voetius (1655: 1195–1217, esp. 1213–7), as well as pp. 1217–28, where it is argued that, although the blessed are generally considered to see God 'in His essence' (*per essentiam*), or *in se*, this is a position Voetius is "not in the least" willing to defend (p. 1227). See also the next footnote.

[191] Matthias Hoë von Hohenegg (1580–1645) was a fierce defender of Lutheranism. He published a friendly, though critical letter to the Czech Protestant nobleman who was later to become a martyr for the Winterking and the Protestant cause, Baron Václav Budovec z Budova (or Wenceslas von Budowa, in Latin Wenceslaus Budowitz a Budowa, 1551–1621). Cf. Hoë van Hohenegg (1617: 8–9). On Matthias Hoë, see ADB 12: 541–9, NDB 9: 300–1, and DBE 5: 85. Note that Voetius was interested in the 1620 reaction to Hoë by the Polish theologian Julian Poniatowski, or Julianus Poniatovius (†1628), which, however, he had not been able to consult. Cf. Voetius (1655: 1215).

[192] Aristotle, *Ethica Nicomachea* I 2 and I 3, 1095a20–22 and 1095b14–22, resp. Note, however, that although 'goods of the soul' rather than external goods contribute to the 'end' of human life, Aristotle does not himself deny the relevance of external goods to happiness; *Ethica Nicomachea* I 8 and I 9, 1098b20–21 and 1099a31–1099b8, resp.

[193] Although Aristotle's ethics is centred around the notion of *hexis* (habit, or disposition) classical and early modern debates on the Aristotelian and Stoic interpretations of the *summum bonum* attributed to Aristotle the idea that virtue was in essence an *actio* and to the Stoics the idea that virtue was a *habitus*. Cf. Aristotle, *Ethica Nicomachea* I 8, 1098a29–1099a7.

accuratus sane vir, Eckardus Leichnerus Medicus Erphordiensis tr. de apodictica scholarum reformatione et alibi. [*207*]

14. Quin et voluptatem animi pro S. B. habendam censet Laurentius Valla in lib. de Vero Bono, et ejus Apologia ad Eugenium IV Pontificem Maximum, ac P. Gassendus in Ethica Epicuri, idque et Aristoteli excidisse VII, Nicomach. 12 et 13 observavit CI. Thomasius Tab. Phil. Pract. XXX, lin. 58. Ad voluptatem animi gloriam, id est triumphum animi internum, sua laude sibi placentis, reducit Th. Hobbes initio librorum de Cive. Fuere qui contemplationem actioni praeferrent, alii contra, alii utramque aequali loco posuere. Breviter quotquot bonorum

of Erfurt, has aligned himself with the Stoics in his treatise *On the Apodictic-Philosophical Correction of the Schools*, and in other works.[194]

14. Indeed, in his book *On the True Good*, and his *Apology to Pope Eugenius IV*, Lorenzo Valla reckons that spiritual pleasure should be held to be the Highest Good,[195] as does Pierre Gassendi in his *Ethics of Epicurus*;[196] and the distinguished Thomasius, in his *Practical Philosophy*, XXX, line 58, observed that this also occurs in Aristotle, *Nicomachean Ethics*, Book VII, chapters 12 and 13.[197] Thomas Hobbes, early on in his *De Cive*, reduces glory to spiritual pleasure, that is, the internal exultation of a mind pleased with itself at being praised.[198] There were some who preferred contemplation to action, others the reverse, others still who put them both on the same level.[199] In short, we must account as many

---

[194] The expression *de apodictica scholarum reformatione* or *emendatione* occurs in the title of a number of works in which the Erfurt professor of medicine Eckhard Leichner (1612–90) pleaded for methodological reforms in science and education. A general introduction was published as Leichner (1652). In Leichner (1664), medical subjects are treated with a critical stance to both ancient and scholastic, as well as to modern traditions. Later adding disputations while using the same title, Leichner would continue critically to discuss recent ideas by the likes of Harvey, Sylvius, and De Graaf, combining his own anatomical observations with vitalistic notions and a rationalist ('apodictic') conception of science that precluded him from accepting their views. Applying a similar critical dogmatism to logical, physical, and theological subjects discussed in contemporary education, Leichner also came to criticize Aristotelian ethics for having held that human well-being would be found in the transient acts of virtue, rather than in virtue itself. Cf. Leichner (1669: 95). Leichner here promises to discuss this point elsewhere in further detail ("Wovon anderswo mit mehren"), but we have not been able to find a positive reference to the Stoics in extant copies of Leichner's rather pompous works. On Leichner, see also: *Zedler* 16, cols. 1562–8.

[195] Cf. Valla (2004), an English translation of which may be found in Valla (1977). Where Valla had offered an Epicurean defence of pleasure in general in *De voluptate*, it is rather in his apology to the Pope that he focused on *spiritual* pleasure. Cf. Valla (1540: 795–800, esp. 797–9a).

[196] Arguing from the premiss that *Felicitas*, as far as it can be had in this life, is the sole factor to be taken as an end in itself, Gassendi's discussion of the *Summum Bonum* develops into a protracted defence of the Epicurean notion of pleasure (*Voluptas*) against those who maliciously interpret the notion of pleasure in "corporeal", "obsceen" and "infamous" ways, contrary to what it had meant to Epicurus himself, namely "tranquillity of mind (*Tranquillitas animi*) and freedom from pain (*Indolentia corporis*)". Cf. Gassendi (1658, vol. 2: 639–735). Quotation from p. 724.

[197] Thomasius (2005), sig. Cr, lines 35 ff. and "Annotationes ad Tabulam XXX, De Voluptate", pp. 41–2, esp. p. 42, where the full reference to Aristotle occurs, here offering Chapters 12 and 14, however. Cf. Aristotle, *Ethica Nicomachea*, VII 12 and 13, 1152b1–1153a35 and Plato, *Republic*, 580d ff.

[198] Hobbes (1839b: 158–61, esp. 160–1). Cf. Hobbes (1841: 2–6, esp. 5): "But all the mind's pleasure is either glory, (or to have a good opinion of one's self), or refers to glory in the end"—from which Hobbes, however, characteristically draws a distinctively egoistic conclusion: "All society therefore is either for gain, or for glory; that is, not so much for love of our fellows, as for the love of ourselves".

[199] Though Christian authors were prone to accept Aristotle's identification of the highest good with contemplative happiness, Aristotle's analysis of both practical and contemplative virtues in the *Nicomachean Ethics* would continue to give rise to elaborate discussions on the relative importance of practical vis-à-vis contemplative virtue and to a division of philosophical positions endorsing either the one or the other.

imae sunt species, quotquot ex illis complexiones, tot sunt summi boni possibiles sectae numerandae.

15. Ex hoc ipso problemate origo est numeri personarum in singulis gradibus Arboris Consanguinitatis, eum nos, ne nimium a studiorum nostrorum summa divertisse videamur, eruemus. Computationem autem canonica neglecta civilem sequemur.

16. Duplex Personarum in singulis gradibus enumeratio est, una generalis altera specialis. [52] In illa sunt tot personae quot diversi flexus cognationis eadem tamen distantia. *Flexus* autem *cognationis*, voco ipsa velut itinera in arbore consanguineitatis, lineas angulosque, dum modo sursum deorsumve modo in latus itur. In hac non solum flexus cognationis varietatem facit, sed et sexus tum intermediarum, tum personae cujus distantia quaeritur a data. In illa enumeratione Patruus, Amita; id est Patris frater sororve: Avunculus, Matertera; id est Matris frater sororque, habentur pro eadem persona, et convenientissime intelliguntur in voce *Patrui*, quia masculinus dignior foemininum comprehendit. Sed in enumeratione speciali habentur pro 4. diversis personis. Igitur illic *cognationes*, hic *personae* numerantur: (Sic tamen ut plures fratres, vel plures sorores, quia ne sexu quidem variant, pro una utrobique persona habeantur). Illa generalis computatio est *Caji in l. 1. et 3.* (quanquam specialis nonnunquam mixta est), haec specialis Pauli in grandi illa *l. 10.*

possible schools of thought regarding the highest good as there are basic species of goods, and complexions of them.

15. Finally, the number of persons in each degree of a Family Tree is part of this same problem, and I shall deal with it now, aiming not to stray too much from the main concern of our studies. Without taking into account the canonical calculation, I will stick to the civil one.[200]

16. There are two ways of enumerating the Persons in each degree, one general, the other special. In the first, there are as many persons as there are different branches of cognation, though they may be at the same distance. I call *Branches of Cognation* the pathways on a family tree, that is, the lines and angles whereby one goes up and down, or sideways. In the second, not only does the branch produce a variety, but also the gender, whether of intermediate persons or of a person whose distance from a given person is required. In the first kind of enumeration, the Paternal Uncle and Paternal Aunt (that is, the Father's brother or sister, respectively), and the Maternal Uncle and Maternal Aunt (that is, the Mother's brother and sister respectively) are taken to be the same person, and they are most conveniently embraced in the term *Paternal Uncle*; the masculine gender, being of greater dignity, standing also for the feminine gender. But in the special enumeration, they are taken to be 4 different persons. Accordingly, *Cognations* are numbered with the former, *persons* with the latter (though in such a way that several brothers or several sisters are taken to be one person standing for both, because we are not considering variation in gender). The former, general calculation is that of Gaius, in his *Institutions of Civil Law*, Books 1 and 3 (though the special is sometimes mixed with it),[201] the latter, special calculation that of Julius Paulus, in his *Digest*, Book 10, *Of Degrees and Affinities*.[202] And though the former is based on Problems

---

[200] Two different ways of calculating the number of persons in a Family Tree are distinguished here: a canonical (in the sense of canon law) and a civil one.

[201] The *Institutiones* of the Roman lawyer Gaius (c. 110–80) found their way into the *Corpus Juris Civilis* as the prime material for the *Institutiones* (or 'Elements of Law') drawn up under emperor Justinian I (c. 482–565). Part 1 covered persons and their status before the law, part 3 the laws of succession. Cf. *Institutiones*, pp. 1–56. *Institutiones* III 6 is a special paragraph on the degrees of cognation, presenting both manners of enumeration Leibniz here identifies. Cf. *idem*, pp. 32–3. A modern edition of Gaius's *Institutiones* after the palimpsest discovered by Barthold Georg Niebuhr (1776–1831) in Verona in 1816, was published in 1877.

[202] The *Digesta* of Julius Paulus Prudentissimus (fl. 230) were to make up a sixth of Justinian's sixth-century *Digest* (or *Pandects*). An extended excerpt of Paulus's Book *De gradibus et adfinibus et nominibus eorum* occurs at *Digesta* XXXVIII 10, *Corpus Iuris Civilis*, pp. 580–5.

*D. de Grad. et Affinibus.* Etsi autem prior fundata est in prob. 1 et 2, quia tamen posterioris fundamentum est, quae huc pertinet, praemittemus.

17. *Cognatio* est forma lineae vel linearum a cognata persona ad datam ductarum; ratione rectitudinis et inflexionis, et harum alternationis. *Persona* h. l. est Persona datae cognationis, et dati gradus, sexusque tum sui, tum *intermediarum*, inter cognatam scilicet et datam. *Datum* a[utem] voco personam, eum eamve, de cujus cognatione quaeritur, ut appellant JCti veteres; Joh. Andreae *Petrucium* nomine sui Bidelli fertur nominasse: Fr. Hottomannus lib. de Gradib. Cognationum, ὑποθετικόν, latine *Propositum*.

18. *Terminus* est persona vel cognatio, quae est de conceptu complexae, v. g. *frater* est Patris filius. Igitur *Patris* et *Filius*, sunt Termini ex quibus conceptus Fratris componitur. Termini autem sunt vel primi, tales accurate loquendo sunt hi solum: Pater et filius, nos tamen commodioris computationis causa, omnes personas lineae rectae vel supra vel infra, supponemus pro primis; vel *orti*: accurate loquendo omnes qui plus uno gradu remoti [208] sunt a Dato; laxius tamen, omnes transversales tantum. Omnes a[utem] [53] transversales componuntur ex duobus terminis lineae rectae; hinc et facillimum prodit artificium data qu[a]cunque cognata numerum gradus complexae invenire, v. g. in simplicissima transversalium persona. *Fratre* seu Patris filio, quia Pater est in 1, filius etiam in gradu 1 + 1 f. 2, in quo est Frater.

19. Caeterum Schemate opus est. Est igitur hoc

I and II, I shall deal with this calculation first, because it is the basis of the latter calculation, which we are concerned with here.

17. *Cognation* has the form of a line or lines drawn from a cognate person to a given person; its principles being straightness and obliquity, and alternation of these two. A *Person* as understood here is a Person of given cognation and given degree, and of gender sometimes his own, sometimes that of intermediate persons — intermediate between, I mean, the cognate and the given. I call a person (whether he or she) whose cognation is required, the *Given*, following the usage of the ancient Jurists. Johannes Andreae is said to have called this person the *Petrucius* after the name of his Beadle.[203] François Hotman, in his book *On the Degrees of Cognation*, [called it] *hypothetikon*; in Latin, the *Proposed*.[204]

18. A *Term* is a person or cognation that belongs to the concept of a complex, for example, a *brother* is the son of one's Father. In this example, *Father* and *Son* are the Terms of which the concept of *Brother* is composed. Further, Terms are either *primitive*, such as (in the strict sense) these alone, Father and son, though for the sake of convenience of calculation I shall take all persons on a straight line, whether above or below, as primitive; or *derived*, (in the strict sense) all persons who are removed by more than one degree from the Given; more loosely, however, all those who are removed merely collaterally. All those who are removed collaterally are composed of two terms of a straight line. Hence, given a cognate person, there is a very easy method of calculating the number of the degree of some complex person. For example, in the case of the simplest person among collaterals: given a Brother, or Son of one's Father, because the Father is in degree 1 and the Son again in degree 1, this makes 1 + 1 f. 2, which is the degree of the Brother.

19. It is time we had a Diagram. I therefore propose this:[205]

---

[203] Andreae 1517, sig. [Qvv°]. The *Arbor consanguinitatis*, or *Lectura super arboribus consanguinitatis et affinitatis* of the Italian lawyer and diplomat Giovanni d'Andrea (c. 1270–1348) received a wide diffusion especially in Germany. Cf. Eis (1965: 5–6, 7).

[204] Note, however, that *Propositum* is Leibniz's own translation, since the French Protestant lawyer and humanist Hotman (1524–90) uses only the Latinised form *hypotheticus* in Hotman (1547), *passim*. In his *Disputations on Civil Law*, moreover, Hotman shows himself to be quite willing to stick to "Titius / Titia" and "Caius / Caia" as the names of the *Is de quo quaeritur*-person in *Thema*-type (in contrast to *Stemma* or *familia*-type) family-trees, i.e. in family trees that carry abstract relations ('themata') rather than proper names. Moreover, Hotman here offers *proposita* as Cicero's word for *themata*, associating *thesis* with *thema*, and giving *hypothesis* as the parallel equivalent of *Stemma* ('family-tree'). Cf. Hotman (1599: 438) and *Disputatio de Gradibus Cognationis, una cum figuris seu schematibus ad graduum enumerandorum rationem adcommodatis*, in *idem*, pp. 810–45.

[205] Leibniz was to propose a very similar tree in a later paper on elimination theory, see: Knobloch (1974: 142–73).

| Gr. Cognationes | | | | Patris | DATUS | Filius | | | | Personae  Gr. |
|---|---|---|---|---|---|---|---|---|---|---|
| | | | | | FRATER 1.1 | | | | | 4 Filius  1. |
| 1.Patris 2 | | | | | | Pa-tru-elis 1.2 | Patru-elis parvus 1.3 | Pro-patru-elis 1.4 | Ab-patru-elis 1.5 | |
| 2. Avi  3 | | | | Patru-us Magnus 2.1 | Con-sobri-nus 2.2 | Sub-conso-brinus 2.3 | Prosub-conso-brinus 2.4 | | | 12 Nepos 2. |
| 3. Proavi 4 | | | Pro-pa-truus 3.1 | Subpa-truus Ma-gnus 3.2 | Prosub-patruus Magnus vel* 3.3 | | | | | 32 Pronepos 3. |
| 4. Abavi 5 | | Ab-pa-truus 4.1 | Subpro-patruus 4.2 | | | | | | | 80 Abnepos 4. |
| 5.Atavi 6 | Subpro-pa-truus 5.1 | | | | | | | | | 192 Atnepos 5. |
| 6.Tritavi 7 | | | | | | | | | | 448 Trinepos 6. |

* Consobrin. secundus

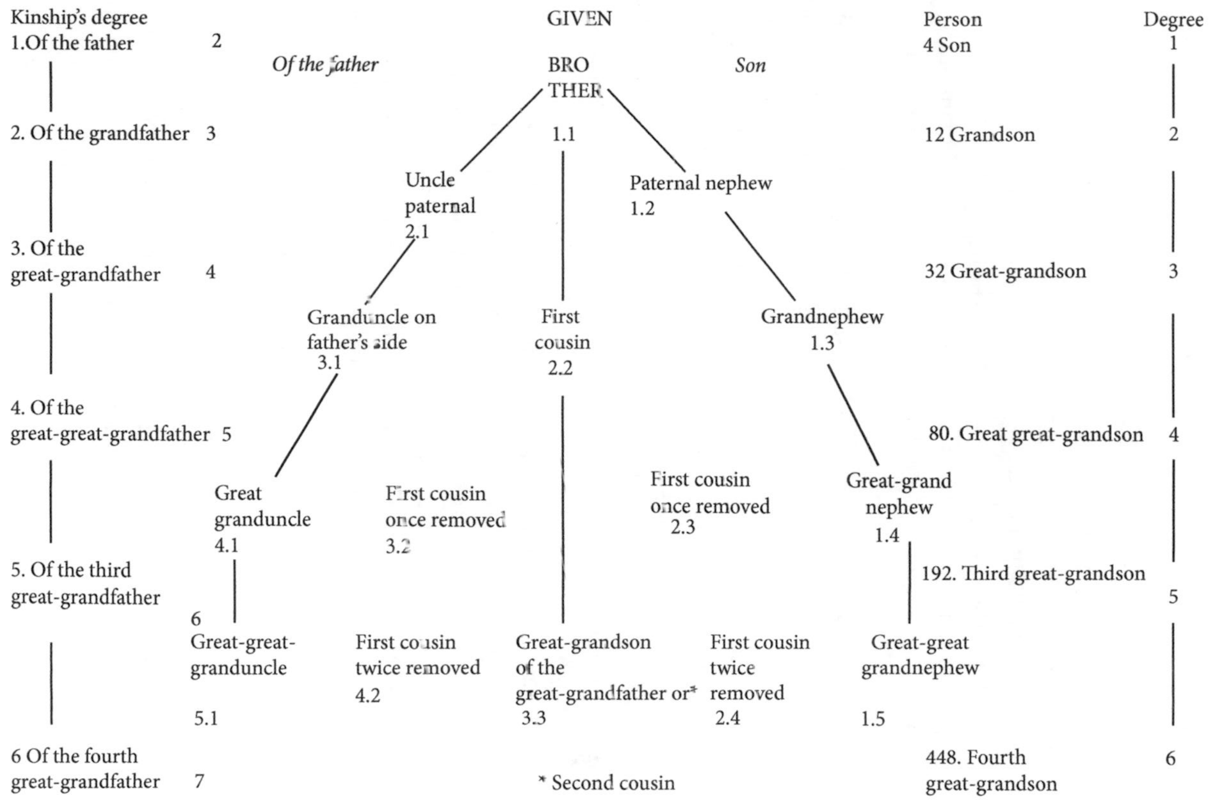
Kinship's degree
1.Of the father 2
2. Of the grandfather 3
3. Of the great-grandfather 4
4. Of the great-great-grandfather 5
5. Of the third great-grandfather 6
6 Of the fourth great-grandfather 7
Of the father
GIVEN
BRO THER
Son
1.1
Uncle paternal 2.1
Paternal nephew 1.2
Granduncle on father's side 3.1
First cousin 2.2
Grandnephew 1.3
Great granduncle 4.1
First cousin once removed 3.2
First cousin once removed 2.3
Great-grand nephew 1.4
Great-great-granduncle 5.1
First cousin twice removed 4.2
Great-grandson of the great-grandfather or* 3.3
First cousin twice removed 2.4
Great-great grandnephew 1.5
* Second cousin
Person
4 Son
12 Grandson
32 Great-grandson
80. Great great-grandson
192. Third great-grandson
448. Fourth great-grandson
Degree
1
2
3
4
5
6

[*209*] 20.    [54] Sunt in hoc Schemate infinita propemodum digna observatione. Nos pauca stringemus. Personae eo loco intelligantur, ubi puncta sunt. Numeri puncta includentes, designant terminos, seu gradus lineae rectae (antecedens ascendentis, sequens descendentis), ex quibus datus gradus transversalis componitur. In eadem Linea transversa dìrecta sunt ejusdem gradus cognationes: obliquae a summo, ad imum dextrorsum ordinem generationis; at sinistrorsum complectuntur cognationes homogeneas gradu differentes. Linea perpendicularis unica a vertice ad basin, triangulum dividens, continet cognationes quarum terminus et ascendens et descendens sunt ejusdem gradus; tales voco *aequilibres*, et dantur solum in gradibus pari numero signatis, in uno non nisi unus.

21.    Nam si libra esse fingatur, cujus Trutina sit linea gradus primi; brachia vero sint: dextrum quidem, linea perpendicularis a summa persona descendentium; sinistrum vero, perpendicularis a summa ascendentium ducta ad terminum vel ascendentem vel descendentem datam cognationem componentem; tum brachiis aequalibus, si utrinque 3, 3 aut 2, 2 etc., cognatio erit aequilibris et ponenda in medio trianguli; in inaequalibus, cognatio talis ponenda in eo latere, quod lineae rectae vel ascendenti vel descendenti, ex qua brachium longius sumtum est, est vicinum.

22.    Hic jam complexionum vis apertissime relucet. Componuntur enim omnes personae transversae ex 2 terminis, una cognatione recta ascendenti altera descendenti. Semper autem sic, ut ascendens in casu obliquo, descendens in casu recto conjungantur, v. g. frater, id est patris filius. At si contra, redibit persona data, nam qui patrem filii sui nominat se nominat. Quia unus pater plures filios habere potest, non contra.

[23].    Ex his jam datur *proposito quocunque gradu cognationum tum numerum, tum species reperire: Numerus* transversalium semper erit unitate minor gradu (numerus omnium semper unitate major, quia addi debent duae cognationes lineae rectae, una sursum altera deorsum), cujus ratio ex inventione *specierum* patebit. "Nam com2nationes partium, oder Zerfallungen in zwey Theil, dati numeri

20.   In this diagram there are almost infinitely many things worthy of notice: I shall confine myself to a few. Persons must be understood in place of points. Numbers containing points designate terms, or the degrees on the straight lines (before the point, of an ancestor, after the point, of descendant) of which a given collateral degree is composed. On the same direct collateral Line are cognations of the same degree: running obliquely from top to bottom right in order of generation; but running from top to bottom left we find homogeneous cognations differing in degree. The single vertical line running from top to bottom, dividing the triangle, contains cognations whose terms, both ancestral and descending, are of the same degree. I call such cognations *equally balanced*, and they exist only in degrees designated by an even number, except in the single case of one.

21.   For if we imagine a pair of scales, whose Pivot is the line of first degree, and whose arms are, on the right, a vertical line from the topmost person of the descendants, and on the left, a vertical line from the topmost person of the ancestors, connecting the terms, whether ancestral or descending, making up the given cognation, then if with the arms equal, it is up and down 3.3 or 2.2 etc., the cognation will be equally balanced, and belongs in the middle of the triangle. With unequal arms, such a cognation belongs on the side that is closer to the straight line, whether ancestral or descending, from which the longer of the arms is measured.

22.   Now the power of complexions is most clearly displayed here. For all collateral persons are composed of 2 terms, with one cognation directed at an ancestor, the other at a descendant; and always in such a way that the ancestor, in the genitive case, and the descendant, in the nominative case, are joined together; for example, a brother, that is the son of one's father. But if we proceed in the opposite direction, we return to the given person, for someone who describes himself as the father of his son describes himself. This is because one father can have several sons, but not the other way round.

[23.]   From this we are now able to find, given *a certain proposed degree of cognation, first their number, then their species*. The *number* of collateral cognations will always be one less than the degree (the number of all cognations always one more than the degree, since two cognations, one at the top of a straight line, the other at the bottom, must be added). The reason for this will become clear when we come to finding *species*. "For the number of com2nations of parts, *or divisions in two parts*,[206] of

[206] Leibniz again (see note 33) uses the German word *Zerfällungen* to designate the *partitions* considered as subsets of combinations.

cujuscunque sunt tot quot unitates habet numeri dati pa[55]ris dimidium, imparis demta unitate dimidium, v. g. 6 habet has: 5, 1, 4, 2, 3, 3. Ejusque rei ratio manifesta est, quia semper numerus antecedens proximus dato cum remotissimo, pene proximus cum pene remotissimo complicatur, etc." Sed cum hic non solum complexioni, sed et situs habenda ratio sit, v.g. alia cognatio est 5, 1, nempe Abpatrui, quam 1, 5. nempe [A]bpatruelis, hinc cum 2 res situm varient 2 vicibus, ergo duplicentur discerptiones, redibit numerus datus si par fuerat; sed cum in ejus discerptionibus detur una homogenea, v. g. 3, 3 in qua nihil dispositio mutat, hinc subtrahatur de numero dato, seu duplo discerptionum, iterum: 1. Si vero numerus datus fuerat impar, redibit numerus unitate minor.

24. Ex hoc manifestum est generaliter: (1) Subtrahatur de numero gradus unitas, productum erit numerus cognationum transversalium. (2) Duo numeri, qui sibi sunt comple[210]mento ad datum, seu quorum unus tantum distat ab 1 quantum alter a dato, complicati dabunt *Speciem* cognationis, si quidem praecedens intelligatur significare ascendentem, sequens descendentem sui gradus.

25. Hac occasione obiter explicandum est, quae sint dati numeri discerptiones, Zerfallungen, possibiles. Nam omnes quidem Discerptiones sunt Complexiones, sed Complexionum eae tantum Discerptiones sunt, quae simul toti sunt aequales. Instigari similiter possunt tum com2nationes tum con3nationes, tum discerptiones simpliciter, tum dato exponente. Quot factores, vel divisores exactos numerus aliquis datus habeat, scio solutum vulgo. Et hinc est quod Plato numerum civium voluit esse 5040, quia hic numerus plurimas recipit divisiones civium pro officiorum generibus, nempe 60, lib. 5 de Legib. fol. 845. Et hoc quidem in multiplicatione et divisione, sed qui additione datum numerum producendi varietates, et subtractione discerpendi collegerit, quod utrumque eodem recidit, mihi notus non est. Viam

any given number is half of the number when it is even, and half of an odd number less one. For example, 6 has these [com2nations]: 5.1, 4.2, and 3.3.[207] The reason for this is obvious, because the nearest antecedent number is always com2ned with the furthest number, the next nearest with the next furthest, and so on."[208] However, we must here take into account not only complexions, but also situs. For example, the cognation 5.1, namely that of Great-great-granduncle, is distinct from the cognation 1.5, namely that of Great-great-grandnephew. Hence, since 2 things can vary in situs in 2 ways, the divisions may be doubled, and the given number will reappear if it was even. But since the divisions contain one homogeneous division, for example, 3.3, in which the disposition does not change anything, this should be subtracted from the given number, that is from the twofold number of divisions. But if the given number was odd, the number [of divisions] will stay at one less than the given number.

24. From this it is clear that generally: (1) 1 is to be subtracted from the number of the degree, and the result will be the number of collateral cognations. (2) Two numbers that are mutually complementary to the given number, that is, of which one is distant from 1 by the same amount as the other is distant from the given number, will, when combined, give the *Species* of cognation, the preceding number being understood to signify the ancestor of that degree, and the following number the descendant of that degree.

25. This is, by the way, a suitable time to explain what the divisions of a given number are, the possible *partitions*.[209] For though all Partitions are Complexions, among Complexions the only Partitions are those that add up to the whole. Com2nations and con3nations can likewise be generated, sometimes as the total number of partitions, sometimes as partitions with a given exponent. I know that the problem of calculating how many factors, or exact divisors a given number can have, has been solved and is quite well known. This is also why Plato wanted the number of his citizens to be 5040, because this number is capable of so many divisions of citizens for the purposes of various kinds of office, that is, 60. (*Laws*, Book 5, folio 845.)[210] So much for

---

[207] Given a number $n$, there are $\frac{n}{2}$ *bipartitions* of it, for $n$ even and $\frac{n-1}{2}$ for $n$ odd, if the order of summands is disregarded (cf. Knobloch 1974: 415).

[208] As in § 27, and elsewhere, Leibniz here uses quotation marks to indicate his solution of a problem.

[209] Again, for '*partitions*' Leibniz uses the German word *Zerfällungen*. See also note 207.

[210] In the 1602 Frankfurt Greek-Latin edition of Plato's works, the view that the right number of citizens is related to a fitting distribution into a variety of sections, as well as to the right distribution of land and houses within Plato's ideal city state, is presented on folio 845, with the exact number of 5040 for the total population, however, only occurring at the top of folio 846. Plato (1602: 845–6). Cf. Plato, *Laws* V, 737c–738b. Note that Plato also discusses questions of distribution between the 5040 citizens at *Laws* V, 739e–741d, 745b–e and 746d ff., as well as *Laws* VI, 771a–d.

autem colligendi com2nationes discerptionum ostendimus proxime. At ubi plures partes admittuntur, ingens panditur abyssus discerptionum. In qua videmur nobis aliquod fundamentum computandi agnoscere, nam [56] semper discerptiones in 3 partes oriuntur ex discerptionibus in 2, praeposita una; exequi vero hujus loci fortasse, temporis autem non est.

26. Caeterum antequam in Arbore nostra a computatione generali ad specialem veniamus, unum hoc admonendum est Definitiones cognationum a nobis assignatas in populari usu non esse. Nam v. g. Patruum nemo definit avi filium, sed potius patris fratrem. Quicunque igitur has definitiones ad popularem efformare morem velit, si quidem persona transversalis ascendit, in termino descendenti loco filii substituat, fratrem; nepotis patruum etc. loco Descendentem ponat uno gradu minorem. Sin descendit, contra.

27. Nunc igitur cum ostendimus cognationes in quolibet gradu, gradus numero unitate majores esse: age et Personas cognationum numeremus. Quae est *Specialis Enumeratio.* Diximus autem in eadem cognatione diversitatem facere tum Sexum cognatae, tum intermediarum inter cognatam et datam personarum. Sexus autem 2plex est. Igitur semper continue numerus personarum est duplicandus, v. g. non solum et pater et mater sexu variant, 2, sed iterum pater habet patrem vel matrem. Et mater quoque. Hinc 4 Avus quoque a patre habet patrem vel matrem, et avia a patre; et avus a matre aviaque similiter: hinc 8 etc. Igitur regulam colligo: "2 ducatur toties in se, quotus est gradus cujus personae quaeruntur, vel quod idem est, quaeratur numerus progressionis geometricae duplae, cujus exponens sit numerus gradus. Is ducatur in numerum cognationum dati gradus: Productum erit numerus personarum dati gradus."

28. Et hac methodo eundem numerum personarum erui, quem Paulus JCtus in d. l. 10, excepto gradu 5, Gr. I, 2 ∩ 2 f. 4. Consentit Paulus

Multiplication and division; but if anyone has calculated all the various ways of producing a given number by addition, and (what comes to the same thing) of dividing up a given number by subtraction, his name is not known to me. I shall shortly show how to calculate com2nations of divisions; but when several parts are allowed, a vast abyss of divisions opens up. It seems to me that I can see some basis for calculating them, as divisions into three parts always arise from divisions into two parts, with one of the parts left constant; but though it may be the right place to go into this further, it is not the right time.

26.   Next, before I pass from general calculations concerning our Tree to particular calculations, I have to issue this one warning, that the Definitions of cognations assigned by me are not those in common use. For example, nobody defines 'Uncle' as 'grandfather's son', but rather as 'father's brother'. Accordingly, if you want to make these definitions conform to common usage, then in the case of a collateral person who is an ancestor, you should substitute 'brother' for 'son', 'uncle' in place of 'grandson', etc. in the descendant term, and place the Descendant one degree lower. But if the collateral person is a descendant, then the opposite.

27.   Therefore, now that I have shown that the number of cognations in any degree is 1 more than the number of the degree, we can go on to enumerate the Persons in these cognations, which constitutes the *Special Enumeration*. I have remarked that within the same cognation diversity arises sometimes from the Sex of the cognate person, sometimes from the sex of persons intermediate between the cognate and the given person. Moreover, as there are two Sexes, the number of persons always needs to be doubled and redoubled. For example, a father and a mother vary not only in sex in 2 ways, but the father has again a father and a mother, and the mother also. Hence, they vary in 4 ways. A grandfather also has through his father, a father or a mother, and a grandmother through her father; and similarly, a grandfather and grandmother through their mother. Hence, they vary in 8 ways, etc. Therefore, I infer the following rule: "multiply 2 by itself as many times as the degree whose persons are required, or what comes to the same thing, look for the term of the geometrical progression of 2 whose exponent is the number of the degree. This needs to be multiplied by the number of cognations of the given degree: the product will be the number of the persons of the given degree."

28.   By this method I have arrived at the same number of persons as Paulus the Jurist in his *Digest*, Book 10, with the exception of Degree 5. Degree 1: 2 ∩ 2f. 4. In his *Digest*, Book 10, § 12, Paulus agrees. Degree 2: 2 ∩ 2 f.

[*211*]    d. l.10, §. 12. Gr. II, 2 ∩ 2 f. 4 ∩ 3 f. 12.§. 13 Gr. III, 2, 2, 2 ∩ f. 8 ∩ 4 f. 32.
§. 14. Gr. IV, 2, 2, 2, 2 ∩ f. 16 ∩ 5 f. 80. §. 15. Gr. V, 2, 2, 2, 2, 2 ∩ f. 32 ∩ 6
f. 192. Dissentit Paulus §. 16 et ponit: 184, cujus tamen calculo errorem
inesse necesse est. Gr. VI, 2, 2, 2, 2, 2, 2 ∩ f. 64 ∩ 7 f. 448. Consentit
Paulus §. 17. Gr. VII, 2, 2, 2, 2, 2, 2, 2 ∩ f. 128 ∩ 8 f. 1024. §. fin. 18. [57]

## Probl. IV.
## DATO NUMERO RERUM VARIATIONES
## ORDINIS INVENIRE.

Solutio: "Ponantur omnes numeri ab unitate usque ad Numerum rerum,
inclusive, in serie naturali: factus ex omnibus continue, erit quaesitum." Ut:
esto Tabula Π quam ad 24. usque continuavimus.

[*212*]

ה

| | |
|---:|:---|
| 1 | 1 |
| 2 | 2 |
| 6 | 3 |
| 24 | 4 |
| 120 | 5 |
| 720 | 6 |
| 5040 | 7 |
| 40320 | 8 |
| 362880 | 9 |
| 3628800 | 10 |
| 39916800 | 11 |
| 479001600 | 12 |
| 6227020800 | 13 |
| 87178291200 | 14 |
| 1307874368000 | 15 |
| 20922789888000 | 16 |
| 355687428096000 | 17 |
| 6402373705728000 | 18 |
| 12164510040883200 | 19 |
| 2432902008176640000 | 20 |
| 51090942171709440000 | 21 |
| 1124000727777607680000 | 22 |
| 25852016738884976640000 | 23 |
| 620448401733239439360000 | 24 |

4 ∩ 3 f. 12. § 13. Degree 3: 2.2.2 ∩ f. 8 ∩ 4 f. 32. § 14. Degree 4: 2.2.2.2 ∩ f. 16 ∩ 5 f. 80. § 15. Degree 5: 2.2.2.2.2 ∩ f. 32 ∩ 6 f. 192. In § 16, Paulus disagrees in favour of 184, but there must be some mistake in his calculations. Degree 6: 2.2.2.2.2.2 ∩ f. 64 ∩ 7 f. 448. In § 17, Paulus agrees. Degree 7: 2.2.2.2.2.2.2 ∩ f. 128 ∩ 8 f. 1024. § 18, finally.[211]

## Problem IV GIVEN A NUMBER OF THINGS, TO FIND THE VARIATIONS OF ORDER

1. Solution: "Place all the numbers from 1 up to the Number of things inclusive in their natural order: their product will be what is required," as in Table ה, which I have continued as far as 24.[212]

ה

| | |
|---:|:---|
| 1 | 1 |
| 2 | 2 |
| 6 | 3 |
| 24 | 4 |
| 120 | 5 |
| 720 | 6 |
| 5040 | 7 |
| 40320 | 8 |
| 362880 | 9 |
| 3628800 | 10 |
| 39916800 | 11 |
| 479001600 | 12 |
| 6227020800 | 13 |
| 87178291200 | 14 |
| 1307874368000 | 15 |
| 20922789888000 | 16 |
| 355687428096000 | 17 |
| 6402373705728000 | 18 |
| 121645100408832200 | 19 |
| 2432902008176640000 | 20 |
| 51090942171709440000 | 21 |
| 1124000727777607680000 | 22 |
| 25852016738884976640000 | 23 |
| 620448401733239439360000 | 24 |

---

[211] Justinianus, *Digesta* XXXVIII 10, 'De gradibus et adfinibus et nominibus eorum', in *Corpus Iuris Civilis*, vol. 1 (1894, esp. XXXVIII 10, §§ 12–8: 581–5).

[212] That is: given, for instance, $n$ things, the number of permutations (variations) that can be made from them equals $n \cdot (n-1) \cdot (n-2) \cdot (n-3) \ldots 3 \cdot 2 \cdot 1 = n!$ (the factorial of $n$). The solution was surely known to Leibniz from Lantz (1616: 180); see Knobloch (1973: 38).

Latus dextrum habet exponentes, seu numeros rerum, qui hic coincidunt; in medio sunt ipsae Variationes. Ad sinistrum posita est *differentia* variationum duarum proximarum, inter quas est posita. Quemadmodum expo[58]nens in latere dextro est ratio variationis datae ad antecedentem. Ratio solutionis erit manifesta, si demonstraverimus *Exponentis dati variationem, esse factum ex ductu ipsius in variationem exponentis antecedentis*, quod est fundamentum Tabulae: ה.

ז

| A | b | cd |
|---|---|----|
| . | . | dc |
| . | c | bd |
| . | . | db |
| . | d | bc |
| . | . | cb |
| B | a | cd |
| . | . | dc |
| . | c | ad |
| . | . | da |
| . | d | ac |
| . | . | ca |
| C | b | ad |
| . | . | da |
| . | a | bd |
| . | . | db |
| . | d | ba |
| . | . | ab |
| D | b | ca |
| . | . | ac |
| . | c | ba |
| . | . | ab |
| . | a | bc |
| . | . | cb |

[2] In hunc finem esto aliud Schema ז. In eo 4 rerum ABCD 24 variationes ordinis, oculariter expressimus. Puncta significant rem praecedentis lineae directe supra positam. Methodum disponendi secuti sumus, ut primum quam minimum B variaretur, donec paulatim omnia. Caeterum quasi limitibus distinximus Variationes exponentis antecedentis ab iis quas superaddit sequens. Breviter igitur: Quotiescunque varientur, res datae, v. g. tres 6 Mahl; addita una praeterea poni poterit servatis variationibus prioris numeri jam initio, jam 2do, jam 3tio, jam ultimo seu 4to loco; seu toties poterit prioribus varie adjungi, quot habet unitates: Et quotiescunque prioribus adjungetur, priores variationes omnes ponet. Vel sic: quaelibet res aliquem locum tenebit semel, cum interim reliquae habent variationem antecedentem inter se, conf. problem. 7. Patet igitur variationes priores in exponentem sequentem ducendas esse.

The right-hand side shows exponents, that is, numbers of things, which here coincide. In the centre are the [numbers of] variations themselves. On the left, and between pairs of adjacent variations, is placed their *difference*.[213] An exponent on the right-hand side is the ratio of the given variation to the preceding variation.[214] The reason for the solution will be clear if we demonstrate that *the variation of a given Exponent is the product of that same exponent and the preceding variation*, which is the basis of Table ח.[215]

[2]. To this end, let us draw up another Diagram ד, in which are shown the 24 variations of order of 4 things, ABCD. The points signify the thing on a preceding line placed directly above them. I have adopted this arrangement so that the first position does not vary until the others have all been varied in turn. Next, I have marked off by means of horizontal lines the Variations with a preceding exponent from those added on by a succeeding exponent. To sum up, therefore: however many given things have to be varied (for example, three things in 6 ways), one thing will then be placed in addition to the already tabulated variations of the preceding number [of things], first at the beginning, then 2nd, then 3rd, and finally last, or in 4th place; that is, it will be adjoined to the preceding things in as many positions as their number has unities. And however often a thing is to be adjoined to the preceding things, it will be set beside all the preceding variations. To put it another way, everything will occupy some place only once, while the rest have all been varied among themselves (see Problem VII). From this it is clear that preceding variations must be multiplied by the succeeding exponent.

|   |   | ד |
|---|---|---|
| A | b | cd |
| . | . | dc |
| . | c | bd |
| . | . | db |
| . | d | bc |
| . | . | cb |
| B | a | cd |
| . | . | dc |
| . | c | ad |
| . | . | da |
| . | d | ac |
| . | . | ca |
| C | b | ad |
| . | . | da |
| . | a | bd |
| . | . | db |
| . | d | ba |
| . | . | ab |
| D | b | ca |
| . | . | ac |
| . | c | ba |
| . | . | ab |
| . | a | bc |
| . | . | cb |

---

[213] As the editors of the Akademie Edition already remarked, the left side of the diagram, containing the list of differences, is missing in the original. Cf. A VI, 1: 212.

[214] That is: 24/6 = 4; 720/120 = 6, etc.

[215] Consider, for example, the variation corresponding to the number 6 (the exponent), i.e. 120. The latter is the product of 5 for the number of the immediately preceding variation (24).

[3.] *Theoremata* hic observo sequentia: (1) omnes numeri variationum sunt pares; (2) omnes vero quorum exponens est supra 5 in cyphram desinunt, imo in tot cyphras, quoties exponens 5narium continet; (3) omnes summae variationum (id est aggregata variationum ab 1 aliquousque) sunt impares; et desinunt in 3 ab exponente 4 in infinitum; (4) quaecunque variatio antecedens, ut et exponens ejus omnes sequentes variationes metitur. [59] (5) Numeri variationum conducunt ad conversionem progressionis arithmeticae in harmonicam. Esto enim progressio arithmetica 1, 2, 3, 4, 5, convertenda in harmonicam; Maximi numeri, h. l. 5, quaeratur variatio: 120; ea dividatur per singulos, prodibunt: 120, 60, 40, 30, 24 termini harmonicae progressionis. Per quos si dividatur idem numerus: 120 numeri progressionis illius arithmeticae redibunt. (6) Si data quaecunque variatio duplicetur, a producto subtrahatur factus ex ductu proxime antecedentis in suum exponentem; residuum erit summa utriusque variationis; v. g. 24 ∩ 2 f. 48 – 6 ∩ 3, 18, f. 30 = 6 + 24 f. 30. (7) Variatio data ducatur in se, factus dividatur per antecedentem, prodibit differentia inter datam et sequentem, v. g. 6 ∩ 6 f. 36 ∪ 2 [*213*] f. 18 = 24 – 6 f. 18. Inprimis autem duo haec postrema theoremata non facile obvia crediderim.

4.  *Usus* etsi multiplex est, nobis tamen danda opera, nec caeteris problematibus omnia praeripiamus. Cumque serias inprimis applicationes Complexionum doctrinae miscuerimus (saepe enim necesse erat Ordinis Varietates in Complexiones duci), erunt hic pleraque magis jucunda, quam utilia.

5.  Igitur quaerunt quoties datae quotcunque personae uni mensae alio atque alio ordine accumbere possint. Drexelius in Phaëthonte orbis, seu de vitiis linguae p. 3, c. 1. ubi de lingua otiosa, ita fabulam narrat: Paterfamilias nescio quis 6 ad coenam hospites invitaverat. Hos cum accumbendi tempus esset, προεδρίαν sibi mutuo deferentes, ita increpat: quid? an stantes cibum capiemus? imo ne sic quidem, quia et stantium necessarius ordo est. Nisi desinitis, tum vero ego vos, ne conqueri

[3]. I note here the following *Theorems*: (1) All of the variations are even in number. (2) And all whose exponent is 5 or more end in zero, in fact in as many zeroes as the exponent contains multiples of 5. (3) All the totals of Variations (that is, the aggregates of variations from 1 up to any number) are odd, and from exponent 4 to infinity end in 3. (4) Every preceding variation, and its exponent, are the factors of all the succeeding variations. (5) The number of variations enables the conversion of an arithmetical progression into a harmonic progression.[216] For suppose that we want to convert the arithmetical progression 1, 2, 3, 4, 5 into a harmonic progression. First, find the Variation of the greatest number, in this case, 5, which is 120;[217] then, divide this by each number in turn, with the result 120, 60, 40, 30, 24, the terms of a harmonic progression; and if we divide the same number 120 into each of them, we re-arrive at the numbers of the former arithmetic progression. (6) If any given variation is doubled, and from the product is subtracted the result of multiplying the immediately preceding variation by its Exponent, the remainder will be the total of both Variations. For example, 24 ∩ 2 f. 48 − 6 ∩ 3 [=18] f. 30 = 6 + 24 f. 30. (7) Multiply a given variation by itself, and divide the result by the preceding variation, and the result will be the difference between the given variation and the succeeding variation. For example, 6 ∩ 6 f. 36 ∪ 2 f. 18 = 24 − 6 f. 18. And these two last theorems in particular are by no means obvious, I would have thought.

4. Although this problem has numerous applications, I must be careful not to steal everything away from the remaining problems. And while earlier on I intermingled serious applications of the doctrine of Complexions [with other matters], since it was often necessary for Variations of Order to be brought into Complexions, here matters will be mostly recreational rather than useful.

5. The enquiry is therefore made into how many ways a given number of persons could sit down at one table in some order or other. Drexel, in his *Phaëthon of the World, or the Vices of Language*, Part 3, Chapter 1, where he treats of idle talk, relates the following tale: The head of a household, I do not know who, had invited 6 guests to dinner, but when the time came to seat them, and each of them in turn yielded the place

---

[216] A harmonic progression is a progression made of the reciprocals of an arithmetic progression.

[217] That is: the factorial of 5 (i.e. 5!).

possitis, toties ad coenam vocabo, quoties variari ordo vester potest. Hic antequam loqueretur, ad calculos profecto non sederat, ita enim comperisset ad 720 variationes (tot enim sunt de 6 exponente, uti Drexelius illic 12 paginis, et in qualibet pagina 3 columnis, et in qualibet columna variationibus oculariter monstravit) totidem coenis opus esse; quae etsi continuarentur, 720 dies, id est 10 supra biennium absument.

6. Harsdörfferus Delic. Math. p. 2. sect. 1. prop. 32, ho[**60**]spites ponit 7; ita variationes, coenae, dies erunt 5040, id est anni 14, septimanae 10. At Georg. Henischius Medicus Augustanus Arithmeticae perfectae lib. 7, pag. 399, hospites vel convictores ponit 12 variationes, coenae, dies prodeunt 479001600 ita absumentur anni 1312333 et dies 5. Imo si quis in hoc Exponente tentare vellet, quod Drexelius in dimidio ejus effecit, nempe variationes oculariter experiri, annos insumeret 110, demto quadrante, et si singulis diebus 12 horis laboraret et hora qualibet 1000 variationes effingeret. Pretium operae si Diis placet!

7. Alii, ut cruditatem nudae contemplationis quasi condirent, versus elaborarunt, qui salvo et sensu et metro, et verbis variis modis ordinari possunt. Tales primus Jul. Caes. Scaliger lib. 2, Poëtices Proteos appellat. Horum alii minus artis habent, plus variationis, ii nempe quorum omnis est a monosyllabis variatio; alii contra, in quibus temperatura est monosyllaborum caeterorumque. Et quoniam in *his* plurimae esse solent inutiles variationes, de quibus problemate 11 et 12 erit contemplandi locus, de *illis* solis nunc dicemus.

of honour to another, he expostulated thus: "What? Are we to dine standing up? In fact, not even that, as even when you stand up you must do so in some order. Unless you give over, to stop you from protesting I shall invite you to dinner as many times as you can be differently arranged."[218] Here he must have spoken before he had actually reckoned up the cost, for in that case he would have ascertained that there are 720 variations (this being the total with 6 as exponent, as Drexel illustrated in his book, giving 12 pages, with 3 columns on each page and 20 variations in each column),[219] and that 720 dinners will be required; which in the end will take 720 days, or two years minus 10 days.

6.  Harsdörffer, in his *Mathematical Recreations*, Part 2, Section 1, Proposition 32, seats 7 guests, so that the variations, dinners, and days will be 5040, that is, 14 years and 10 weeks.[220] But Georg Henisch, a Physician of Augsburg, in his *Complete Arithmetic*, Book 7, p. 399, seats 12 guests or companions, and 479,001,600 variations, dinners, and days result, stretching over 1,312,333 years and 5 days.[221] In fact, if anyone wanted to attempt what Drexel did for half its number with this exponent, that is, to display all the variations, he would require, if he worked for 12 hours every day and set out 1000 variations every hour, 110 years less a quarter of a year. Worth the trouble, if it is pleasing to the Gods!

7.  In order as it were to mitigate the crudity of bare contemplation, some have elaborated verses whose words can be ordered in a variety of ways, while preserving sense and metre. Julius Caesar Scaliger, in Book 2 of his *Poetics*, was the first to call them *Protean*.[222] Some of them, namely those whose variation is all in monosyllables, have less art and more variety; in others, by contrast, modification is both in monosyllables and in other words as well. Since the latter kind are liable to have useless variations, which it will be more convenient to deal with in Problems XI and XII, I shall for the present speak only of the former kind.

---

[218] Cf. Drexel (1631), the quotation occurring at p. 525, with the tables (as if these could not themselves count as a paradigm of futility) ranging from pp. 526–31.
[219] The 1631 edition (see the former footnote) has six pages of five columns and twenty-four rows.
[220] Harsdörffer's text, however, mentions the erroneous result of "5050 mal ( ... ) / das ist bey 14 Jahren." Harsdörffer (1990: 29).
[221] Henisch (1609: 399).
[222] Julius Scaliger (1994: 588–9). The term 'Protean' stems from Proteus, a seagod in Greek mythology who was able to change his form.

8. Bernardus Bauhusius Societatis Jesu, Epigrammatum insignis artifex tali Hexametro Salvatoris nostri velut Titulos μονοσυλλάβους complexus est:

Rex, Dux, Sol, Lex, Lux, Fons, Spes, Pax, Mons, Petra
## CHRISTUS

Hunc Eryc. Puteanus Thaumat. Pietat. Y, pag. 107, aliique ajunt variari posse vicibus 362880, scilicet monosyllabas tantum respicientes, quae 9. Ego numerum prope decies majorem esse [214] arbitror, nempe hunc: 3628800. Nam accedens decima vox CHRISTUS etiam ubique potest poni, dummodo Petra maneat immota, et post Petram vel vox Christus vel 2 monosyllabae ponantur. Erunt igitur variationes inutiles, quibus post Petram ponitur 1 monosyllaba proxime antecedente Petram Christo, id contingit quoties caeterae 8 monosyllabae sunt variabiles, nempe 40320 𝔐𝔞𝔥𝔩. Cum ultima possit esse quaecunque ex illis 9, 40320 ∩ 9 f. 362880 – 3628800 f. 3265920. Qui est numerus utilium versus hujus Bauhusiani variationum. [61]

8.  Bernard Bauhuysen S.J., a celebrated inventor of Epigrams, included something like the monosyllabic[223] Titles of Our Saviour in this Hexameter verse:

Rex, Dux, Sol, Lex, Lux, Fons, Spes, Pax, Mons, Petra,
CHRISTUS[224]

Erycius Puteanus, in his *Wonders of Piety*, p. 107, and others, say that this can be varied in 362,880 ways, I mean, in ways that involve only the monosyllables, of which there are 9. I believe the number to be nearly ten times greater, namely this, 3,628,800.[225] For CHRISTUS, which follows the tenth word, can also be placed in all positions, while 'Petra' remains fixed; and after 'Petra' one can place either the word 'Christus' or 2 of the monosyllables. There will entail useless variations, in which one of the monosyllables is placed after 'Petra', with 'Christus' coming just before 'Petra'. So it turns out that the 8 remaining monosyllables can be varied as many as 40320 times. Since the last word can be any of the 9, [the result is] 40,320 ∩ 9 f. 362,880 – 3,628,800 f. 3,265,920, which is the number of useful variations of this verse of Bauhuysen.

9.  In the preface to his *Consultations*, Thomas Lansius made further advances, devising something like this:

---

[223] Here, as in § 11, Leibniz uses the Greek: *monosyllabous*.
[224] "King, Governor, Sun, Law, Light, Fountain, Hope, Peace, Mountain, Rock, CHRIST." Bernard Bauhuysen, or Bauhuis, S.J. (1575–1619), is the author of a collection of epigrams, including this Protean verse on the various titles of Christ. Cf. Bauhusius (1616: 48). Bauhuysen offers Biblical references for each of Christ's titles: Apoc. 19 for *Rex*, Matt. 2 for *Dux*, Mal. 4 for *Sol*, Mic. 4 for *Lex*, John 1 and 8 for *Lux*, Zech. 13 for *Fons*, I Tim. 1 for *Spes*, Eph. 2 for *Pax*, Ps. 67 and Dan. 2 for *Mons*, and I Cor. 10 for *Petra*, respectively. Cf. Bernardus Bauhusius, *Epigrammatum Selectorum Libri V*, Cologne: Gualtherus, s.d., p. 50. Note that this latter edition of Bauhusius's *Epigrammata*, "Nunc postremum correcti & aucti [Now finally corrected and augmented]", though sometimes dated '1615' because of the date of the approbation of Bauhuysen's work by the Antwerp censor on August 26, 1615, probably postdates the 1618 Gualtherus edition. Leibniz may have known the poem from Puteanus instead of Bauhuysen, since the latter already mentions in numbers the number of 3,628,800 variations (see next footnote).
[225] The Dutch humanist, prolific writer, and Leuven professor of eloquence Hendrick van den Putte, Henri du Puy, or Erycius Puteanus (1574–1646) in fact mentions exactly this number, which Leibniz must have misread. Puteanus (1617: 107): "CHRISTI Proteus, tricies sexies centies millies, vicies octies millies, & octingenties potest [converti]". Cf. Bauhusius (1616: 48).

9. Thomas Lansius vero amplius progressus praefatione Consultationum
tale quid molitus est :

> Lex, Rex, Grex, Res, Spes, Jus, Thus, Sal, Sol(bona) Lux, Laus.
> Mars, Mors, Sors, Lis, Vis, Styx, Pus, Nox, Fex(mala) Crux, Fraus.

Hic singuli versus, quia 11 monosyllabis constant, variari possunt vicibus: 39916800.

10. Horum exemplo Joh. Philippus Ebelius Giessensis Scholae Ulmensis
quondam Rector, primum Hexametrum, deinde Elegiacum Distichon
commentus est. Ille extat praefat. n. 8 hoc, quia et retrocurrit, in ipso
opere pag. 2, Versuum Palindromorum, quos in unum fasciculum
collectos, Ulmae anno 1623, in 12mo edidit. Hexameter ita habet :

> DIs, VIs, LIs, LaVS, fraVs, stlrps, frons, Mars,
> regnat In orbe.

Lex, Rex, Grex, Res, Spes, Jus, Thus, Sal, Sol (bona), Lux, Laus,
Mars, Mors, Sors, Lis, Vis, Styx, Pus, Nox, Fex (mala), Crux, Fraus.[226]

Here, each of the verses, since they consist of monosyllables, can be
varied in 39,916,800 ways.

10. Following their example, Johann Philipp Ebel of Giessen, former Rector
of the Latin School at Ulm, devised first a Hexameter, then an Elegiac
Couplet. The former is found in the preface, Paragraph 8, and the latter,
because it is reversible, on p. 2, of the same work, on *Palindromic
Verses*,[227] which, bound together, he published at Ulm in 12vo, 1623.
His Hexameter runs like this:

DIs, VIs, LIs, LaVs, fraVs, stIrps, frons, Mars,
regnat in orbe,[228]

where the year in which he both composed it, and it was all too true, the
1620th from the birth of Christ, is expressed in the selfsame verse.[229]
Since there are 8 monosyllables, 40320 variations must arise from it.

---

[226] "Law, King, Flock, Thing, Hope, Right, Incense, Salt, Sun (good things) Light, Praise / War,
Death, Fate, Dispute, Force, Styx, Pus, Night, Dregs (bad things), Cross, Deceit". Lansius (ed.)
(1620), 'Praefatio ad Lectorem', 2v ff. The book, also known under the title of *Orationes seu
consultatio de principatu* (etc.), contains a partly printed, partly handwritten, collection of more
than twenty proclamations by German lords and nobles for and against the great states and empires
of Europe, all in the form of 'consultations' addressed to Friedrich Achilles, the Duke of Wittemberg. The verses expressing good things and bad things also occur in Johann Phillip Ebel, on whom
see note 229, below. On Lansius, see DBE 6: 250. Many of the 'bad things' reported here appear
more often in Baroque variations of Protean verse. See, for instance, Julien Waudré's maximally
Protean hexameter verse on the Devil, *Foex, pus, sus, nox, fraus, lis, crux, mors, grips, latro, pix, styx*
('Dregs, Pus, Pig, Night, Fraud, Dispute, Cross, Death, Griffin, Brigand, Tar, Styx)', in Beer, Enenkel,
and Rijser (2009: 285–6). Note that, although Protean verse might come in the form of hymns of
praise like those of Bauhusius and Puteanus, there was also a tendency in early-modern combinatorial poetry to articulate the possibility of deceit, the unreliability of fate and the vanity of all things.
Scaliger's prototypical verse *Perfide sperasti divos te fallere Proteu* ('Deceitfully you hoped to cheat
the Gods, Proteus'), for instance, quoted by Leibniz below, brought up the theme of duplicity, and
the example below, quoted from Harsdörffer (*Ehr/Kunst/Geld/Guth/Lob/Weib und Kind | Man hat/
sucht/fehlt/hofft/und verschwind*; 'Honour, Art, Money, Goods, Praise, Wife and Child / One has,
seeks, lacks, hopes for – and vanishes'), gives a stark voice to the theme of vanity. See also notes 237
and 238, below, and, on the theme of duplicity, Göncz (2013: 14–9), as well as Cramer (2003:
220–1), who mentions both the 'Wheel of Fortune' and the *vicissitudo rerum* theme ('all things are
subject to change') as popular motifs in early-modern art.
[227] A *palindromic verse* is a verse that, as the English word 'madam', can be read with the
same meaning in both 'directions': from left to right as well as from right to left.
[228] "Wealth, Force, Dispute, Praise, Deceit, Lineage, Appearance, War / dominate the world".
The capital letters MDLLVVVIIII seem to be meant as a reference to 1619, "the 1620th [year]
from the birth of Christ", in which Ebel composed his verse as the Thirty Years' War was
starting to spread into German lands.

Ubi eadem opera annus, quo et compositus est, et verissimus erat, a Christo nato 1620mus, exprimitur. Cujus cum monosyllabae sint 8, 40320 variationes necesse est nasci.

11. At Distichon ad Salvatorem tale est:

> Dux mihi tu, mihi tu Lux, tu Lex, Jesule, tu Rex:
> Jesule tu Pax, tu Fax mihi, tu mihi Vox.

Variationes ita computabimus: tituli Salvatoris μονοσίλλαβοι, sunt 7, hi inter se variantur 5040 vicibus. Cumque singulis adjecta sit vox Tu, quae cum titulo suo variatur 2 vicibus, quia jam ante, jam post poni potest, idque contingat vicibus septem, ducatur 2narius septies in se, 2, 2, 2, 2, 2, 2 ∩ 2 f. 128. seu Bissurdesolidum de 2, factus ducatur in 5040 ∩ 128 f. 645120. Productum erit quaesitum.

12. Hos inter nomen suum voluit et Joh. Bapt. Ricciolus legi, ut alieniori in opere Poëtica facultas professoris quondam sui tanto clarius reluceret. Symbola ejus Almagest. nov. P. 1. lib. 6, c. 6. Scholio 1, fol. 4l3, talis:

> Hoc metri tibi me en nunc hic, *Thety*, Protea sacro:
> Sum Stryx, Glis, Grus, Sphynx, Mus, Lynx, Sus, Bos, Cap[er] et Hydrus.

11. As for the Couplet, addressed to our Saviour, it runs like this:

> Dux mihi tu, mihi tu Lux, tu Lex, Jesule, tu Rex,
> Jesule tu Pax, tu Fax mihi, tu mihi Vox.[230]

I shall calculate the variations as follows: There are 7 monosyllabic[231] titles of our Saviour, which can be varied in 5040 ways, and since one may attach to each of them the word 'Tu', which may be varied with its title in 2 ways, as it can be placed either before or after it, and this can happen in seven ways, let 2 be multiplied by itself seven times, 2.2.2.2.2.2 ∩ 2 f. 128, or the seventh power of 2. Multiplying by this factor, 5040 ∩ 128 f. 645,120, the product will be what is required.

12. Giovan Battista Riccioli wanted these verses to be ascribed to himself, in order that in the Poetic work of another the genius of his onetime professor might shine so much the more brightly. His own contribution is to be found in his *New Almagest*, Part 1, Book 6, Chapter 6, Scholium 1, folio 413, namely:

> Hoc metri tibi me en nunc hic, *Thety*, Protea sacro;
> Sum Stryx, Glis, Grus, Sphynx, Mus, Lynx, Sus, Bos, Caper et Hydrus.[232]

Its 9 monosyllables can be varied in 262,880 ways. If in place of the last words, 'et Hydrus', he had substituted monosyllables, for example,

---

[229] Ebel (1623), 'Praeloquium ad Lectorem', § VIII (s.p.). Ebel here also discusses the Protean verses by Scaliger, Kleppisius, Carolus à Goldstein, and Lansius that Leibniz mentions. On Ebel, see *Jöcher-F* II, col. 806 and *Zedler* 8, col. 33.

[230] Ebel (1623: 2): "You my Guide, you my Light, Law, oh Jesus, you King / Oh Jesus, you Peace, my Torch, my Voice".

[231] In this case, too, Leibniz employs the Greek word: *monosyllaboi*.

[232] "Oh Thetis, now and here this Protean verse I devote to you / I am an Owl, a Dormouse, a Crane, a Sphynx, a Mouse, a Lynx, a Pig, an Ox, a Goat and a Dragon". Cf. Riccioli (1651). The verse itself occurs in slightly different form on p. 414. Riccioli had come to discuss Protean verses in this Scholium upon having mentioned Erycius Puteanus's "very crafty verse" *Tot tibi sunt dotes Virgo, quot sidera caelo* on the Blessed Virgin, the Protean number of variations of which added up to 1022, equalling the number of known stars, "hodie mille viginti duae recensentur", which, however, is still thought to be short of the number of the Virgin's gifts; Puteanus (1617: 82–3). Cf. also pp. 95: "Hic numerus stellarum est" and 102: "Tot in Virgine Dotes esse, quot in caelo stellae sunt: non tot in caelo stellas esse, quot in Virgine Dotes sunt (The Virgin has so many Endowments as there are stars in heaven; in heaven there are not so many stars as there are Endowments in the Virgin)". The verse, however, is by Bauhuysen, whom Puteanus praises in his book. Cf. Bauhusius (1616: 47). It is not clear on which grounds Leibniz concludes that Riccioli pillaged either Lansius or Ebel. On Riccioli, see Luigi Campedelli, 'Riccioli, Giambattista', DSB XI: 411–2. See also below, notes 268 and 278.

Cujus 9 monosyllabae variantur 262880 vicibus. Si loco po[62] stremarum vocum: et Hydrus, substituisset monosyllabas, v. g. Lar, Grex, ascendisset ad Lansianas varietates. Hic ad[215]monere cogor, ne me quoque contagio criminis corripiat, primam in *Thety* correptam non legi. Et succurrit opportune Virgilianus ille, Georg. lib. I, v. 3I.

Teque sibi generum *Thetys* emat omnibus undis.

Nam alia Thetys, oceani Regina, Nerei conjux; alia Thetis, nympha marina vilis, Peleo mortali nupta, Achillis parens, nec digna cui se Proteus sacret. Ea sane corripitur:

Vecta est frenato caerula pisce *Thetis*.

Caeterum Ricciolus Scaligerum imitari voluit, utriusque enim de Proteo Proteus est. Hujus autem iste:

Perfide sperasti divos te fallere Proteu.

De cujus variationibus infra probl. fin.

[13.] Ne vero Germani inferiores viderentur, elaborandum sibi Harsdörfferus esse duxit, cujus Delic. Math. P. 3. Sect. 1, prop. 14, distichon extat:

Ehr, Kunst, Geld, Guth, Lob, Weib und Kind
Man hat, sucht, fehlt, hofft und verschwind.

'Lar' or 'Grex',[233] he would have raised the number of varieties to Lansen-like heights. Here I cannot forbear to point out (in case I might be thought to condone the error) that the first syllable of *Thety* should not be read as short, to which this verse of Virgil (in Book 1, verse 31 of the *Georgics*) aptly lends support:

> Teque sibi generum *Thetys* emat omnibus undis.[234]

For one is Thetys, Queen of the ocean, consort of Nereus; the other is Thetis, a lowly sea-nymph, wedded to the mortal Peleus, and mother of Achilles, and not the illustrious one to whom Proteus consecrated himself. And she clearly deserves to be short in this verse:[235]

> Vecta est frenato caerula pisce *Thetys*.[236]

Riccioli next wished to imitate Scaliger, for in each of them there is a Protean verse concerning Proteus, of whom he [Riccioli] says:

> Perfide sperasti divos te fallere Proteu.[237]

[13]. And for the variations of this, see below at the conclusion of this problem. But Harsdörffer took it to be his business that the Germans should not be thought to be inferior, and in his *Mathematical Recreations*, Part 3, Section 1, Proposition 14, this couplet is to be found:

> Ehr/Kunst/Geld/Guth/Lob/Weib *und Kind*
> Man hat/sucht/fehlt/hofft/*und verschwind*,[238]

in whose 11 monosyllables there are 39,916,800 variations. So much for verses; but *Anagrams* are also relevant here, as they are nothing else but

---

[233] That is: 'a deity' or 'a flock'.

[234] "And Tethys shall give you all her waters to be her son-in-law". Virgil, *Georgica* I, 31.

[235] Here Leibniz plays with the name 'Thetis', which deserves to be short (for metric reasons) because Thetis the sea-nymph belongs to a lower rank compared to the other, *Thetys*, the Queen of the Sea.

[236] Tibullus, *Eleg.* I, 5, v. 46: "The sea-blue Thetis is carried on a bridled fish".

[237] Riccioli argues that verses like those given by Puteanus and his own "Vi metri tibi me..." are called 'Protean' on account of their variability. Cf. Riccioli (1651: 413). The verse itself is indeed Scaliger's. Cf. Scaliger (1561: 588). See also notes 222, 226 and 283.

[238] Harsdörffer, (1990: 60). See note 226, above.

Cujus 11 monosyllabae habent variationes 39916800. Tantum de Versibus. Quanquam autem et *Anagrammata* huc pertinent, quae nihil sunt aliud, quam Variationes utiles literarum datae orationis; nolumus tamen vulgi scrinia compilare.

14. Unum e literaria re vel dissensu computantium quaeri dignum est: quoties situs literarum in Alphabeto sit variabilis. Clav. Com. in Sphaer. Joh. de Sacro Bosco cap. 1, pag. 36, 23 literarum linguae latinae dicit variationes esse 25852016738884976640000, cui nostra assentitur computatio. 24 literarum germanicae linguae variationes Laurembergius assignavit 620448397827051993. Erycius Puteanus dicto libello, 62044801733239439360000. At Henricus ab Etten: 62044859343886061336000. Omnes justo pauciores. Numerus verus, ut in Tabula ת manifestum, est hic: 620448401733239439360000. Omnes in eo conveniunt, quod numeri initiales sint: 620448.

useful variations of the letters of a given utterance. However, I have no
wish to dip into the well-known sources.[239]

14. One matter concerning letters, or rather concerning a disagreement
among those making calculations, is worth looking at, namely, how
many permutations there are of the letters of an Alphabet. Clavius,
in his *Commentary on the* De Sphaera *of Johannes Sacrobosco*,
Chapter 1, p. 36, says that there are 25,852,016,738,884,976,640,000
permutations of the 23 letters of the Latin language, which agrees with
my calculation.[240] Lauremberg distinguished 620,448,397,827,051,933
permutations of the 24 letters of the German language. Erycius
Puteanus, in the volume cited, has 62,044,801,733,239,439,360,000,
but Hendrik van Etten, 620,448,593,438,860,613,360,000.[241] These are
all too small.[242] The true number, as is clear from Table ⊓, is this:
620,448,401,733,239,439,360,000.[243] They all agree that the initial

---

[239] A reminiscence of Horace's verse, *Satires* I, 120–1: *Crispini scrinia* (...) *compilasse*, "to
plagiarize Crispino's books".
[240] Clavius (1999: 19). On Clavius and John of Sacrobosco, see also note 23, above.
[241] Leibniz here draws his information exclusively from Harsdörffer (1990: 516). On
Harsdörffer and the background to his work, see also above, notes 21 and 158. Harsdörffer's
own calculation was taken from Lauremberg (1640: 9). On Johann (or Hans) Lauremberg (or
Laurenberg; 1590-1658), see DBE 6: 270. In order to join the Royal Academy at Sorø in
Denmark, Lauremberg in 1624 left his chair in Poetry at Rostock University to his older
brother Peter (1585–1639), who had until then been Professor of Physics and Mathematics at
Hamburg. Besides many books on other subjects, both brothers published didactic works on
arithmetic and corresponded on arithmetical problems. In his *Ocium Soranum*, Lauremberg
refers to the idea in Hegias Olynthius's *Metastasis* (quoted by Hellanicus), that the letters of
the Greek alphabet could be "mutually replaced and transmuted" in so many ways as to
outnumber the total amount of people spread over the world. See also below, note 244.
Lauremberg adds to this that not even if humans filled the whole face of the earth and
successive generations would also be counted, could the total number of men exceed that of
the varieties of letters from 1 to 24. To achieve the number of 620,448,397,827,651,993 that he
gives (and not Leibniz's 620,448,397,827,051,933, which is taken over from Harsdörffer), the
world, in his reckoning, would have to have been filled with human beings for as much as
16,717,941 years—a number for the existence of this "worldly machine" that, according to
Laurenberg, our religion gives us reason to doubt. Cf. Laurenberg (1640: 7-9). Note, however,
the erratum at the end of the book "p. 9. lin. 14: adde numero scripto, 6000", which would
supposedly change the total number of variations in a 24-letter alphabet according to
Lauremberg to 6,204,483,978,276,519,936,000. "H. van Etten", or Hendrick van Etten, referred
to as "Henrich von Etten" in Harsdörffer, is named as the author of the 1624 edition of
*Recreation Mathematique. Composee de plusievrs problemes plaisants & facetieux. En faict
d'Arithmeticque, Geometrie, Mechanicque, Opticque, & autres parties de ces belles sciences*,
Pont-à-Mousson: J. Appier Hanzelet, 1624. The book saw numerous later editions, amongst
which an English edition of 1633, as well as an otherwise anonymous twentieth-century
facsimile edition. It contains an assortment of mathematical, mechanical, and other puzzles,
tricks, and marvels, for which the author more or less excuses himself in a dedication to his
uncle Sir Lambert Verreyken, Lord of Himden and Wolverthem, and Captain in the service of
the King of Spain. The author's name was long held to be a pseudonym for the professor of
philosophy

Puteanaeae computationis error non mentis, sed calami vel typorum esse videtur, nihil aliud enim, quam loco 7mo numerus 4 est omissus.

15. (Aliud autem sunt variationes, aliud nu[63]merus vocum ex datis literis componibilium. Quae enim vox 23 literarum est? Imo quantacunque sit, inveniantur omnes complexiones 23 rerum, in singulas ducantur variationes suae juxta probl. 2, num. 59 productum erit numerus omnium vocum nullam literam repetitam habentium. At habentes reperire docebit problema 6). Porro tantus hic numerus est, ut, etsi totus globus terraqueus solidus circumquaque esset, et cuilibet spatiolo homo insisteret, et quotannis, imo singulis horis morerentur omnes[216]surrogatis novis; summa omnium ab initio mundi ad finem usque multum ab futura sit: ut ait Harsdörff. d. l. Hegiam Olynthium Graecum dudum censuisse. His contemplationibus cum nuper amicus quidam objiceret, ita sequi, ut liber esse possit in quo omnia scripta scribendaque inveniantur. Tum ego: et fateor, inquam, sed legenti grandi omnino fulcro opus est, ac vereor ne orbem terrarum opprimat. Pulpitum tamen

numbers should be 620,448. The error in Puteanus's calculation seems to be not one of understanding, but a slip of the pen or a typographical error, as it involves nothing more than the omission of the number of 4 in the seventh place.

15. (Variations are one thing, the number of words that can be made up out of given letters is another thing. For what word has 23 letters? In any case, whatever the size of the word, we have to find all the complexions of 23 things, and multiply each of them by their variations, in accordance with Problem II, paragraph 59. This product will be the number of words that do not have any letter repeated. Problem VI will explain how to deal with words that do have letters repeated.) I should add that this is a number so great that even if the whole terraqueous globe were solid all over, and a man stationed on every square foot, and given new recruits in the very year, in fact at the very hour when each of them passed away, the total work of all of them from the foundation of the world up to its

and mathematics at Pont-à-Mousson, Jean Leurechon, S.J. (1591–1670). In 1969, however, Trevor Henry Hall argued that there was actually no reason to doubt that the book was indeed compiled by Hendrick van Etten during his studies at Pont-à-Mousson on the basis of a similar publication by the scholar of Greek Claude-Gaspar Bachet (1581–1638). See: Hall (1969). In his Introduction to the as yet unpublished new edition of the seventeenth-century English translation of *Récréations mathématiques*, however, Albrecht Heeffer has made a good case for the idea that the Pont-à-Mousson printer Jean Appier the younger, "dit Hanzelet" (1596–1647), was actually the driving force behind the work. Cf. Heeffer (2004: 11). The permutation problem does not occur in the 1624 edition of *Récréation Mathématique*, but it does occur in later editions, and is included in the 1633 English edition in a passage "Of Changes in Bells, in musicall Instruments, transmutation of places, in numbers, letters, men or such like". Cf. van Etten (1633: 185–9). It also occurs in another work by Jean Leurechon, however, though with the outcome of the permutation of 23 letters only. Cf. Leurechon (1622), Propositiones Arithmeticae, Thesis IV, p. 3: '23 milites, litterae &c. permutari possunt: 25852016738884976640000.' This is in fact the number that Clavius gave and then Leibniz accepted. Note that Leurechon here also offers the *Tot tibi sunt dotes, Virgo* and *Rex, Dux, Sol* examples of protean verse, for which see above, note 232. Although Leibniz seems to imply that he found it there, Puteanus offers no calculation of the permutations of the letters of the alphabet in his *Pietatis thaumata*. Puteanus does, however, present the sum that both Harsdörffer and Leibniz mention in *De Anagrammatismo*. Cf. Puteanus (1643: 37–9), where he says that one can continue calculating the possible transposition of letters of long and fictive names (such as those given by Plautus in *The Persian*, Act 4, Scene 6) until one arrives at names with even more letters than the alphabet, which itself has "xxiv. apud Graecos & nostrates" that may be transposed. Cf. *ibidem*, p. 39. "A lot of work!", Puteanus adds, "I calculate the sum, you pronounce it: 620448401733239439360000".

[242] Note, however, that the number suggested by van Etten actually exceeds the amount offered in Leibniz's solution. On the various attempts made in Leibniz's days to calculate the number of all possible permutations with the letters of a given alphabet, see Knobloch (1973: 42–4).

[243] Leibniz's result is correct.

commodius non inveneris cornibus animalis illius, quo Muhamed in
coelum vectus arcana rerum exploravit, quorum magnitudinem et dis-
tantiam Alcorani oracula dudum tradiderunt.

17. Vocum omnium ex paucis literis orientium exemplo ad declarandam
originem rerum ex Atomis usus est ex doctrina Democriti ipse Aristot.
1, de Gen. et Corr. text. 5 et illustrius lib. I. Metaph. c. 4, ubi ait ex
Democrito: Atomos differre σχήμαθι id est figura, uti literas A et N;
θέσει id est situ, uti literas N et Z, si enim a latere aspicias, altera in
alteram commutabitur; τάξει id est ordine, v. g. syllabae AN et
NA. Lucret. quoque lib. 2 ita canit:

Quin etiam refert nostris in versibus ipsis
*Cum quibus (complexiones) et quali sint ordine*

end would still leave much to be done: as Harsdörffer (in the place cited) says that the Greek Hegias Olynthius already thought.[244]

16. A friend has recently objected to these reflections in this way, that there could exist a book in which one would find everything that ever has been and ever will be written down. I reply: yes, I admit it, but for its reader you would need so altogether vast a lectern as I am afraid would be too heavy even for the world's great globe; and you may not be able to find a pulpit even more commodious than the horns of that beast on which Mohammed was carried up into Heaven to search out the arcana of things, whose grandeur and remoteness have been handed down to us by the oracles of the Koran.[245]

17. From this example of the origin of all words from a few letters, I pass to the origin of things from atoms, the application being taken from the teachings of Democritus. Aristotle himself, in *On Generation and Corruption*, Book 1, Text 5, and more famously in *Metaphysics*, Book 1, Chapter 4, says, following Democritus, that Atoms differ in *schema*, that is, in Figure, like the letters A and N; in *thesis*, that is, in situation, like the letters N and Z, which if you view them sideways on they change into one another; and in *taxis*, that is, in order, like the syllables AN and NA.[246] Lucretius also, in *On the Nature of Things*, Book 2, expressed himself thus:

Is it not also the case in these very verses, where letters are set in place *with certain others (complexions)*, and *in a certain kind of order (variation of situs)*? For the same letters represent the

---

[244] Harsdörffer (1990: 59–60), '*Die XIV. Frage. Von Versetzung der Buchstaben in dem Abc*', where the comparison to countless generations of men alternately filling the surface of the Earth is attributed to Hegias Olynthius. After numerous fruitless efforts to trace the author whom Johann Leuremberg, Georg Philipp Harsdörffer, and Leibniz refer to as 'Hegias Olynthius', the editors were pleased to obtain from Professor Franco Montanari a confirmation of their suspicion that there seems to be no evidence for the figure mentioned by Leibniz and Lauremberg. Though regularly referred to in secondary sources on the early modern history of the art of permutation, a 'Hegias Olynthius' (a Hegias from Olynthos, that is), as Franco Montanari concludes in private correspondence, "seems to be utterly unknown". Nor is there any figure "known from Olynthos who dealt with the matters referred to" in Lauremberg's text. Finally, there is neither a candidate author for the work referred to as *Metastasis*. The only option seems to be that Hegias Olynthius is a later, perhaps medieval, invention, and that the spurious Hegias may have been associated with the name of Hellanicus on the basis of Hellanicus of Mytilene's "attested genealogical and ethnographical interests". The editors are indebted to Franco Montanari for his effort to help them in their search, which, thus far, has not yielded any references to Hegias predating Lauremberg's. For Lauremberg, and his reference to Hegias, see note 241.

[245] On 27 Rajab, Mohammed mounted the steed *Al-burāq* (*Lightning*), according to tradition at the site of the Al-Aqsa mosque in Jerusalem, to make his Night Journey into the heavens. Cf. *Al-Isra* sura, verses 1 and 60.

[246] Aristotle, *De Generatione et Corruptione* I 2, 315b6–11 and 315b32–316a1; and I 9, 327a17–19; *Metaphysica* I 4, 985b13–20.

(*variatio situs*) quaeque locata.

Namque eadem coelum, mare, terras, flumina, Solem
Significant: eadem fruges, arbusta, animantes:
Si non omnia sint, at multo maxima pars est
Consimilis; verum positura discrepitant haec. [64]
Sic ipsis in rebus item jam materiai
Intervalla, viae, connexus, pondera, plagae,
Concursus, motus, ordo, positura, figura
Cum permutantur, mutari res quoque debent.

Et Lactant. Divin. Inst. lib. 3, c. 19, pag. m. 163. *Vario inquit* (Epicurus), *ordine ac positione conveniunt atomi sicut literae, quae cum sint paucae, varie tamen collocatae innumerabilia verba conficiunt.* Add. Pet. Gassend. Com. in lib. 10, Laertii ed. Lugduni anno 1649, fol. 227 et Joh. Chrysost. Magnen. Democrit. redivivo Disp. 2 de Atomis c. 4, prop. 32, p. 269.

18. Denique ad hanc literarum transpositionem pertinet ludicrum illud docendi genus, cujus meminit Hieronymus ad Paulinam, tesserarum usu literas syllabasque puerulis imprimens. Id Harsdörfferus ita ordinat Delic. Math. P. 2, Sect. 13, prop. 3. Sunt 6 cubi, quilibet cubus sex laterum est, eruntque inscribenda 36, haec nempe: I. ɑ.e. i. o. u. ŋ.

heavens, the sea, the dry land, rivers, and the Sun, the same letters represent the fruits of the soil and flourishing groves. And though they are not all the same, yet for the greatest part by far they are the same: it is just that their position is entirely different. So it is that in things themselves there are sometimes voids in their matter, sometimes channels, connections, preponderances, surfaces, assemblies, motions, order, situation, and figure, which when they are altered, the things themselves are bound to be changed.[247]

And Lactantius, in his *Divine Institutions*, Book 3, Chapter 19, page m. 163: *With variety of order and position, he* [Epicurus] *says, atoms come together like letters, which, though they are few in number, yet, when brought together in a variety of ways, make up innumerably many words.*[248] Add to these, Pierre Gassendi, in his *Commentary on Book 10 of Diogenes Laërtius*, folio 227, published at Lyon in 1649,[249] and Jean-Chrysostôme Magnen in *Democritus Revived, or, On Atoms*, Disputation 2, Chapter 4, Proposition 32, p. 269.[250]

18. Finally, also relevant to this transposition of letters is that agreeable method of instruction (which Jerome mentions in a letter to Paulina[251]) using blocks to impress letters and syllables on the minds of the young. Harsdörffer sets it out in his *Mathematical Recreations*, Part 2, Section 13, Proposition 3, as follows. There are 6 cubes, each cube having 6 faces, and 36 things to be written on them, namely: I. *a e i o*

---

[247] Lucretius, *De rerum natura* II· 1013–22 (transl. Martin Wilson). The words between brackets have been interpolated by Leibniz. See also *De rerum natura* II: 688–99.

[248] The quote is actually from *Divinae Institutiones* III, 17.24. Cf. Lactantius (2007: 263–4). Cf. Lactantius (2003: 200). Cf. Lucretius *De rerum natura* I: 196–7, 823–5, 912–4, and II: 688–90.

[249] Gassendi (1649: 227), where Gassendi compares letters, as the "elements of writing" out of which "syllables, expressions, sentences, speeches and books" are composed, with atoms as "the elements of things" composing bodies of various shapes.

[250] Magnen (1643: 269), where atoms are compared to letters, from the transposition of which *flumina* ('sweet-voiced floods of words'), for instance, as well as *fulmina* ('strokes of fire and brimstone') may be formed, and "comedies and tragedies are expressed through the very same signs".

[251] Leibniz here interprets Harsdörffer's "Hieron. ad Paulin." as a reference to a letter from Jerome "to Paulina". A reference to the process of learning how to read by the use of letters occurs neither in a letter to Paulina, however, nor in any of Jerome's letters to Paulinus, though there is such a reference in a letter to Paula. Cf. Jerome (1910–18, vol. 1: 245), where it is stated that, in order to read, one should start with the elements. The obvious source of Harsdörffer's reference, however, is the passage in Jerome's famous letter on the education of girls, the Letter to Laeta, Paula's daughter-in-law, in which the Church Father discusses the education of Laeta's daughter, also called Paula, and suggests giving her wooden or ivory letters to play with. Cf. Jerome (1910–18, vol. 2: 294). See also notes 253, 266 and 267 below.

II. b. c. ð.f. g. h.   III. k. l. m.n. p. q.   IV. r. s. β. t. w. x.   V. v.j.s. r. a. o.
VI. ff. ff. k. fch. ch. z.   Alphabetum autem lusus unius tesserae, syllabas (das Buchstabieren) duarum docebit: inde paulatim voces
orientur. [*217*]

## Probl. V.
## DATO NUMERO RERUM VARIATIONEM SITUS
## MERE RELATI SEU VICINITATIS INVENIRE.

1. "Quaeratur Variatio situs absoluti, seu ordinis, de numero rerum unitate
   minori quam est datus, juxta probl. 4. Quod invenietur in Tab. ה erit
   quaesitum."
2. Ratio Solutionis manifesta est ex Schemate ז quo rationem solutionis
   problematis praecedentis dabamus. V. g. in variationibus vicinitatis,
   variationes hae: Abcd. Bcda. Cdab. Dabc. habentur pro una, velut in
   circulo scripta. Et ita similiter de caeteris, omnes igitur illae 24 variationes
   dividendae sunt per numerum rerum, qui hoc loco est 4, prodibit variatio
   ordinis [65] de numero rerum antecedenti, nempe 6.
3. Finge tibi hypocaustum rotundum in omnes 4 plagas januas habens, et in
   medio positam mensam; (quo casu quis sit locus honoratissimus disputat

*u y*. II. *b c d f g h*. III. *k l m n p q*. IV. *r s*[252] *ß t w x*. V. *v j s r ä ö*. VI. *ff ss tz sch ch z*. Playing with one block will teach the Alphabet, playing with two blocks, syllables (*das Buchstabiren*), from which words will gradually arise.[253]

## Problem V
## GIVEN A NUMBER OF THINGS, TO FIND THE VARIATION OF A PURELY RELATIVE SITUS, OR NEIGHBOURHOOD

1. "We require the Variation of the absolute situs, or order, of the number of things less by 1 than the given number, in accordance with Problem IV, which will be found in Table ה. This will be what is required."[254]

2. The reason for the solution is clear from Diagram ז, in which I gave the reason for the solution of the preceding Problem. For example, in variations of a neighbourhood, the variations *Abcd, Bcda, Cdab, Dabc*, are taken to be one, as if they were written in a circle. And so similarly of the rest, so that the 24 variations must all be divided by the number of things, which in this case is 4. The variation of order will result from that of the preceding number of things, namely, 6.

3. Imagine for yourself a circular dining-room having doors at all four compass-points, and a table set in the midst of the room; and reflect that

---

[252] Namely, the long s: "ſ".

[253] Cf. Harsdörffer (1990: 513), which deals with teaching children how to read and write through the use of an increasing number of dice with letters painted on their sides. It is here that we also find a reference to the Lutheran theologian and poet Johann Michael Dilherr (1604–69), who inaugurated his office of rector at the Nürnberg Gymnasium with an *Oratio* on the topic (also typical of Jerome) of avoiding any morally destructive influence on the young child. Dilherr there refers to Jerome's advice on playful ways of learning the alphabet, quoting the passage from Jerome's Letter to Laeta given in note 251, above. Cf. *idem*, sigg. [B3v]–[B4]. Harsdörffer obviously extracted his reference to "*Hieron. ad Paulin.*" from Dilherr, as Leibniz tapped Harsdörffer's text. On Dilherr, see: *Zedler* 7, cols 924–5, and DBE 2: 545. Note that *Buchstabe* is German for a 'letter' of the alphabet; that *buchstabieren* means 'to spell' and *das Buchstabieren* '[the act of] spelling'.

[254] With "variation of purely relative situs or neighbourhood", Leibniz means a variation of position that preserves some given order. The problem concerns 'cyclic permutations' (cf. Knobloch 1973: 44–5, 1974: 415). Thus, to stick to his own example, suppose that we have a sequence of four letters *abcd*, in this order, and that we change the place of *a*, removing it from the beginning and placing it at the end, leaving the other letters in place. In this case, we have produced a variation 'of purely relative situs'. To calculate the number of all possible variations of this kind out of *k* elements, Leibniz suggests that we first calculate the number of all possible variations, according to the solution of the preceding problem (i.e. *k!*), and then divide the resulting number by the number of the elements $\frac{k!}{k}$. This, however, as Leibniz explains, amounts to the same thing as calculating the value of $(k-1)!$

Schwenter, et pro janua orientem spectante decidit, e cujus regione collocandus sit honoratissimus hospes. Delic. Math. sect. VII. prop. 28) atque ita hospitum situm variari cogita prioritatis posterioritatisque consideratione remota.

4. Hic obiter aliquid de Circulo in demonstratione perfecta dicemus. Ejus cum omnes Propositiones sint convertibiles, prodibunt syllogismi sex, circuli tres. Ut esto demonstratio: I. O. rationale est docile. O. homo est rationalis. E. O. homo est docilis. II. O. homo est docilis. O. rationale est homo. E. O. rationale est docile. 2. III. O. homo est rationalis. O. docile est homo. E. O. docile est rationale. IV. O. docile est rationale. O. homo est docilis. E. O. homo est rationalis. 3. V. O. homo est docilis. O. rationale est homo. E. O. rationale est docile. VI. O. rationale est docile. O. homo est rationalis. E. O. homo est docilis.

## Probl. VI.
### DATO NUMERO RERUM VARIANDARUM, QUARUM ALIQVA VEL ALIQUAE REPETUNTUR, VARIATIONEM ORDINIS INVENIRE.

1. "Numerentur res simplices et ex iisdem repetitis semper una tantum; et ducantur in variationem numeri numero variationum dato unitate minoris; productum erit quaesitum", v. g. sint sex: a, b, c, c, d, e sunt simplices 4 + 1 (duo illa c. habentur pro 1) f. 5 ∩ 120 (120 autem sunt variatio numeri 5 antecedentis datum 6) f. 600.[*218*]

the situs of the guests may in this way be varied without consideration of priority or posteriority. (Schwenter, in *Mathematical* Recreations, Section VII, Proposition 28, debates which in this case is the place of honour, before coming down in favour of the door with an eastern aspect, on which side the most honourable guest is to be placed).[255]

4. I shall say something in passing here about the role of the Circle in perfect demonstration. When all its Propositions are convertible, six syllogisms and three circles will result, the demonstration being like this: I. Every rational being is teachable, Every man is rational, Therefore every man is teachable. II. Every man is teachable, Every rational being is a man, Therefore every rational being is teachable. III. Every man is rational, Every teachable being is a man, Therefore every teachable being is rational. IV. Every teachable being is rational, Every man is teachable, Therefore every man is rational. V. Every man is teachable, Every rational being is a man, Therefore every rational being is teachable. VI. Every rational being is teachable, Every man is rational, Therefore every man is teachable.

## Problem VI
## GIVEN A NUMBER OF THINGS TO BE VARIED, OF WHICH ONE OR SOME ARE REPEATED, TO FIND THE VARIATION OF ORDER.[256]

1. "Count the number of simple things, and one only of the same things repeated. Then multiply them by the variation of the number that is one less than the number of things to be varied. The product will be what is required." For example, suppose that there are six things, *a b c c d e*. The simple ones are 4 + 1 (the two c's being taken to be one) f. 5 ∩ 120 (120 being the variation of the number 5 that precedes the given number 6) f. 600.[257]

[255] Schwenter (1991: 333–4).
[256] Given a set S of *n* elements, the problem amounts to calculating the number of sequences of length *k*, possibly containing repetitions, over S. The formula to perform the calculus is $P(^n_k) = n^k$. Leibniz, however, gives a false rule, which amounts to determine the number k(n - 1)! (see Knobloch 1973: 45).
[257] If among the elements to be varied, repetitions of the same items are allowed, Leibniz suggests that to calculate the total number of variations, one must proceed as follows: (1) calculate the number of simple, non-duplicated elements; (2) add to this number the number of those elements which have been duplicated, without considering its replicas; (3) multiply the number thus obtained with the number of all possible variations out of the total number of elements (including replicas), diminished with 1. Thus, suppose we have 6 elements, as in Leibniz's example: *a, b, c, c, d, e*; the total number of variations is 5 × 120 (= 5 × 5!) = 600. In case we increase by 1 the number of replicas of *c* (thus giving rise to the sequence: *a, b, c, c, c, d, e*), we must multiply 5 by the factorial of 6. The rule stated by Leibniz, however, is false; see the previous note.

2. Ratio manifesta est, si quis intueatur Schema ⁊, corruent enim omnes variationes quibus data res pro se ipsa ponitur. Usum nunc monstrabimus.

3. Esto propositum: dato textu omnes melodias possibiles invenire. Id Harsdörfferus quoque Delic. Math. sect. 4, prop. 7 tentavit. Sed ille in textu 5 Sylla[66]barum melodias possibiles non nisi 120 esse putat, solas variationes ordinis intuitus. At nobis necessarium videtur etiam complexiones adhibere, ut nunc apparebit.

4. Sed altius ordiemur: Textus est vel simplex, vel compositus. Compositum voco in lineas, Reimʒeilen, distinctum. Et compositi textus variationem discemus melodiis simplicium in se continue ductis per probl. 3. Textus simplex vel excedit 6 syllabas, vel non excedit. Ea differentia propterea necessaria est, quia 6 sunt voces: Ut, Re, Mi, Fa, Sol, La (ut omittam 7mam: Bi, quam addidit Eryc. Puteanus in Musathena). Si non excedit, aut sex syllabarum, aut minor est.

5. Nos in exemplum de Textu hexasyllabico ratiocinabimur, poterit harum rerum intelligens idem in quocunque praestare. Caeterum in omnibus plus quam hexasyllabicis necesse est vocum repetitionem esse. Porro in textu hexasyllabico capita variationum sunt haec:

I. ut, re, mi, fa, sol, la. Variatio ordinis est              720

II. ut, ut, re, mi, fa, sol. Variatio ordinis est 720 – 120. f. 600. Non solum autem ut, sed et quaelibet 6 vocum potest repeti 2 Mahl. E. 6 ∩ 600 f. 3600. Et reliquarum 5 vocum semper 5 Mahl aliae 4 possunt poni post ut ut; nempe: re mi fa sol. re mi fa la. re mi sol la. re fa sol la. mi fa sol la. Seu

2. The reason for this is clear if you consider diagram ꜟ, where all the variations in which a given thing is put in place of itself will fall away. I shall now show an application.

3. Let us propose the following question: given a text, to find all its possible melodies. Harsdörffer attempted this also in his *Mathematical Recreations*, Section 4, Proposition 7. But he thinks that for a text of 5 syllables there can only be 120 melodies, having looked only at variations of order.[258] However, it seems to me that one must also take complexions into account, as we shall now see.

4. But I shall begin by making a distinction: a Text is either simple or composite. I call a text composite when it is divided into lines, or *Reimzeilen*.[259] And I shall look at the variation of a composite text by means of melodies of simple texts set to them one after another in accordance with [Problem IV].[260] A simple text either exceeds 6 syllables or it does not. That distinction is necessary, because there are 6 notes, Ut Re Mi Fa Sol La (omitting the seventh, Bi, which Erycius Puteanus added in his *Musathena*).[261] If it does not exceed 6, then it is either of 6 syllables or fewer.

5. I shall consider an example of a six-syllabled Text, as anyone who understands these will be able to apply the same knowledge to anything. Observe that with more than six syllables, some notes must necessarily be repeated. Now in a six-syllabled text, these are the heads of variations:

I. ut re mi fa sol la. The variation of order is..................................720.[262]

II. ut ut re mi fa sol. The variation of order is

720 – 120 f. 600. And not only 'ut', but any six of the notes can be the one repeated twice. Therefore,

6 ∩ 600 f. 3600. And of the remaining 5 notes, another 4 can be placed after ut ut in 5 ways, namely, re mi fa sol, re mi fa la, re mi sol la, re fa sol la, and mi fa sol la;

---

[258] Harsdörffer (1990: 142). As Knobloch remarks, Leibniz is quite reticent about the fact that Harsdörffer's inquiry rests on Mersenne 1625 (as Harsdörffer openly recognizes). For a comparison between Leibniz's method for solving the problem and Mersenne's, see Knobloch (1973: 45–7).

[259] *Reimzeilen* is German for 'rhyming lines' or 'lines of rhyme', i.e. 'verses'.

[260] The 1666 edition has "probl. 3".

[261] Puteanus (1602: 35): "Ego adiungo, et ( ... ) Notarum numerum augeo: & senis receptis, ut MUSATHENA constituatur comitem unam adiicio, ex eodem illo Hymno: BI." The 'BI' directly follows the LA in the verse Guido d'Arezzo (c. 991–after 1033) was thought to have introduced as a means to remember the series: "VT queant laxis REsonare fibris / MIra gestorum FAmuli tuorum, / SOLve pollutis LAbij reatum / Sancte Ioannes". Cf. Puteanus, *ibidem*, and Guido d'Arezzo (1999: 19–20 esp.), and (1999: 615) under *Ut queant laxis*.

[262] This number is obtained through the application of the false rule that Leibniz presents as the solution of PROBLEM II.

| | | |
|---|---|---:|
| | 5 res habent 5 con4nationes: 5 ∩ 3600 f | 18000 |
| III. | ut ut re re mi fa. 480 ∩ 15 f. 7200 ∩ 6 f | 43200 |
| IV. | ut ut re re mi mi. 360 ∩ 20 f | 7200 |
| V. | ut ut ut re mi fa. 360 ∩ 6 f. 2160 ∩ 20 f | 43200 |
| VI. | ut ut ut re re mi. 360 ∩ 6 ∩ 5 ∩ 4 f | 43200 |
| VII. | ut ut ut re re re. 240 ∩ 15 f | 3600 |
| VIII. | ut ut ut ut re mi. 360 ∩ 6 ∩ 10 f | 21600 |
| IX. | ut ut ut ut re re. 240 ∩ 6 ∩ 5 f. | 7200 |
| | Summa | 187920 |

6. Quid vero si septimam vocem Puteani Bi, si pausas, si inaequalitatem celeritatis in notis, si alios characteres musicos adhi[67]beamus computationi; si ad Textus plurium syllabarum quam 6, si ad compositos progrediamur, quantum erit mare melodiarum, quarum pleraeque aliquo casu utiles esse possint? [*219*]

7. Admonet nos vicinitas rerum posse cujuslibet generis carminum possibiles species seu flexus, et quasi Melodias inveniri, quae nescio an cuiquam hactenus vel tentare in mentem venerit.

8. Age in Hexametro conemur. Cum hexametro sex sint pedes, in caeteris quidem dactylus spondaeusque promiscue habitare possunt, at penultimus non nisi dactylo, ultimus spondaeo aut trochaeo gaudet. Quod igitur 4 priores attinet, erunt vel meri dactyli, 1; vel meri spondaei, 1; vel tres dactyli unus spondaeus, vel contra: 2; vel 2 dactyli 2 spondaei, 1; et ubique variatio situs, 12, 2 + 1 f. 3 ∩ 12 f. 36 + 1 + 1 f. 38. In singulis autem his generibus ultimus versus vel spondaeus vel trochaeus est, 2 ∩ 38 f. 76. Tot sunt genera hexametri si tantum metrum spectes.

9. Ut taceam varietates quae ex vocibus veniunt, v. g. quod vel ex monosyllabis vel disyllabis etc. vel his inter se mixtis constat; quod vox modo cum pede finitur, modo facit caesuram eamque varii generis; quod crebrae intercedunt elisiones aut aliquae aut nullae.

that is, the 5 things have 5 con4nations. And 5 ∩ 3600 f.......................18000.

III.  ut ut re re mi fa. 480 ∩ 15 f. 7200 ∩ 6 f.................................43200.

IV.  ut ut re re mi mi. 360 ∩ 20 f................................7200.

V.  ut ut ut re mi fa. 360 ∩ 6 f. 2160 ∩ 20 f.................................43200.

VI.  ut ut ut re re mi. 360 ∩ 6 ∩ 5 ∩ 4 f.................................43200.

VII.  ut ut ut re re re. 240 ∩ 15 f................................3600.

VIII. ut ut ut ut re mi. 360 ∩ 6 ∩ 10 f................................21600.

IX.  ut ut ut ut re re. 240 ∩ 6 ∩ 5 f................................7200.

Total     187920.

6.  But what if we admit Puteanus's seventh note Bi, if we admit pauses, if we admit inequality of length in notes, if we admit other musical qualities into the calculation? What if we go on to Texts of more than 6 syllables, if we go on to composite Texts? What an ocean of melodies there will be, most of which could be useful in certain cases!

7.  This circumstance reminds us that for any class of poem we can find possible kinds or tropes, and, as it were, Melodies, such as I do not know whether it has yet come into anyone's mind to test.

8.  Let us then venture upon the Hexameter.[263] Now there are six feet in a hexameter, all but two of which may be either a dactyl or a spondee, while the next to last may be only a dactyl, and the last either a spondee or a trochee.[264] Hence, so far as the first 4 are concerned, they may be all pure dactyls (1 case) or pure spondees (1 case), three dactyls and one spondee, or the other way round (2 cases), or two dactyls and two spondees (1 case), giving, with overall variation of situs in 12 ways, 2 + 1 f. 3 ∩ 12 f. 36 + 1 + 1 f. 38. Finally, in each kind of hexameter, the last foot is either a spondee or a trochee, giving 2 ∩ 38 f. 76. And this is the number of different kinds of hexameter if you consider only metre.

9.  Thus, I say nothing about variations that arise from words, for example, that the hexameter consists of monosyllables or disyllables, etc., or of mixtures of them; that sometimes the end of a word coincides with the end of a foot; that sometimes it has a caesura, and in various ways; and that elisions intervene, whether frequently, sometimes, or not at all.

[263] In ancient Greek and Latin poetry, the *hexameter* was a metrical line of verses composed, as Leibniz explains, of six feet. A foot is composed of two long syllables, giving rise to a *spondee*, or one long and two short syllables: a *dactyl*. A long syllable, according to prosodic conventions, is denoted by means of the sign '–', whereas a short syllable is denoted by means of the symbol 'ᴗ'. Thus, a dactyl is represented as '–ᴗᴗ' and a spondee as '– –'.

[264] In Greek and Latin, a *trochee* is a metrical foot made of a long syllable followed by a short one.

10. Caeterum et multitudine literarum hexametri differunt, quam in rem
    extat carmen Publilii Porphyrii Optatiani (quem male cum Porphyrio
    Graeco, philosopho, Christianorum hoste, Caesar Baronius confudit) ad
    Constantinum Magnum 26 versibus heroicis constans, quorum primus
    est literarum, caeteri continue una litera crescunt, usque ad 26tum qui
    habet 50. Ita omnes organi Musici speciem exprimunt. Meminere
    Hieron. ad Paulinam, Firmicus in myth. Rab. Maurus, Beda de re
    metrica. Edidit Velserus ex Bibliotheca sua Augustae cum figuris An.

10. Next, hexameters differ also in quantity of letters, in which connection there has come down to us a poem of Publilius Optatianus Porphyrius (whom Cesare Baronius wrongly confused with the Greek philosopher Porphyry, an opponent of the Christians), addressed to Constantine the Great, and consisting of 26 heroic verses, of which the first is of 25 letters, and the others increase by one letter at a time up to the 26th, which has 50 letters.[265] In this way, they jointly depict a kind of Musical organ. Those who remembered him were Jerome in his letter to Paulina, [Fulgentius] in his *Mythologies*, Rabanus Maurus, and Bede in *On Metre*.[266] Welser published

---

[265] Leibniz's reference to Baronius may have been taken from Vossius's *De Poetis Latinis*. Vossius took his information regarding the confusion of Porphyrius, or Porfyrius († c. 335), the poet, and the philosopher Porphyry, from Joseph Scaliger, who scolded the 'founder of the Church Annals' for having confused the two men. Cf. Scaliger (1606a: 229). On the reference to Baronius, see also note 269, below. On Porphyrius and his *Carmina Figurata*, see also: Optatianus Porfyrius (1973); Barnes (1975: 173–86); Edwards (2005: 447–66).

[266] The four authors mentioned are bracketed together with the Porphyrius example because they mention Porphyrius; not for similar attempts at varying the number of letters within a verse, or because of the images they themselves may have made from their verses—as Raban Maur, for instance, did. Leibniz seems to have derived these names from his reading of Gerhard Johannes Vossius, who in turn based his information on Joseph Scaliger. He could also have found them in Welser (see the next footnote), although it is unclear why he would then have missed or simply skipped the further references available in that work. He mistakenly mentions "Firmicus", however, instead of Fulgentius as the author of the *Mythologies*. The mistake itself is understandable in view of the fact that, like Porphyrius, Julius Firmicus Maternus (* c. 305 AD) was a contemporary of Constantine the Great and someone known for having wavered between paganism and Christianity. Firmicus was the author of a book on astrology, the *Mathesis*, that was written in 337, before his conversion. The book is the most complete work of ancient astrology that survives. Cf. P. Monat, 'Introduction', in Maternus (1992: 7–43). Fabius Planciades Fulgentius (fl. c. 500 or 550), on the other hand, was a late-classical author of influential works on mythography, as well as of an interesting commentary on the *Aeneid* entitled *Expositio Virgilianae continentiae* or *The Exposition of Virgil's Modesty*. Fulgentius refers only to Porphyrius's *Epigrams* and does so in the *Mythologiae* and the *Expositio continentiae Virgilianae secundum philosophos moralis*, respectively. Cf. Fulgentius (1997: 72–3, 98–9, n.98, resp.). See also: Withbread (1971: 67, 131, 149, 151). The poet Porphyrius was further remembered by Jerome in his addenda to Eusebius's *Chronology*, of which a variety of enlarged sixteenth- and seventeenth-century versions were known, sometimes covering AD dates up to the present, sometimes BC dates as well. See, e.g. Aubertus Miraeus's edition of the *Hieronymi Presbyteri Chronicon* in: Eusebius, Hieronymus, Sigebertus Gemblacensis, Anselmus Gemblacensis, Aubertus Miraeus, et al. (1608), s.p., 332. XXIII. See also Scaliger (1606: 181), as well as Donalson (1996: 40, 62, n. 232e, 101). Leibniz's reference to the 'Letter to Paulina' is a slip of the pen, probably due to the fact that he had already made this particular reference to Jerome quoting a similar reference ("*Hieron. ad Paulin.*") in Harsdörffer (see above, note 251). Raban Maur (or Hrabanus Maurus, c. 780–856) was the author of *In honorem Sanctae Crucis*, the first part of which consists of twenty-eight *carmina figurata*, poems ordered in such a way that the images and symbols outlined in its verses contain new verses, thereby offering a second level of reading. Cf. Rabanus Maurus (1997). In the 'Prologus', Maur, "le Porfyrius de l'époque carolingienne" as his editor Perrin aptly calls him, compares his own method to that of his Constantinian example. Cf. p. 19, and *idem*, 'Introduction', pp. ix–x. Finally, the Venerable Bede (c. 673–735) in his book *On Meter* mentions the alternative meters in "the volume of the poet [Porphyrius], which when he sent it to the Emperor Constantine earned him release from exile". Bede, however, chose not to discuss these verses "because of their pagan nature". Cf. Bede (1991: 161).

1591. Adde de eo Eryc. Puteanum in Thaum. Pietatis lit. N. qui ait hoc carmine revocari ab exilio meruisse; Gerh. Joh. Vossium syntag. de Poet. Latinis v. Optatianus, item de Historicis Graecis, I. 16, Casp. Barthium Commentariolo de Latina Lingua, et Aug. Buchnerum Notis in Hymnum Venantii Fortunati (qui vulgo Lactantio ascribitur) de Resurrect. ad v. 29, pag. 27. Qui observat Hexametros fistulis, Versum per me[68] dium ductum: *Augusto victore*, etc. regulae organi, jambos anacreonticos dimetros omnes 18 literarum, epitoniis respondere. Versus ipsos quia ubique obvii non sunt expressimus.

them from his Augsburg publishing house, with illustrations, in [1595].[267] Add in this connection also, in his *Wonders of Piety*, Letter N, Erycius Puteanus, who says that for this poem he [Optatian] deserved to be recalled from exile;[268] Gerardus Johannes Vossius in his *Syntagma of Latin Poets* (see under "Optatian"), and also in his *Greek Historians* l. 16;[269] Kaspar von Barth in his *Brief Commentary on the Latin Language*;[270] and August Buchner in his *Notes on the Hymn to the Resurrection of Venantius Fortunatus* (which is usually ascribed to Lactantius), p. 27, the note on verse 29.[271] As one can see, the Hexameters correspond to the pipes, the Verse *Augusto Victore etc.*, written down the middle, corresponds to the windchest of the organ, and the anacreontic iambic dimeters, of 18 letters each, correspond to the keyboard. I have reproduced the verses, as they are not widely available.

---

[267] In 1595, Paul Welser (1555–1620), the elder brother of the Augsburg merchant-diplomat, renowned humanist scholar, publisher, and expert of Roman remains Marcus Welser (1558–1614), published an edition of the Porphyrius poem as Paullus Velserus (1595). The book, which is now very rare, received a wider diffusion by being included in the 1682 edition of Paul's brother's *Opera*: Marcus Velserus (1682). Leibniz, however, probably consulted neither of these works, since he seems to have taken his information from Vossius, making two mistakes, however, along the way. The first concerns the year of publication of Paul Welser's work. Vossius gives the right date, namely 1595, where Leibniz's text has "1591". Secondly, Leibniz refers to "Hieron. ad Paulinam.", whereas the reference to Jerome concerns the latter's addenda to Eusebius's *Chronology* (see the former footnote). Leibniz may have taken the text of the poem itself from Bücher's edition of Fortunatus, for instance, but most probably not from any of the Welser editions, or he would have mentioned the other sources named there, rather than copying Vossius's list. Besides Paul and Marcus, two further brothers, namely Anton (1551–1618) and Matthäus (1553–1633), also owned extensive libraries. On the Welser family, see Häberlein et al. (2002) and Johnson (2008). On Marcus Welser, see also *DBE* 10. 42*f*.

[268] Puteanus (1617: 89–93, esp. 93), where Puteanus argues Bauhuisen has deserved even more: having shown piety besides candour, he merits not just his homeland, but heaven. See also above, notes 224 and 232.

[269] According to Vossius, the type of poetry represented by Optatian was praiseworthy for the effort put into it, rather than for its "poetical spirit". Cf. Vossius (1654: 54), which includes a series of later references to Optatian by Jerome, Fulgentius, Beda and Rhabanus Maurus, and mentions the Welser edition, as well as the confusion of the two Porphyrios. Vossius also mentions 'the very learned Baronius', whom Leibniz refers to, above. As an instance of the confusion in ecclesiastical chronology, see, for instance, Abraham Bzowski (1617), col. 441, XIX, nn. m–n. Note that Vossius elsewhere refers to Joseph Scaliger as his own source on the confusion of the poet Optatian and the philosopher Prophyry. Cf. Vossius (1623: 198–9).

[270] Barthius (1600: 365–415). Von Barth mentions Optatian along with the 1595 edition by Paul Welser on p. 410, for which see also note 266, above.

[271] Fortunatus (1627: 27). The poem, or 'Publilii Optatiani Porphyrii Organon' itself, is depicted on a separate fold inserted between pp. 27 and 28. Fortunatus's editor August Buchner discusses the question of the conventional attribution to Lactantius right at the start of his commentary, pp. 1–2 (second series). Verses 29 and 30 of the *Hymn*, associating the sweet song of birds in spring with the attunement of organs through their reeds, prompted Buchner's observations on Porphyrius's poem. The poet and Bishop of Poitiers Venantius Honorius Clementius Fortunatus (*c.* 536–610), still known today for having inspired well-known Christian hymns, wrote his poem in honour of St. Felix of Nantes. Cf. Fortunatus (1880: 59–62), 'Ad Felicem episcopum de pascha'.

AUGUSTO VICTORE JUVAT RATA REDDERE VOTA

| | |
|---|---|
| 25 Post martios labores | ‖ O si diviso metiri limite Clio |
| 26 Et Caesarum parantes | ‖ Una lege sui uno manantia fonte |
| 27 Virtutibus, per orbem | ‖ Aonio versus heroi jure manente |
| 28 Tot laureas virentes, | ‖ Ausuro donet metri felicia texta |
| 29 Et Principis trophaea | ‖ Augeri longo patiens exordia fine |
| 30 Felicibus triumphis | ‖ Exiguo cursu parvo crescentia motu |
| 31 Exultat omnis aetas, | ‖ Ultima postremo donec vestigia tota |
| 32 Urbesque flore grato, | ‖ Ascensus jugi cumulato limite cludat |
| 33 Et frondibus decoris | ‖ Uno bis spatio versus elementa prioris |
| 34 Totis virent plateis. | ‖ Dinumerans cogens aequali lege retenta |
| 35 Hinc ordo veste clara | ‖ Parva nimis longis et visu dissona multum |
| 36 In purpuris honorum | ‖ Tempore sub parili metri rationibus isdem |
| 37 Fausto precantur ore, | ‖ Dimidium numero Musis tamen aequiparantem |
| 38 Feruntque dona laeti. | ‖ Haec erit in varios species aptissima cantus |
| 39 Iam Roma culmen orbis | ‖ Perque modos gradibus surget fecunda sonoris |
| 40 Dat munera et coronas | ‖ Aere cavo et tereti calamis crescentibus aucta |
| 41 Auro ferens coruscas | ‖ Quis bene suppositis quadratis ordine plectris |
| 42 Victorias triumphis, | ‖ Artificis manus innumeros clauditque aperitque |
| 43 Votaque iam theatris | ‖ Spiramenta probans placitis bene consona rythmis |
| 44 Redduntur et Choreis. | ‖ Sub quibus unda latens properantibus incita ventis |
| 45 Me sors iniqua laetis | ‖ Quas vicibus crebris iuvenum labor haud sibi discors |
| 46 Solemnibus remotum | ‖ Hinc atque hinc animaeque agitant augetque reluctans |
| 47 Vix haec sonare sivit | ‖ Compositum ad numeros propriumque ad carmina praestat |
| 48 Tot vota fronte Phoebi, | ‖ Quodque queat minimum admotum intremefacta frequenter |
| 49 Versuque comta solo | ‖ Plectra adaperta sequi aut placitos bene claudere cantus |
| 50 Augusta rite seclis. | ‖ Iamque metro et rythmis praestringere quicquid ubique est |

Ex quibus multa circa scripturam Veterum observari possunt, inprimis Diphthongum *AE* duabus literis exprimi solitam; qui tamen mos non est cur rationem vincat, unius enim soni una litera esse debet. Sed de hoc Optatiano vel propterea fusius diximus, ut infra dicenda praeoccuparemus; ubi versus Proteos ab eo compositos allegabimus.

AUGUSTO VICTORE JUVAT RATA REDDERE VOTA

| | |
|---|---|
| 25 Post martios labores | ‖ O si diviso metiri limite Clio |
| 26 Et Caesarum parantes | ‖ Una lege sui uno manantia fonte |
| 27 Virtutibus, per orbem | ‖ Aonio versus heroi jure manente |
| 28 Tot laureas virentes, | ‖ Ausuro donet metri felicia texta |
| 29 Et Principis trophaea | ‖ Augeri longo patiens exordia fine |
| 30 Felicibus triumphis | ‖ Exiguo cursu parvo crescentia motu |
| 31 Exultat omnis aetas, | ‖ Ultima postremo donec vestigia tota |
| 32 Urbesque flore grato, | ‖ Ascensus jugi cumulato limite cludat |
| 33 Et frondibus decoris | ‖ Uno bis spatio versus elementa prioris |
| 34 Totis virent plateis. | ‖ Dinumerans cogens aequali lege retenta |
| 35 Hinc ordo veste clara | ‖ Parva nimis longis et visu dissona multum |
| 36 In purpuris honorum | ‖ Tempore sub parili metri rationibus isdem |
| 37 Fausto precantur ore, | ‖ Dimidium numero Musis tamen aequiparantem |
| 38 Feruntque dona laeti. | ‖ Haec erit in varios species aptissima cantus |
| 39 Iam Roma culmen orbis | ‖ Perque modos gradibus surget fecunda sonoris |
| 40 Dat munera et coronas | ‖ Aere cavo et tereti calamis crescentibus aucta |
| 41 Auro ferens coruscas | ‖ Quis bene suppositis quadratis ordine plectris |
| 42 Victorias triumphis, | ‖ Artificis manus innumeros clauditque aperitque |
| 43 Votaque iam theatris | ‖ Spiramenta probans placitis bene consona rythmis |
| 44 Redduntur et Choreis. | ‖ Sub quibus unda latens properantibus incita ventis |
| 45 Me sors iniqua laetis | ‖ Quas vicibus crebris iuvenum labor haud sibi discors |
| 46 Solemnibus remotum | ‖ Hinc atque hinc animaeque agitant augetque reluctans |
| 47 Vix haec sonare sivit | ‖ Compositum ad numeros propriumque ad carmina praestat |
| 48 Tot vota fronte Phoebi, | ‖ Quodque queat minimum admotum intremefacta frequenter |
| 49 Versuque comta solo | ‖ Plectra adaperta sequi aut placitos bene claudere cantus |
| 50 Augusta rite seclis. | ‖ Iamque metro et rythmis praestringere quicquid ubique est |

From these verses one can deduce many things about Classical orthography, and in particular, that the Diphthong 'Æ' was customarily represented by two letters; not, however, that such a custom should override reason, as there should be only one letter for each sound. I have discussed this poem of Optatian at rather greater length in order to foreshadow what I shall have to say later on, when I come to address the question of Protean verses composed in imitation of it.

# [69] Probl. VII.
## DATO CAPITE VARIATIONES REPERIRE.

1. Hoc in Complexionibus solvimus supra. De situs variationibus nunc: Sunt autem diversi casus. Caput enim Variationis hujus aut constat una re, aut pluribus: si una, ea[*220*]vel monadica est, vel dantur inter Res (variandas) alia aut aliae ipsi homogeneae. Sin pluribus constat, tum vel intra caput dantur invicem homogeneae vel non, item extrinsecae quaedam intrinsecis homogeneae sunt vel non.

2. "Primum igitur capite variationis fixo manente numerentur res extrinsecae; et quaeratur variatio earum inter se (et si sint discontiguae seu caput inter eas ponatur) praeciso capite, per prob. 4, productum vocetur A. Si caput multiplicabile non est, seu neque pluribus rebus constat, et una ejus res non habet homogeneam, *productum A erit quaesitum.*"[*221*]

3. "Sin caput est multiplicabile, et constat 1 re habente homogeneam, productum A multiplicetur numero homogenearum aeque in illo capite ponibilium, et *factu erit quaesitum.*"

4. "Si vero caput constat pluribus rebus, quaeratur variatio earum inter se (etsi sint discontiguae seu res extrinsecae interponantur), per probl. 4, ea ducatur in productum A, quodque ita producitur dicemus B. Jam

# Problem VII GIVEN A HEAD, TO FIND THE VARIATIONS

1. I solved this Problem earlier for Complexions.[272] But as regards variations of situs, there are various cases, as the head of a variation may consist either of one thing or of many. If of one thing, either it is monadic, or there exist among the Things to be varied one or several things homogeneous with it. But if it consists of several things, then either there exist within the head mutually homogeneous things, or there do not. And again, there may be certain extrinsic things homogeneous with the intrinsic things, or there may not.

2. "First, then, with the head of the variation remaining fixed, count up the extrinsic things, and look for their variation among themselves (even if they are discontiguous, that is, if the head is scattered among them) with the head removed, according to Problem IV. Call this result A. If the head is not a manifold, i.e. if it does not consist of several things, and if there is nothing homogeneous with the single thing of which it is composed, *the result A will be what is required.*[273]

3. But if the head is a manifold, and consists of a single thing having a homogeneous thing, multiply the result A by the number of homogeneous things equally capable of being placed in the head; and *this product will be what is required.*[274]

4. But if the head consists of several things, look for their variation among themselves (even if they are discontiguous, that is, if extrinsic things are interpolated) according to Problem IV, and multiply that by result

[272] Cf. above, the Applications of PROBLEMS I and II. For the notion of 'head' see above Introduction, § 13. In the case of *variations*, Leibniz calls 'head' the items (or item) which remain invariant with respect to a set of variations (permutations). At the very beginning of the DAC (see above, 'Definitions'), Leibniz defines a homogeneous thing as "something that is equally disposable in a given place, with the only exception of the head". A given head is 'monadic' if it has no homogeneous elements; 'multipliable' if its elements can be permuted.

[273] Suppose as given a set of 4 things: $a, b, c, d$; and suppose that $a$ is the head. This means that $a$ is invariant in respect to the other things. Thus, keep $a$ separate from the remaining items and simply calculate the number of permutations that can be made out of them, i.e. 3! = 6. If $a$ "does not consist of several things" and there are no other things homogeneous with it, then add $a$ to each permutation out of '$b, c, d$'. As a result, we have 6 permutations (variations) of the 4 items $a, b, c, d$, with $a$ as the head.

[274] If $a$ is the only head and there is one thing $a_1$ that is homogeneous with $a$, then multiply the previous result A (6 in our example; see the preceding footnote) by the number of homogeneous things, i.e. the number of permutations is $A(a_1 + 1)$. Cf. Knobloch (1973: 49) and Introduction.

si res capitis nullam habet homogeneam extra caput, *productum B erit quaesitum.*"

[5.] "Si[**70**]res capitis habet homogeneam tantum extra caput, non vero intra, productum B multiplicetur numero rerum homogenearum, et si saepius sunt homogeneae, factus ex numero homogenearum priorum multiplicetur numero homogenearum posteriorum continue, et *factus erit quaesitum.*"

6. "Sin res capitis habet homogeneam intra caput et extra, numerentur primo res homogeneae intrinsecae et extrinsecae simul, et supponantur pro Numero complicando; deinde res datae homogeneae tantum intra caput supponantur pro exponente. Dato igitur numero et exponente quaeratur complexio per probl. 1 et si saepius contingat homogeneitas, ducantur complexiones in se invicem continue. Complexio vel factus ex complexionibus ducatur in productum B. Et *factus erit quaesitum.*"

7. Hoc problema casuum multitudo operosissimum efficit, ejusque nobis solutio multo et labore et tempore constitit. Sed aliter sequentia problemata ex artis principiis nemo solvet. In illis igitur usus hujus apparebit.

A. I shall call this product B. Now if there is no thing outside the head homogeneous to things within the head, *the result B will be what is required.*[275]

[5]. If a thing in the head has a homogeneous thing only outside the head, and not inside it, multiply the product B by the number of homogeneous things. If the homogeneous things are scattered, multiply the product resulting from the number of earlier homogeneous things one after another by the number of later homogeneous things, and *this result will be what is required.*[276]

6. But if a thing in the head has a homogeneous thing both within the head and outside it, first count the intrinsic and extrinsic things together, and take this to be the Number for the purposes of complication. After this, take as the exponent the number of given homogeneous things within the head only. Then, look for the complexion with the given number and exponent according to Problem I, and if the homogeneity happens more than once, multiply the complexions together one after another. Multiply this complexion, or this product of the complexions, by product B, *and the result of it will be what is required.*"[277]

7. This Problem is made up of a lot of quite difficult cases, and its solution cost me a great deal of labour and time. But the following Problems will not otherwise be solved on the basis of the Principles of the art. It is in those Problems, therefore, that its application will become clear.

[275] If the head consists of $k$ things, first calculate all the permutations of $k$ elements (= $k!$); then, to continue Leibniz's example, multiply the result by A: $A(k!)$ and call the result '$B$'. Cf. Knobloch (1973: 49) and Introduction.

[276] If outside the head there are elements homogeneous with the elements $a_1$, $a_2$....$a_k$ of the head (whereas the elements belonging to the head are not homogeneous among them), the number of permutations is $B(a_1 + 1) (a_2 + 1)....(a_k + 1)$. Cf. Knobloch (1973: 49) and Introduction.

[277] If the head contains $k$ 'homogeneous' things and if ouside the head there are $h$ 'homogeneous' things as well, first add together $k$ and $h$. Then, consider $k + h$ as the number of things to be combined and $k$ as the exponent. In other words, calculate all complexions of $k$ elements out of $k + h$ things. As a final step, multiply the number of complexions of $k$ elements for the number corresponding to $B$ above. Cf. above Introduction and Knobloch (1973: 48–9).

## Probl. VIII.
## VARIATIONES ALTERI DATO CAPITI COMMUNES REPERIRE.

8. "Utrumque caput ponatur in eandem Variationem quasi esset unum caput compositum (etsi interdum res capitis compositi sint discontiguae) et indagentur variationes unius capitis compositi per probl. 10, *productum erit quaesitum.*" [222]

## Probl. IX.
## CAPITA VARIATIONES COMMUNES HABENTIA REPERIRE.

9. "1. Si plura capita in variatione ordinis in eundem locum incidunt vel ex toto vel ex parte, non habent variationes communes. 2. Si eadem res monadica in plura capita incidit, ea non habent variationes communes. Caetera omnia habent variationes communes."[71]

## Probl. X.
## CAPITA VARIATIONUM UTILIUM AUT INUTILIUM REPERIRE.

10. *Capita in universum reperire expeditum est.* Nam quaelibet res per se, aut in quocunque loco per se, aut cum quacunque alia aliisve, quocunque item loco cum alia aliisve, breviter omnis complexio aut variatio proposita minor et earundem rerum, seu quae tota in altera continetur, est caput. Methodus autem in disponendis capitibus utilis, ut a minoribus ad majora progrediamur, quando v. g. propositum nobis est omnes variationes oculariter proponere, quod Drexelius loco citato, Puteanus et Kleppisius et Reimerus citandis factitarunt.

## Problem VIII
## TO FIND THE VARIATIONS IN COMMON WITH ANOTHER GIVEN HEAD

8. "Place each head into the same Variation as though it were a single, composite head (and even if the things in the composite head are discontiguous), and look for the variations under the single, composite head according to Problem X. *The result will be what is required.*"

## Problem IX
## TO FIND THE HEADS THAT HAVE COMMON VARIATIONS

9. "1. If several heads in a variation of order occur in the same position, either in whole or in part, they do not have common variations. 2. If the same monadic thing occurs in several heads, they do not have common variations. [3]. All other heads have common variations."

## Problem X
## TO FIND THE HEADS OF USEFUL OR USELESS VARIATIONS

10. *To find all the heads in general is an easy matter.* For anything by itself, or in some position by itself, or with some other thing or other things, or in some position with another thing or other things, in short, every complexion or variation smaller than the one proposed, and of the same things, that is, which is wholly contained in another variation, may be a head. And this is a useful method of dealing with heads, that is, to advance from smaller heads to greater heads, when, for example, we have to display all the variations, as Drexel (in the work cited), Puteanus, Kleppis, and [Reimer] (in works to be cited) have frequently done.[278]

---

[278] Enumerations of all variations of a series of elements, such as in Drexel's variations of *a b c d e f*), include lists in which '*a*', '*a b*', '*a b c*', etc. may be seen as so many heads of different lengths. Puteanus (1617: 13–50), offers 1022 variations of the verse *Tot tibi sunt dotes, Virgo, quot sidera caelo*, for which see also above, notes 232 and 285. For Kleppis, see note 294, below. The 1666 original has 'Reiner' for 'Reimer', i.e. Reimarus or 'Reimerus', on whom see below, note 298.

11. Caeterum ut *Capita utilia vel inutilia reperiantur*, adhibenda disciplina est ad quam res variandae, aut totum ex iis compositum pertinet. Regulae ejus inutilia quidem elident, utilia vero relinquent. Ibi videndum quae cum quibus et quo loco conjungi non possint, item quae simpliciter quo loco poni non possint v. g. primo, tertio, etc. Inprimis autem primo et ultimo. Deinde videndum quae res potissimum causa sit anomaliae (v. g. in versibus hexametris Proteis syllabae breves). Ea ducenda est per omnes caeteras, omnia item loca, si quando autem de pluribus idem judicium est, satis erit in uno tentasse.

**Probl. XI.**
**VARIATIONES INUTILES REPERIRE.**

12. Duae sunt viae. (1) Per probl. 12 hoc modo: "Inventa summa variationum utilium et inutilium per probl. 4 subtrahatur summa utilium per probl. 12 viam secundam; Residuum erit quaesitum." (2) Absolute hoc modo: "Inveniantur capita variationum inutilium per probl. 10. Quaerantur singulorum capitum variationes per probl. 7. Si qua capita communes habent variationes per probl. 9, numerus earum inveniatur per probl. 8 et in uno solum capitum variationes Communes habentium relinquatur, de caeterorum variationibus subtrahatur; aut si hunc laborem subtrahendi subterfugere velis, initio statim ca[72]pita quam maxime composita pone, conf. probl. 8. Aggregatum omnium variationum de omnibus complexionibus, subtractis subtrahendis, erit *quaesitum*." [223]

**Probl. XII.**
**VARIATIONES UTILES REPERIRE.**

13. Solutio est ut in proxime antecedenti, si haec saltem mutes, in via 1, loco problem. 12 pone 11, etc. et subtrahatur summa inutilium per probl. 11 viam secundam. In via 2. inveniantur capita variationum utilium, caetera ut in probl. proximo.

*Usus Problem. 7, 8, 9, 10, 11, I2.*

11. Next, *to find useful or useless heads*. We must apply a set of rules, to which the things to be varied, or the whole composed of them, are subject. These rules will get rid of the useless things, and the useful things will remain. They will show us what cannot be joined with what, and in what position, and also what simply cannot be placed in what position, such as first and third place, and so on, and especially in first and last position. They will further show us, above all, which thing might be the cause of an anomaly (for example, short syllables in protean hexameter verses). Such a thing must be tested with all the others, and similarly in all positions. But in many cases, when the judgement is similar, it will suffice to have tried it in one case.

## Problem XI
### TO FIND THE USELESS VARIATIONS

12. There are two ways of proceeding, the first according to Problem XII, like this: "from the total of useful and useless variations found according to Problem IV subtract the total of useful variations according to the second way of Problem XII: the Difference will be what is required". The second way is direct, like this: "Find the heads of the useless variations according to Problem X, and look for the variations of each head according to Problem VII. If these heads have common variations according to Problem IX, find the number of each according to Problem VIII, retain the number from only one of the heads having Common variations, and subtract it from the variations of the others. Or if you want to avoid all this trouble of subtracting, then right from the start make the heads as composite as possible, in accordance with Problem VIII. The Aggregate of all the variations of all the complexions, with the necessary subtractions, will be *what is required*."

## Problem XII
### TO FIND THE USEFUL VARIATIONS

13. The solution is as in the last Problem, at least if in the first way of proceeding you make this change, putting XI in place of Problem XII etc., and subtract the total of useless variations as in the second way of Problem XI. In the second way, find the heads of the useful variations. The rest is as in the last Problem.

14. Si cui haec problemata aut obvia aut inutilia videntur, cum ad praxin superiorum descenderit, aliud dicet. Rarissime enim vel natura rerum vel decus patitur omnes variationes possibiles utiles esse. Cujus specimen in argumento minus fortasse fructuoso, in exemplum tamen maxime illustri daturi sumus.

15. Diximus supra *Proteos* versus esse *pure* Proteos, id est in quibus pleraeque variationes possibiles utiles sunt, ii nimirum qui toti propemodum monosyllabis constant; vel *mixtos*, in quibus plurimae incidunt inutiles, quales sunt qui polysyllaba, eaque brevia continent.

16. In hoc genere inter veteres, qui mihi notus sit tentavit tale quiddam idem ille, de quo probl. 6, Publilius Porphyrius Optatianus. Et Erycius Puteanus Thaumat. Piet. lit. N. pag. 92 ex aliis ejus de Constantino versibus hos refert:

> Quem divus genuit Constantius Induperator
> Aurea Romanis propagans secula nato.

Ex illis primus est Torpalius, vocibus continue syllaba crescentibus constans; alter est Proteus sexiformis, si ita loqui fas est.

> Aurea Romanis propagans secula nato
> Aurea propagans Romanis secula nato
> Secula Romanis propagans aurea nato
> Secula propagans Romanis aurea nato
> Propagans Romanis aurea secula nato
> Romanis propagans aurea secula nato.

# The Applications of Problems VII, VIII, IX, X, XI, and XII

14. Anyone who finds these Problems either obvious or useless will think otherwise when he has got down to the practice of what has been said above. It is not so often that either the nature or appearance of things allows all the possible variations to be useful. I am going to give you an example of this in a subject of perhaps no great consequence, but as a model, of the utmost clarity.

15. I said earlier that *Protean verses* are either *purely* protean, that is, those in which most of the possible variations are useful, those, in fact, which consist almost entirely of monosyllables; or *mixed*, in which many useless variations occur, such as those that contain polysyllables, and short ones at that.[279]

16. Among the Ancients, the same Publilius Porphyrius Optatianus whom I noted in Problem VI attempted something of the latter kind.[280] Erycius Puteanus, among others, in his *Wonders of Piety*, Letter N, p. 92, refers to these verses of his about Constantine:

> Quem divus genuit Constantius Induperator
> Aurea Romanis propagans secula nato.[281]

The first of these verses is Rhopalic, consisting of words whose syllables increase gradually in number; the second is Protean in six forms [*sexiformis*] (if that is the right way to put it).

> Aurea Romanis propagans secula nato
> Aurea propagans Romanis secula nato
> Secula Romanis propagans aurea nato
> Secula propagans Romanis aurea nato
> Propagans Romanis aurea secula nato
> Romanis propagans aurea secula nato

---

[279] See above § 11 of Problem III.
[280] Among his panegyric verses to Constantine the Great, Publilius Optatianus Porphyrius offers an example of fifteen lines, the first of which contains only disyllables, the second only trisyllables, the third only tetrasyllables, the fourth pentasyllables, the fifth a monosyllable, disyllable, trisyllable, tetrasyllable, and pentasyllable, and the sixth five terms, the first four of which, says Welser, may be transposed. These last two lines, lines five and six, are the ones Leibniz here refers to quote from Puteanus. Cf. Publilius Optatianus Porphyrius, *Panegyricus Dictus Constantino Augusto*, number III in Paul Welser's 1595 edition, on which see notes 266 and 267 above.
[281] Puteanus (1617: 92): "To whom the divine Constantius Imperator gave birth, born to increase golden times for the Romans".

[73] 17.  Verum plures habet primus ille Virgilianus:

> Tityre tu patulae recubans sub tegmine fagi

quem usus propemodum in jocum vertit. Ejus variationes sunt hae: pro *tu sub* 2; pro patulae recubans 2; et *Tityre* jam initio, ut nunc; jam *tegmine* initio; jam *Tityre tegmine,* fine; jam *tegmine Tityre,* fine, $4 \cap 2 \cap 2$ f. 16. Verum in Porphyrianaeis non singuli Protei, sed omnes, neque unus versus, sed carmen totum talibus plenum admirandum est. Ejusmodi versus composituro danda opera, ut voces consonis aut incipiant aut finiant. [*224*]

18.      Alter qui et nomen Protei indidit, est Jul. Caes. Scaliger, vir si ingenii ferocia absit, plane incomparabilis, Poet. lib. 2, c. 30, pag. 185. Is hunc composuit, formarum, ut ipse dicit, innumerabilium, ut nos 64:

> Perfide sperasti divos te fallere Proteu.

Plures non esse facile inveniet, qui vestigia hujus nostrae computationis leget. Pro *Perfide fallere* 2, pro *Proteus divos* $2 \cap 2$ f. 4. *Sperasti divos te,* habet variationes $6 \cap 4$ f. 24. *Divos perfide Te sperasti,* habet var. 2. *Divos Te sperasti perfide,* habet $6 + 2 + 2$ f. $10 \cap 4$ f. $40 + 24$ f. 64. Observavimus ex Virgilio, aeque, imo plus variabilem, Aen. lib. 1, v. 282. Queis (pro: His) ego nec metas rerum nec tempora pon[o]. Nam *perfide* una vox est; *queis ego* in duas discerpi potest.

17. And indeed, that opening line of Virgil,

> Tityre tu patulae recubans sub tegmine fagi,

has several variations, whose application makes it almost comic.[282] Its variations are these: for *tu sub*, 2 variations; for *patulae recubans*, 2 variations; with *Tityre* sometimes at the beginning, as now; with *tegmine* sometimes at the beginning, with *Tityre tegmine* sometimes at the end, with *tegmine Tityre* sometimes at the end, $4 \cap 2 \cap 2$ f. 16. As for the Porphyrian verses, they are not Protean individually, but all taken together; and the whole poem, not just one verse, is marvellously full of such devices. To anyone who would like to compose such verses in his spare time, I recommend words either beginning or ending with consonants.

18. Someone else who called his verses 'Protean' is Julius Caesar Scaliger (a man quite incomparable, were it not for the savagery of his temper) in his *Poetics*, Book 2, Chapter 30, p. 185. He composed this verse of (as he himself says) innumerable forms (64 according to my calculation):

> Perfide sperasti divos te fallere, Proteu.[283]

Anyone who follows this calculation of mine closely will have difficulty in finding any more. For *Perfide fallere*, 2 variations; for *Proteu divos* $2 \cap 2$ f. 4; *Sperasti divos te* has variations $6 \cap 4$ f. 24; *Divos perfide te sperasti* has 2 Variations; *Divos te sperasti perfide* has $6 + 2 + 2$ f. $10 \cap 4$. f $40 + 24$ f. 64. I have also noticed that a certain verse of Virgil in the *Aeneid*, Book 1, v. 282, *Queis* (in place of '*His*') *ego nec metas rerum nec tempora* [*pono*],[284] is even more variable, as *perfide* is one word, while *queis ego* can be divided into two.

---

[282] Virgil, *Eclogue* I, 1, which literally reads: "You, Tityrus, stretched out beneath a wide beech canopy"; the comic element presumably being engendered by altering sequences in the combination of references to size, a reclining position, and a cover of leaves.

[283] "Deceitfully you hoped to cheat the Gods, Proteus." See also notes 222, 226 and 237 above.

[284] Virgil, *Aeneid* I, 282: *His ego nec metas rerum nec tempora pono*, "For these [the Romans], I set no limits in power or time." In John Dryden's classic translation: "To them no bounds of empire I assign / Nor term of years to their immortal line")—a line that, through its sequence in *Aeneid* I, 283 (*imperium sine fine dedi*) has developed into the English expression 'Empire without end'. Note that *Queis* is an archaic form for *Quibus*.

19. Venio ad ingeniosum illum Bernhardi Bauhusii Jesuitae Lovaniensis, qui inter Epigrammata ejus extat; utque superior, v. probl. 4, de Christo, ita hic de Maria est:

> Tot tibi sunt dotes virgo, quot sidera coelo.

Dignum hunc peculiari opera esse duxit vir doctissimus Erycius Puteanus libello, quem *Thaumata Pietatis* inscripsit, edito Antverpi anno 1617. forma 4.ta, ejusque variationes utiles omnes enumerat a pag. 3 usque ad 50 inclusive, quas autor, etsi longius porrigantur, intra cancellos numeri 1022 continuit, tum quod totidem vulgo stellas numerant Astronomi, ipsius autem institutum est ostendere dotes non esse pauciores quam stellae sunt; tum quod nimia propemodum cura omnes illos evitavit, qui dicerc videntur, tot sidera coelo, quot Mariae [74] dotes esse, nam Mariae dotes esse multo plures. Eas igitur variationes si assumsisset (v.g. Quot tibi sunt dotes virgo, tot sidera coelo), totidem, nempe 1022, alios versus ponendo *tot* pro *quot,* et contra, emersuros fuisse manifestum est. Hoc vero etiam in praefatione Puteanus annotat pag. 12, interdum non sidera tantum, sed et dotes coelo adherere, ut coelestes esse intelligamus, v. g.

> Tot tibi sunt coelo dotes, quot sidera virgo.

Praeterea ad variationem multum facit, quod ultimae in *Virgo,* et *Tibi* ambigui quasi census et corripi et produci patiuntur, quod artificium quoque infra in Daumiano illo singulari observabimus.

19. I come now to that ingenious verse of Bernard Bauhuysen, a Jesuit of Louvain, which is found among his *Epigrams*; where the verse above, in Problem IV, is about Christ, here it is addressed to Mary:

> Tot tibi sunt dotes virgo, quot sidera coelo.[285]

That most learned of men, Erycius Puteanus, thought this verse worthy of an essay of its own in the book that he entitled *Wonders of Piety*, published in quarto at Antwerp, 1617, and on pp. [13]–50[286] inclusive he lists its variations; which the author, though they could have gone on longer, kept within the bounds of the number 1022, perhaps because Astronomers traditionally number the stars at this total, and his intention was to show that her endowments are at least as numerous as the stars; or perhaps because he was almost excessively concerned to avoid all the variations that seem to say that there are as many stars in the heavens as endowments of Mary, the endowments of Mary being many more than that. It is clear that if he had accepted those variations (for example, *Quot tibi sunt dotes virgo, tot sidera coelo*), then by putting *tot* in place of *quot*, and the other way round, just as many other verses, namely, 1022, would have emerged. In the course of his Preface, p. 12, Puteanus even noted that not only the stars but also her endowments belong to the heavens, in order that we may understand them to be heavenly; for example,

> Tot tibi sunt coelo dotes, quot sidera virgo.[287]

Besides, what contributes to variation, is that the last syllables in *Virgo* and *Tibi*, as ambiguous quantities, permit themselves to be shortened and lengthened, an artifice that we shall also encounter below in the singular verse of Daum.[288]

---

[285] "You have so many endowments, Virgin, as there are stars in heaven". The verse, according to Hoyt Hudson, would become "perhaps the most famous Protean line in literature ( … )". Cf. Hudson (1966: 155). See also notes 232, 268 and 298.

[286] Cf. Puteanus (1617: 13–50). Leibniz's (1666) text erroneously refers to pp. 3–50: "à pag. 3. usque ad 50".

[287] Cf. Puteanus (1617: 12). Note that the verse "You have so many endowments, Virgin, as there are stars in heaven" is here transposed into "You, Virgin, have so many endowments in heaven, as there are stars".

[288] See § 31 below in this chapter.

20. Meminit porro Thaumatum suorum et Protei Bauhusiani aliquoties Puteanus in apparatus Epistolarum cent. I. ep. 49 et 57 ad Gisbertum Bauhusium Bernardi Patrem; add. et ep. 51, 52, 53, 56 ibid. Editionem autem harum Epistolarum habeo in 12 Amstelodami anno 1647, nam in editione Epistolarum in 4to, quia jam anno 1612 prodiit, frustra quaeres.

21. Caeterum Job. Bapt. Ricciol. Almag. nov. P. I. lib. 6, c. 6, schol. 1, f. 413 peccato μνημονικῷ Versus Bauhusiani Puteanum autorem praedicavit his verbis: *quoniam vero vetus erat opinio a Ptolemaeo usque propagata, stellas omnes esse 1022, Erycius Puteanus pietatis et[225]ingenii sui monumentum posteris reliquit, illo artificiosissimo carmine, Tot tibi etc.,* qui tamen non autor sed commentator, commendatorque est.

22. Denique similem prorsus versum in Ovidio, levissima mutatione observavimus hunc Metam. XII. fab. *7,* v. 594:

> Det mihi se, faxo triplici quid cuspide possim
> Sentiat etc.

Is talis fiet :

> Det mihi se faxo trina quid cuspide possim.

Nam etiam ultima in mihi et faxo anceps est.

23. Extat in eodem genere Georg. Kleppisi nostratis Poëtae laureati versus hic:

> Dant tria jam Dresdre, ceu sol dat, lumina lucem.

20. Puteanus further recalls his *Wonders of Piety* and Bauhuysen's Protean verse several times in his collection of *Epistles*, Century I, Epistles 49 and 57 to Gisbert Bauhuysen, Father of Bernard Bauhuysen, and in addition, Epistles 51, 52, 53, and 56 in the same work. I refer you to the edition of the *Epistles* published in Amsterdam, 1647, as you will now look in vain for the edition of the *Epistles* in quarto, which came out in 1612.[289]

21. Next, Giovanni Battista Riccioli, in his *New Almagest*, Part 1, Book 6, Chapter 6, Scholium 1, folio 413, described Puteanus, by a slip of memory, as the author of Bauhuysen's verse in these words: .... *and since the prevalent opinion of the ancients from Ptolemy onwards was that there are altogether 1022 stars, Erycius Puteanus left us a monument of his piety and great intellect in that most ingenious poem, Tot tibi etc.* However, he was not the author, only a commentator and eulogiser.[290]

22. Finally, I have noticed this verse in Ovid's *Metamorphoses*, XII, Fable 7, v. 594, which, with a very slight change, is quite similar:

Det mihi se, faxo triplici quid cuspide possim Sentiat etc.[291]

It can be changed into this:

Det mihi se, faxo trina quid cuspide possim.

For the last foot can also be either 'mihi' or 'faxo'.

23. [Gregor][292] Kleppis, our native Poet laureate, has left us this verse of the same kind:

Dant tria iam Dresdae, ceu sol dat, lumina lucem.[293]

---

[289] Erycius Puteanus's four books *Epistolarum Selectarum Apparatus Miscellaneus, et Novus* were published in a bound edition as Puteanus (1647), all with separate page numbers and separate title pages dated '1646'. The term *apparatus* here refers to the collection of letters as such. Leibniz's references are to Book 1, pp. 62–4, 65–6, 67, 67–8, and 70–1, respectively. Meanwhile, the 1612 edition published by Jean-Christophe Flavius has not been lost. Cf. Puteanus (1612–13).

[290] Cf. above, note 232.

[291] Ovid, *Metamorphosis* XII, 594: "Let he [i.e., Achilles] but come within my reach. [I'll make him feel] what I can do with my three-forked spear". The words are Neptune's, the translation from Ovid (1984: 223).

[292] The 1666 original has "Georg".

[293] "Three lights now give to Dresden as the Sun gives light".

cujus variationes peculiari libro enumeravit 1617: occasionem dedere tres soles qui anno 1617 in coelo fulsere, quo tempore Dresdae convenerant tres soles terrestres ex Austriaca domo: Matthias Imperator, Ferdinandus Rex Bohemiae, et Maximilia[75]nus Archidux, supremus ordinis Teutonici Magister. Libellum illis dedicatum titulo Protei Poëtici eodem anno edidit, quem variationum numerus signat.

24. Omnino vero plures sunt variationes quam 1617. Quod ipse tacite confitetur autor dum in fine inter Errata ita se praemunit: fieri potuisse, ut in tanta multitudine aliquem bis posuerit, supplendis igitur lacunis novos aliquot ponit, quos certus sit nondum habuisse. Nos ut aliquam praxin proximorum problematum exhibeamus, Variationes omnes utiles computabimus. Id sic fiet, si inveniemus omnes inutiles. Capita variationum expressimus notis quantitatis, sic tamen ut pro pluribus transpositis unum assumserimus, v. g. – –. –. –. ∪ ∪. etiam continet hoc: –. – –. –. ∪ ∪ etc. Punctis designamus et includimus unam vocem.

25. Summa omnium Variationum utilium et inutilium            362880.
Catalogus Variationum inutilium:

    l. ∪ ∪. v.g. *tria* dant jam Dresdae ceu sol dat lumina lucem.   40320

    2. – –. ∪ ∪. *Dresdae tria* dant jam ceu sol etc.   10080

    3. –. –. ∪ ∪. *dant jam tria.*   14400

    4. – –. –. –. ∪ ∪. *Dresdae dant jam tria.*   28800

    5. – –. – –. ∪ ∪. *Dresdae lucem tria.*   1440

    6. –. –. –. –. ∪ ∪. Dant jam ceu sol tria.   2880

    7. – –. – –. –. –. ∪ ∪. *Dresdae lucem ceu sol tria.*   28800

    8. – –. –. –. –. –. ∪ ∪. *Dresdae dant jam ceu sol tria.*   7200

    9. – –. – –. –. –. –. –. ∪ ∪. *Dresdae lucem dant jam ceu sol tria.*   7200

    10. in fine ∪ ∪. v. g. Tria   40320

He calculated the number of its variations at 1617 in his book on the subject, which celebrated the occasion when in the year 1617 three Suns shone in the heavens, on which date there had been a conjunction in Dresden of those three earthly Suns of the House of Austria: Emperor Mathias, Ferdinand, King of Bohemia, and Archduke Maximilian, Grand Master of the Teutonic Order. Under the title of *Poetic Proteus*, he published the book dedicated to them in the selfsame year indicated by the number of variations.[294]

24. But in truth there are far more variations than 1617, something that the author himself concedes when at the end, among the Errata, he defends himself by saying that among so many verses he might have included a verse twice over, and so includes some new ones (which he is certain that he has not yet mentioned) in order to make up for some of the deficiencies.[295] To show how the last few Problems work in practice, I shall calculate all the useful variations, which can be done by finding all the useless ones. I have denoted the heads of the variations by means of marks of quantity, but on the understanding that to represent several transposed variations I shall take only one. For example, – –. –. –. UU also includes this, –. – –. –. UU, and so on. I indicate and enclose each word with points.

25. The total of all the variations, useful and useless                     362880

A list of the useless Variations:

| | |
|---|---:|
| 1. UU. For example, *tria dant iam Dresdae ceu sol dant lumina lucem* | 40320 |
| 2. – –. U U. *Dresdae tria dant iam ceu sol etc.* | 10080 |
| 3. –. –. U U. *dant iam tria* | 14400 |
| 4. – –. –. –. U U. *Dresdae dant iam tria* | 28800 |
| 5. – –. – –. U U. *Dresdae lucem tria* | 1440 |
| 6. –. –. –. –. U U. *dant iam ceu sol tria* | 2880 |
| 7. – –. – –. –. –. U U. *Dresdae lucem ceu sol tria* | 28800 |
| 8. – –. –. –. –. –. U U. *Dresdae dant iam ceu sol tria* | 7200 |
| 9. – –. – –. –. –. –. –. U U. *Dresdae lucem dant iam ceu sol tria* | 7200 |
| 10. At the end, UU. For example, *tria* | 40320 |

[294] To commemorate the gathering of the three princes, Imperial Poet Laureate Gregor Kleppis (fl. *c.* 1620) indeed published a book simply containing no less than 1617 variations of the verse. Cf. Kleppisius (1617) and, on Kleppis, *Jöcher-FIII*, cols 488–9.

[295] The 'Ad Spectatorem', right at the end of Kleppis's book, is accordingly a list of further additions rather than the list of *errata* it looks like.

[*226*] 26. Summa Variationum ob vocem *Tria* inutilium, quae exacte constituit
dimidium summae Variationum possibilium      181440

    11. ab initio: –. – ∪ ∪. *dant lumina.*      18000

    12. –. – – –. – ∪ ∪. *dant Dresdae lumina.*      9600

    [76]13. –. –. –. – ∪ ∪. *dant jam ceu lumina.*      4320

    14. –. –. –. –. –. – ∪ ∪. *dant jam ceu sol dat lumina.*      240

    15. –. – – –. – – –. – ∪ ∪. *dant Dresdae lucem lumina.*      2160

    16. –. –. –. – – –. – ∪ ∪. *dant jam ceu lucem lumina.*      5700

    17. –. –. –. –. –. – – –. – ∪ ∪. *dant ceu jam sol dat lucem lumina.*      0

    18. –. –. –. – – –. – – –. – ∪ ∪. *dant ceu jam Dresdae lucem lumina.*      1200

    19. –. –. –. –. –. – – –. – – –. – ∪ ∪. *dant ceu jam sol dat lucem*
        *Dresdae lumina.*      0

    20. fine – ∪ ∪. v. g. *lumina.*      11620

27. Summa Variationum ob solam vocem: *lumina* inutilium      52900

    21. ubicunque: – ∪ ∪. ∪ ∪. *lumina tria.*      40320

    22. – ∪ ∪. –. –. ∪ ∪. *lumina Dresdae tria.*      14440

    23. – ∪ ∪. –. –. ∪ ∪. *lumina ceu jam tria.*      4800

    24. – ∪ ∪. –. –. –. –. ∪ ∪. *lumina ceu jam sol dat tria.*      1440

    25. – ∪ ∪. – – –. – – –. ∪ ∪. *lumina Dresdae lucem tria.*      480

    26. – ∪ ∪. –. –. – – –. ∪ ∪. *lumina ceu jam Dresdae tria.*      4800

    27. – ∪ ∪. –. –. – – –. – – –. ∪ ∪. *lumina ceu jam Dresdae lucem tria.*      4080

    28. – ∪ ∪. –. –. –. –. – – –. ∪ ∪. *lumina ceu jam dat sol lucem tria.*      532

    29. – ∪ ∪. –. –. –. –. – – –. – – –. ∪ ∪. *lumina ceu jam dat sol*
        *lucem Dresdae tria.*      2978

28. Summa Var. inut. ob complicationem *Lumina* et *Tria*, illo
            praeposito      59870

    30. –. ∪ ∪. –. –. ∪ ∪. *dant tria jam lumina.*      2400

    31. –. ∪ ∪. – –. –. ∪ ∪. *Dant tria jam Dresdae lumina.*      3840

    32. –. ∪ ∪. –. –. –. – ∪ ∪. *ceu sol.*      1440

    33. –. ∪ ∪. –. –. –. – – –. – ∪ ∪. *dant tria jam ceu sol lucem*
        *lumina.*      5760

    34. –. ∪ ∪. –. –. –. – – –. – – –. – ∪ ∪. *dant tria jam ceu sol lucem*      9360
        *Dresdae lumina.*      [*227*]

26. The number of Variations useless on account of the word *tria*, which
    constitutes exactly half of the total of possible variations     181440
    11. From the beginning: −. − ∪ ∪. *dant lumina*     18000
    12. −. − −. − ∪ ∪. *dant Dresdae lumina*     9600
    13. −. −. −. − ∪ ∪. *dant iam ceu lumina*     4320
    14. −. −. −. −. −. − ∪ ∪. *dant iam ceu sol dat lumina*     240
    15. −. − −. − −. − ∪ ∪. *dant Dresdae lucem lumina*     2160
    16. −. −. −. − −. − ∪ ∪. *dant iam ceu lucem lumina*     5760
    17. −. −. −. −. −. − −. − ∪ ∪. *dant ceu iam sol dat lucem lumina*     0
    18. −. −. −. − −. − −. − ∪ ∪. *dant ceu iam Dresdae lucem lumina*     1200
    19. −. −. −. −. −. − −. − −. − ∪ ∪. *dant ceu iam sol dat lucem*     0
        *Dresdae lumina*
    20. At the end, − ∪ ∪. For example, *lumina*     11620
27. The number of Variations useless on account of the word
    *lumina*     52900
    21. Everywhere: − ∪ ∪. ∪ ∪. *lumina tria*     40320
    22. − ∪ ∪.− −. ∪ ∪. *lumina Dresdae tria*     14440
    23. − ∪ ∪. −. −. ∪ ∪. *lumina ceu iam tria*     4800
    24. − ∪ ∪. . −. −. −. ∪ ∪. *lumina ceu iam sol dat tria*     1440
    25. − ∪ ∪.− −. − −. ∪ ∪. *lumina Dresdae lucem tria*     480
    26. − ∪ ∪. −. −. − −. ∪ ∪. *lumina ceu iam Dresdae tria*     4800
    27. − ∪ ∪. −. −. − −. − −. ∪ ∪. *lumina ceu iam Dresdae lucem tria*     4080
    28. − ∪ ∪. −. −. −. −. − −. ∪ ∪. *lumina ceu iam dat sol lucem tria*     532
    29. − ∪ ∪. −. −. −. −. − −.[− −.]∪ ∪. *lumina ceu iam dat sol*
        *lucem Dresdae tria*     2978
28. The number of Variations useless on account of the complication
    *lumina* and *tria*, with the former placed first     59870
    30. −.∪ ∪. −. −. − ∪ ∪. *dant tria iam lumina*     2400
    31. −.∪ ∪. −. − −. − ∪ ∪. *dant tria iam Dresdae lumina*     3840
    32. −.∪ ∪. −. −. −. − ∪ ∪. *dant tria iam ceu sol lumina*     1440
    33. −.∪ ∪. −. −. −. − −. − ∪ ∪. *dant tria iam ceu sol lucem lumina*     5760
    34. −.∪ ∪. −. −. −. − −. − −. − ∪ ∪. *dant tria iam ceu sol lucem*     9360
        *Dresdae lumina*

[77]Summa Var. inut. ob complic. *Tria* et *Lumina*, illo praeposito

|  | |
|---|---:|
| | 22800 |
| | 59870 |
| | 52900 |
| | 181440 |
| Summa summarum Var. inut. | 317010 |
| subtrahatur de summa Universali | 362880 |

Remanet: Summa utilium Variationum versus Kleppisi
admissis spondaicis                                                45870
Spondaicos reliquimus ne laborem computandi augeremus, quot tamen
inter omnes variationes utiles et inutiles existant spondaici, sic invenio:

| | |
|---|---:|
| 1. si in fine ponitur –. – –. v. g. dant lucem. | 100800 |
| 2. – –. – –. v.g. Dresdae lucem. | 10080 |
| 3. –. –. –. v.g. dant ceu sol. | 43200 |
| Summa omnium spondaicorum util. et inut. | 154080 |

Extat praeterea versus nobilissimi herois Caroli a Goldstein:

Ars non est tales bene structos scribere versus,

in arte sibi neganda artificiosus, qui 1644 variationes continere dicitur.
Aemulatione horum, Kleppisi inprimis, prodiit Henr. Reimerus Lüne-
burgensis, Scholae Patriae ad D. Johannis Collega, Proteo instructus tali:

Da pie Chrlste Vrbl bona paX slt teMpore nostro.

The number of Variations useless on account of the complication

| | |
|---|---:|
| *tria* and *lumina*, with the former placed first | 22800 |
| | 59870 |
| | 52900 |
| | 181440 |
| The grand total of useless Variations | 317010 |
| Subtract from the Global total | 362880 |

29. There remains the number of useful Variations of Kleppis's
    verse with spondaic verses admitted          45870

I have left the spondaic verses in order not to increase the
labour of calculation. But I find the number of spondaic
verses that may exist among the useful and useless
variations as follows:

| | |
|---|---:|
| 1. If a spondee is placed last: –. – –. For example, *dant lucem* | 100800 |
| 2. – –. – –. For example, *Dresdae lucem* | 10080 |
| 3. –. –. –. For example, *dant ceu sol* | 43200 |
| The total of all the spondaic verses, useful and useless | 154080 |

30. There has come down to us also a verse by the most noble hero Carolus
    à Goldstein.[296]

Ars non est tales bene structos scribere versus,[297]

ingenious in the art that it denies to itself, and which is said to contain
1644 variations. In emulation of these examples, especially those of
Kleppis, Heinrich Reimarus of Lüneburg, commissioned by a Colleague
in the Elementary School at the Hospital of St John, came forward with
this Protean verse:

Da pIe ChrIste VrbI bona paX sIt teMpore nostro,

---

[296] Carolus à Goldstein may have been the military man Carl Goldstein (1570–1628), who
waged war against the Swedes in Polish and Lithuanian service, later became courtier of
Christian II, Elector of Saxony, and would end his life in the service of the Principality of
Braunschweig-Wolfenbüttel during the Thirty Years' War. Cf. *Zedler* 11, col. 144. The verse
"Carolus Goldsteins / 1644. numeris absolutis" occurs in Ebel (1623) *Praeloquium ad Lectorem*,
§ VIII, s.p.

[297] "The art is not to write verses so well structured".

qui idem annum 1619, quo omnes ejus variationes uno libello in 12 Hamburgi edito, inclusae prodierunt, continet.

3I. Laboriosissimus quoque Daumius, vir in omni genere poëmatum exercitatus, ne hoc quidem intentatum voluit a se relinqui. Nihil de ejus copia dicam, qua idem termillies aliter carmine dixit (hic enim non alia verba, sed eorundem verborum alius ordo esse debet), quod in hac sententia: fiat justitia aut pereat mundus, Vertumno poëtico Cygneae anno I646, 8, edito praestitit. Hoc saltem adverto, quod et autori annotatum, in Millenario I num. 219 et 220 versus Proteos esse. Hi sunt igitur: [78]

> v. 2I9. Aut absint vis, fraus, ac jus ades, aut cadat aether.
>
> v. 220. Vis, fraus, lis absint, aequum gerat, aut ruat orbis.

Nacti vero nuper sumus, ipso communicante, alium ejus versum invento sane publice legi digno, quem merito *plus quam Protea* dicas, neque enim in idem tantum, sed alia plurima carminis genera convertitur. Verba enim haec: *O alme* (se. Deus) *mactus Petrus* [*228*] (sponsus) *sit lucro duplo*: varie transposita dant Alcaicos 8, Phaleucios 8, Sapphicos 14, Archilochios 42, in quibus omnibus intercedit elisio. At vero sine elisione facit Pentametros 32, Jambicos senarios tantum 20, Scazontes tantum 22,

which itself contains the year 1619 in which all its variations appeared included in a book published in duodecimo in Hamburg.[298]

31. Nor did the indefatigable Daum, a man practised in every kind of poem, wish to leave this kind of poem unattempted by him. I shall say nothing of his copiousness in saying the same thing three thousand times over in a poem (since in the present work the order must be not other words, but another order of the same words), the same thing as is said in this sentence: *fiat iustitia, aut pereat mundus.* He presented it in his *Poetic Calendar*, published at Zwickau in octavo, 1646.[299] I simply draw attention to the fact (which was also noted by the author) that the verses in Millenary 1, nos. 219 and 220, are Protean verses. Here they are:

v. 219: Aut absint vis, fraus, ac ius [adest], aut cadat aether.
v. 220: Vis, fraus, lis absint, aequum gerat, aut ruat orbis.[300]

32. But I have recently happened upon another verse from the same source, of an ingenuity that deserves to be better known, and which you would be justified in calling *more than Protean*, as it converts not only into the same kind of metre, but into many kinds of metre. For these words, *O alme* (meaning God) *mactus Petrus* (the bridegroom) *sit lucro duplo*, variously transposed, yield 8 Alcaics, 8 Phaleucians, 14 Sapphics, and 42 Archilochians, in all of which elision plays a part. But even without elision it produces 32 Pentameters, as many as 20 six-foot Iambics, as many as 22 Scazons, and 44 Scazons that are simultaneously Iambics (so

---

[298] Reimarus (1619). "DA, PIE CHRISTE, URBI BONA PAX SIT TEMPORE NOSTRO" ("Holy Jesus, grant our city a favourable peace in our days") occurs in 2150 variations. The letters MDCXVIIII stand for the year. A reference to the otherwise virtually unknown author whom Leibniz here refers to as "Reimerus" also occurs in Burton's *Anatomy of Melancholy* 2.2.4.1, as one of the many examples of mental exercises with which to divert the mind and cure it of its melancholy thoughts. Cf. Burton (1994: 94): "[...] and rather then doe nothing, vary a verse a thousand waies with Putean, so torturing his wits, or as Rainnerus of Luneburge, 2150 times in his Proteus poeticus, or Scaliger, Chrysolitus, Cleppissius, and others have in like sort done". Cf. *idem* (2000), vol. 6, p. 409. Note that the example of varying "a verse a thousand waies" that Burton here refers to, is Puteanus's record of the 1022 variations of *Tot tibi sunt dotes virgo, quot sidera coelo*; for which see above, notes 232, 278, and 285.

[299] Daum (1646). The work contains a thousand and then another thousand variations in all types of verse of the idea that justice should prevail. A third collection of a thousand varieties was published separately later that year (1646). On the Zwickau Schoolmaster and polyglot Christian Daum (1612–87), see *Jöcher* II, cols 53–4.

[300] Daum (1646: 17). The lines read: "Either power and fraud are missing and there is law, or heaven will fall upon us" and "Power, fraud and dispute begone, let fairness rule, or the world will collapse".

Scazontes et Jambos simul 44 (et ita Jambos omnes 64, Scazontes omnes 66), si syllabam addas fit Hexameter, v. g.

Fac duplo Petrus lucro sit mactus, o alme!

variabilis versibus 480.

33. Caeterum artificii magna pars in eo consistit, quod plurimae syllabae, ut prima in duplo, Petrus, lucro, sunt ancipites. Elisio autem efficit ut eadem verba, diversa genera carminis syllabis se excedentia, efficiant. Alium jam ante anno 1655 dederat, sed variationum partiorem, nempe Alcaicum hunc:

Faustum alma sponsis da Trias o torum!

convertibilem in Phaleucios 4, Sapphicos 5, Pentametros 8, Archilochios 8, Jambicos senarios 14, Scazontes 16.

Sed jam tempus equum spumantia solvere colla.

34. Si quis tamen prolixitatem nostram damnat, is vereor ne, cum ad praxin ventum erit, idem versa fortuna de brevitate conqueratur.

# FINIS

that the total of Iambics is 64, and of Scazons 66). If you add a syllable, you get a Hexameter. For example,

> Fac duplo Petrus lucro sit mactus, o alme![301]

can be varied 480 times.

33. Moreover, a great deal of the ingenuity consists in the fact that most of the syllables, such as the first in 'duplo', 'Petrus', and 'lucro', are ambiguous.[302] Also, elision makes the same words give rise to different kinds of metre, some of which exceed others in the number of syllables. Just before 1655, he [Daum] had given another verse, though of fewer variations, namely, this Alcaic:

> Faustum alma sponsis da Trias o torum![303]

which is convertible into 4 Phaleucians, 5 Sapphics, 8 Pentameters, 8 Archilochians, 14 six-foot Iambics, and 16 Scazons.

> Sed iam tempus equum spumantia solvere colla.[304]

34. But if anyone condemns me for being prolix, I am afraid that when it comes to practice, he may complain in these changed circumstances that I have been too brief.

# THE END

---

[301] "Make that the sacrificed Peter is good for a twofold advantage, O Lord!"
[302] That is: ambiguous as to their metrical value.
[303] "Grant these fiancés, oh feeding Trinity, a favourable spond!"
[304] Virgil, *Georgica* II, 542, the final line of book 2: *Et iam tempus equum spumantia solvere colla*: "It is time to loosen the necks of our foaming horses".

# Bibliography and Abbreviations

## A Leibniz's Works

A           G. W. Leibniz, *Sämtliche Schriften und Briefe*, herausgegeben von der Deutschen Akademie der Wissenschaften zu Berlin (Darmstadt 1923 ff., Leipzig 1938 ff., Berlin 1950 ff.).

GP          G. W. Leibniz, *Die Philosophischen Schriften*, (ed.) C. I. Gerhardt, 7 vols. Berlin: Weidmannsche Buchhandlung. 1875–90.

GM          G. W. Leibniz, *Mathematische Schriften*, (ed.) C. I. Gerhardt, 7 vols. Berlin: A. Asher/ Halle: H. W. Schmidt. 1849–63.

L           G. W. Leibniz, *Philosophical Papers and Letters*. A Selection Translated and Edited, with an Introduction by Leroy E. Loemker. Dordrecht: D. Reidel (second edition) 1969.

*Essais*    G. W. Leibniz, *Essais scientifiques et philosophiques*. Les articles publiés dans les journaux savants recueillis par Antonio Lamarra et Roberto Palaia, 3 vols. Hildesheim, Zürich, and New York: Georg Olms. 2005.

*Logical Papers*   G. W. Leibniz, *Logical Papers*. A Selection Translated and Edited with an Introduction by G. H. R. Parkinson (Oxford: Clarendon Press) 1966.

*New Essays*   G. W. Leibniz, *New Essays on Human Understanding*, Translated and Edited by Peter Remnant and Jonathan Bennett. Cambridge: Cambridge University Press. 1981.

*Philosophical Texts*   G. W. Leibniz, *Philosophical Texts*. Transl. by Richard Francks and R. S. Woolhouse. With Introduction and Notes by R. S. Woolhouse. Oxford-New York: Oxford University Press, 1998.

*Schriften*   G. W. Leibniz, *Schriften zur Syllogistik*. Herausgegeben, übersetzt und mit Kommentaren versehen von Wolfgang Lenzen. Hamburg: Felix Meiner Verlag, 2019.

## B Primary Sources

Agrippa, Henricus Cornelius (1533). *In Artem Brevem Raymundi Lullij Commentaria*. Cologne: Ioannes Soter.

Alcinous (1993). *The Handbook of Platonism*, (ed.) John Dillon. Oxford: Clarendon Press.

Alsted, Johann Heinrich (1609). *Clavis Artis Lullianae, et verae Logices duos in libellos tributa, Id est, Solida Dilucidatio Artis magnae, generalis, & ultimae, quam Raymundus Lullius invenit, ut esset quarumcunque artium & scientiarum clavigera & serperastra: edita in usum & gratiam eorum, qui impendiò delectantur compendijs, & confusionem sciolorum, qui juventutem fatigant dispensijs.* Strasbourg: Lazari Zetzneri, reprinted (1983) Hildesheim, Zürich, and New York: Georg Olms.

Alsted, Johann Heinrich (1612). *Trigae Canonicae, Quarum Prima est Dilucida Artis Mnemologicae, vulgò Memorativae, traditae à Cicerone, Quintiliano, aliisque oratoribus quà priscis, quà recentibus, explicatio & applicatio. Secunda, est Artis Lullianae, A Multis neglecta & nescio quo edicto proscripta, architectura, & usus locupletissimus. Tertia, est Artis Oratoriae Novum magisterium, quo continetur utilis introductio ad copiam rerum camparandam per tres rotas sive circulos Generis demonstrativi, Deliberativi, & Iudicialis: itemque ad comparandam copiam verborum per Triangulum aliasque figuras.* Frankfurt: Wolfgang Richter.

Andreae, Ioannes (1517). *Arbor Consanguineitatis*, in: Bartholus [de Sassoferrato], Joannes Andree, *et al.*, *In utriusque iuris libros introductorium: Modus legendi abbreviaturas in utroque iure. Tractatus iudiciorum Bartholi. Tractatus renunciationum beneficiorum in publicis intrumentis. Processus Sathanae infernalis contra genus humanum. Ars notariatus. Summa Joannis Andreae super secundo Decretalium. Summa Joannis Andreae super quarto Decretalium. Arbor consanguinatis [et] affinitatis Joannis Andreae. [Arbor] cognationis spiritualis. [Arbor] cognationis legalis.* Basel: Adam Petrus de Lange.

Aristotle (1984). *The Complete Works of Aristotle*, The Revised English Translation, (ed.) Jonathan Barnes, Bollingen Series LXXI. Princeton, NJ: Princeton University Press.

Augustinus, Aurelius (1955). *De Civitate Dei*, in *Opera*, XIV, 2, Corpus Christianorum Series Latina XLVIII. Turnhout: Brepols.

Bacon, Francis (1605). *The Twoo Bookes of Francis Bacon. Of the Proficience and Advancement of Learning, Divine and Humane* London: [Thomas Purfoot and Thomas Creede] for Henrie Tomes.

Bacon, Francis (1857). *De Augmentis scientiarum, The Works of Francis Bacon*, ed. James Spedding, Robert Leslie Ellis, and Douglas Denon Heath, vol. 1. London: Longman, et al., reprinted (1962), Stuttgart-Bad Canstatt: Frommann-Holzboog.

Bacon, Francis (1858). 'Of the Dignity and Advancement of Learning', in *The Works of Francis Bacon*, (eds) James Spedding, Robert Leslie Ellis, and Douglas Denon Heath, vol. 4. London: Longman, et al., reprinted (1962), Stuttgart-Bad Canstatt: Frommann-Holzboog.

Bacon, Francis (1861). *The Works*. London: Longman, et al., reprinted (1963), Stuttgart-Bad Canstatt: Frommann-Holzboog.

Barrow, Isaac (1655). *Euclidis Elementorum libri XV, breviter demonstrati*. Cambridge: Coll. Trin. Soc.

Barthius, Caspar (1600). *De latina lingua et scriptoribus latinis commentatio*, in Dilherr (1660).

Bauhusius, Bernardus (1616). *Epigrammatum Selectorum Libri V*. Antwerp: Plantijn.

Becher, Johann Joachim (1661). *Clavis convenientiae linguarum. Character, Pro Notitia Linguarum Universali: Inventum Stenographicum hactenus inauditum, quo quilibet suam Legendo vernaculam diversas imò omnes Linguas, unius etiam diei informatione, explicare ac intelligere potest.* Frankfurt: Ammonius and Serlinus.

Bede (1991). *Libri II De Arte Metrica et De Schematibus et Tropis / The Art of Poetry and Rhetoric* Calvin B. Kendall (ed.), Bibliotheca Germanica Series Nova, vol. 2. Saarbrücken: AQ-Verlag.

Bisterfeld, Johann Heinrich (1657a). *Phosphorus Catholicus; Seu Artis meditandi Epitome, Ex rerum naturâ, bonorumque authorum analysi educta, & per praecipua mentis munia deducta. Cui subjunctum, Consilium de Studiis foeliciter instituendis*, in Johann Heinrich Bisterfeld, *Elementorum Logicorum Libri tres: ad praxin exercendam apprimè utiles. Atque ita instituti, ut Tyro, trimestri spatio, fundamenta Logices, cum fructu jacere possit.* Leiden: H. Verbiest.

Bisterfeld, Johann Heinrich (1657b). *Philosophiae Primae Seminarium.* Leiden: Daniel and Abraham Gaasbeck.

Bisterfeld, Johann Heinrich (1658). *Elementorum logicorum libri tres.* Accedit *Phosphorus Catholicus, seu artis meditandi epitome.* Leiden: H. Verbiest.

Bruno, Giordano (1890). *Opera latine conscripta,* (eds) F. Tocco and H. Vitelli. Florence: Le Monnier heirs. Reprinted (1962) Stuttgart-Bad Cannstatt: Frommann-Holzboog.

Boxhornus, Marcus Zuerius (1668). *Institutiones Politicae cum Commentariis ejus dem, et Observationibus Georgii Hornii.* Amsterdam: C. Crommelin.

Budowitz a Budowa (1617). *Epistola Quam ad Generosum et Illustrem Dominum Wenceslaum Budowetz, Baronem à Budowa, Monachogrecii et Zasadeci Dominum, Sacrae Caesareae Maiestatis Consiliarum, De Circulo Horologii Lunaris et Solaris nuperrimè ab isto Typis Wechelianis Hanoviae edito, modestè, & solius veritatis divinae vindicandae gratiâ, exaravit Matthias Hoe ab Hoenegg (etc.).* Leipzig: Lamberg and Glück.

Burton, Robert (1994). *The Anatomy of Melancholy,* ed. Thomas C. Faulkner, Nicolas K. Kiessling, Rhonda L. Blair, J. B. Bamborough, and Martin Dodsworth, Vol. 2. Oxford: Clarendon Press.

Bzowski, Abraham ed. (1617). *Historiae Ecclesiasticae ex Illustriss. Cesaris Baronii ( . . . ) Annalibus, Aliorumque Virorum Illvstrorum Ecclesiasticis Historicisque monumentis.* Cologne: Antonius Boetzerus.

Caramuel, Juan (1681). *Critica philosophica artium scholasticarum cursum exhibens.* Vigevano: C. Conrada.

Carbo, Ludovicus (1597). *Introductionis in Logicam, Sive Totius Logicae Compendii absolutissimi Lib. VI.* Venice: Baptist and Jo. Sessa.

Cardanus, Hieronimus C. (1539). *Practica arithmetice, et mensurandi singularis.* Milan: A. Castellioneus and B. Caluscus.

Carpzovius, Benedictus (1638). *Iurisprudentia forensis Romano-Saxonica secundum ordinem constitutionum D. Augusti electoris Saxoniæ in part. IV. divisa: rerum et quæstionum in foro, præsertim Saxonico, ut plurimum occurrentium et in dicasterio septemvirali Saxonico celeberrimo, quod vulgo Scabinatum Lipsiensem appellitant, ex iure civili, Romano, imperiali, canonico, Saxonico & provinciali tractatarum ac decisarum* : *definitiones iudiciales succinctas et nervosas, placitisque & sententiis dominorum Scabinorum corroboratas exhibens.* Frankfurt: Clemens Schleichius.

Champier, Claude (1579). *Hippocratis aphorismi ex nova Claudii Campensii* (...) *interpretatione.* Lyon: Cl. Ravot.

Cicero, Marcus Tullius (1979). *De Natura Deorum / Academica,* (ed.) H. Rackham, *Cicero in Twenty-Eight Volumes,* vol. XIX, Loeb Classical Library, vol. 268. Cambridge, MA, and London: Harvard University Press and Heinemann.

Clavius, Christophorus (1585). *In Sphaeram Joannis de Sacro Bosco Commentarius.* Rome: Domenico Basa.

Clavius, Christophorus (1999). *In Sphaeram Ioannis de Sacro Bosco Commentarius,* (ed.) Eberhard Knobloch. Hildesheim, Zürich, and New York: Olms-Weidmann.

Daum, Christian (1646). *Vertumnus Poeticus ad Scitum illud Imperatorium Fiat Justitia, aut Pereat Mundus, bis millies transformatus, Justitiae Cultum salutarem, Neglectum πανόλεθρον, Varijs Versuum Generibus detinens, Exercitio tralatitio.* Zwickau: Melchior Göpner.

David, Iohannes (1601). *Veridicus Christianus.* Antwerp: Plantijn.

Descartes, René (1637). *Discours de la Methode Pour bien conduire sa raison, & chercher la verité dans les sciences. Plus La Dioptrique. Les Meteores. La Geometrie. Qui sont des essais de cette Methode.* Leiden: I. Maire. (AT: VI).

De Soto, Domingo (1554). *Summulae.* Salamanca: Andreas a Portomaris.

Digby, Kenelm (1655). *Demonstratio immortalitatis animae rationalis, sive Tractatus duo philosophici, in quorum priori natura et operationes corporum, in posteriori vero, natura animae rationalis, ad evincendam illius immortalitatem, explicantur.* Paris: Fredericus Leonard.

Dilherr, Johann Michael (1642). *Oratio, de recta liberorum educatione.* Nuremberg: W. Endter.

Dilherr, Johann Michael (1660). *Apparatus Philologiae, Sive Justi Lipsii Orthographia.* Nuremberg: Johann Tauber.

Drexel, Hieremia S. J. (1631). *Orbis Phaëton: Hoc est De Universis Vitiis Linguae.* Cologne: Cornelis ab Egmond.

Duns Scotus, Johannes (1639). *Opera omnia,* (ed.) Lucas Wadding et al. Lyon: L. Durand. Reprinted (1968). Hildesheim: Georg Olms.

Duns Scotus, Johannes (1891). *Opera omnia editio nova.* Paris: L. Vivès.

Ebel, Johann Philip (1623). *Epigrammata palindroma, qua antrorsum ac retrorsum eodem plane sensu legi possunt.* Ulm: Eberken.

Eggefeld, Johann Chrisostomus (1661). *Nova detecta Veritas: seu In Veterem Ratiocinandi Artem ab Aristotele Peripatus Authore inventam nova Methodo instituta Philosophica Animadversio pro invenienda breviori, et veriori Philophandi doctrina.* [s.l.]: A. Verus.

Etten, Hendrik van (1624). *Recreations mathématiques: composées de plusieurs problemes plaisans & facetieux d'arithmetique, geometrie, astrologie, optique, perspective, mechanique, chymie, & d'autres rares & curieux secrets: plusieurs desquels n'ont iamais esté imprimez... le tout representé par figures.* Pont-à-Mousson: Jean Appier Hanzelet.

Etten, Hendrik van (1633). *Mathematicall Recreations. Or a Collection of sundrie Problemes, extracted out of the Ancient and Moderne Philosophers, as secrets in nature, and experiments in Arithmeticke, Geometrie, Cosmographie, Horologiographie, Astronomie, Navigation, Musick, Opticks, Architecture, Staticke, Mechanicks, Chimestrie, Waterworkes, Fireworks, &c.* London: T. Cotes and R. Hawkins.

Fabry, Honoré (1646). *Philosophiae tomus primus, qui complectitur scientiarum Methodum sex libris explicatam: Logicam Analyticam, duodecim Libris demonstratam, et aliquot Controversias logicas, breviter disputatas. Auctore Petro Mosnerio Doctore Medico. Cuncta Excerpta ex Praelectionibus R. P. Hon. Fabry Soc. Iesu..* Leiden: J. Champion.

*Fama Fraternitatis* (1614). *Deß Löblichen Ordens des Rosencreutzes, an all Gelehrte und Häupter Europae geschrieben.* Kassel: Wilhelm Wessell.

*Fama Fraternitatis* (1998). *Het oudste manifest der Rozenkruizers Broederschap, bewerkt aan de hand van teruggevonden manuscripten, ontstaan vóór de eerste druk van 1614/ Fama Fraternitatis: Das Urmanifest der Rosenkreuzer Bruderschaft zum ersten Mal nach den Manuskripten bearbeitet, die vor dem Erstdruck von 1614 entstanden sind,* (ed.) Carlos Gilly and Pleun van der Kooij. Haarlem: Rozenkruis Pers.

Fortunatus, Venantius Honorius Clementius (1627). *Hymnus de resurrectione Domini,* (ed.) August Buchner. Wittenberg: Haeredes Selfischiani.

Fortunatus, Venantius Honorius Clementius (1881) *Opera Poetica,* (ed.) Friedrich Leo. *Monumenta Germaniae Historica* IV–I. Berlin: Weidmann. Reprinted (1991). Munich: Monumenta Germaniae Historica.

Frey, Jano Caecilio (1628). *Via ad divas scientias, artesque, linguarum notitiam, sermones extemporaneos nova & expeditissima.* Paris: D. Langlois.

Fulgentius (1997). *Commento all'Eneide,* (ed.) Fabio Rosa, Bibliotheca Medievale 5. Milan and Trento: Luni.

Galenus, Claudius (1821). *Opera Omnia,* in *Medicorum Graecorum Opera Quae Exstant,* (ed.) Karl Gottlob Kühn. Leipzig: K. Knobloch.

Galenus, Claudius (1996). *On the Elements According to Hippocrates,* (ed.) Phillip de Lacy, in *Corpus Medicorum Graecorum* V 1, 2. Berlin: Akademie Verlag.

Gassendi, Pierre (1649). *Animadversiones in decimum librum Diogenis Laertii, Qui est De Vita, Moribus, Placitisque Epicuri.* Lyon: Guillelmus Barbier. Reprinted (1987) *Greek and Roman Philosophy,* vol. 19. New York and London: Garland.

Gassendi, Pierre (1658). *Opera Omnia,* in 6 vols. Lyon: Anisson and Devenet. Reprinted (1964) (ed.) Tullio Gregory. Stuttgart-Bad Cannstatt: Frommann-Holzboog.

Grégoire, Pierre [Gregorius Tholusanus] (1575a). *Syntaxes Artis Mirabilis, in Libros Septem Digestae. Per quas de omni re proposita, multis & propè infinitis rationibus disputari, aut tractari, omniúmque summaria cognitio haberi potest.* Lyon: Ant. Gryphius.

Grégoire, Pierre (1575b). *Commentaria in Prolegomena Syntaxeon Artis Mirabilis.* Lyon: Ant. Gryphius.

Grégoire, Pierre (1585). *Syntaxeon Artis Mirabilis, Alter Tomus.* Lyon: Ant. Gryphius.

Guido d'Arezzo (1999). *Regule Rithmicae, Prologus in Antiphonarium, and Epistola ad Michaelem: A Critical Text and Translation,* (ed.) Dolores Pesce. Ottawa: The Institute of Mediaeval Music.

Guthrie, Kenneth Sylvan (ed.) (1987). *The Pythagorean Sourcebook and Library*. Grand Rapids, MI: Phanes Press.

Harsdörffer, Georg Philipp, and Daniel Schwenter (1990). *Deliciae physico-mathematicae, oder, Mathematische und philosophische Erquickstunden*, in 3 vols, (ed.) Jörg Jochen Berns. Frankfurt: Keip.

Henischius, Georg (1609). *Arithmetica perfecta, et demonstrata, doctrinam de numero triplici, vulgari, cossico & astronomico nova methodo per propositiones explicatam*. Augsburg: D. Franck.

Herodotus (1998). *The Histories*. Oxford: Oxford University Press.

Hobbes, Thomas (1839a). *Elementorum Philosophiae Sectio Prima De Corpore*, in *Opera Philosophica Omnia*, (ed.) William Molesworth, vol. 1. London: John Bohn.

Hobbes, Thomas (1839b). *Elementorum Philosophiae Sectio III De Cive*, in *Opera Philosophica Omnia*, (ed.) William Molesworth, vol. 1. London: John Bohn.

Hobbes, Thomas (1839c). *Elements of Philosophy: The First Section, Concerning Body*, in *The English Works of Thomas Hobbes*, (ed.) William Molesworth, vol. 1. London: John Bohn.

Hobbes, Thomas (1841). *Philosophical Rudiments Concerning Government and Society*, in *The English Works of Thomas Hobbes*, ed. William Molesworth, vol. 2. London: John Bohn.

Hobbes, Thomas (1999). *De Corpore*, (ed.) Karl Schuhmann. Paris: Vrin.

Hoë von Hohenegg (1617). *Epistola Quam ad Generosum et Illustrem Dominum Wenceslaum Budowetz, Baronem à Budowa, Monachogrecii et Zasadeci Dominum, Sacrae Caesareae Maiestatis Consiliarum, De Circulo Horologii Lunaris et Solaris nuperrimè ab isto Typis Wechelianis Hanoviae edito, modestè, & solius veritatis divinae vindicandae gratiâ, exaravit Matthias Hoe ab Hoenegg (etc.)*. Leipzig: Lambergius and Glück.

Hospinianus, Johannes [Wirth, Johannes] (1543). *Quaestionum Dialecticarum Libri Sex*. Basel: [s.n.].

Hospinianus, Johannes [Wirth, Johannes] (1560). *Non esse tantum triginta sex bonos malosque categorici syllogismi modos, ut Aristoteles cum interpretibus docuisse videtur: sed quingentos et duodecimo quorum quidem probentur triginta sex, reliqui verò omnes reiiciantur*. Basel: [s.n.].

Hospinianus, Johannes [Wirth, Johannes] (1576). *De Controuersijs Dialecticis Liber*. Basel: Sebastianus Henricpetrus.

Hotman, François (1547). *De gradibus cognationis et affinitatis libri duo*. Paris: Jean Bogard.

Hotman, François (1599). *Disputatio de Gradibus Cognationis, Iurisc. Disputationvm Iuris Civili, vol. 1, in: Idem, Iurisconsulti Operum, Tomus Primus, Iacobus Bongarsius Bodrianus*(ed.) [Genève]: Eustathius Vignon heirs and Iacobus Stoer.

Jerome (1910–18). *Sancti Eusebii Hieronymi Epistulae*, (ed.) Isidorus Hilberg. Vienna and Leipzig: Tempsky and Freytag.

Kepler, Johannes (1619). *Harmonices Mundi libri V*. Linz: Godefredus Tampachus and Ioannes Plancus. Reprinted (1969). Bologna: Forni.

Kepler, Johannes (1997). *The Harmony of the World*, (ed.) E. J. Aiton, A. M. Duncan, and J. V. Field, *Memoirs of the American Philosophical Society*, vol. 209. Philadelphia: American Philosophical Society.

Kircher, Athanasius (1646). *Ars Magna Lucis et Umbrae In decem Libros digesta. Quibus Admirandae Lucis et Umbrae in mundo, atque adeò universa natura, vires effectusque uti nova, ita varia novorum reconditiorumque speciminum exhibitione, ad varios mortalium usus, panduntur.* Rome: Hermannus Scheus and Ludovicus Grignani.

Kircher, Athanasius (1663). *Polygraphia nova et universalis ex combinatoria arte detecta. Quà Quivis etiam Linguarum quantumvis imperitus triplici methodo Prima, Vera & reali, sine ulla latentis Arcani suspicione, manifestè; Secunda, per Technologiam quondam artificiosè dispositam; Tertia, per Steganographiam impenetrabili scribendi genere adornatam, unius vernaculae linguae subsidio, omnibus populis & linguis clam, apertè; obscure, & dilucidè scribere & respondere posse docetur, & demonstratur.* Rome: Varesius.

Kircher, Athanasius (1669). *Ars Magna Sciendi, In XII Libros Digesta, Qua Nova & Universali Methodo Per Artificiosum Combinationum contextum de omni re proposita plurimis & prope infinitis rationibus disputari, omniumque summaria quaedam cognitio comparari potest.* Amsterdam: Johannes Janssonius à Waesberge and the Widow of Elizeus Weyerstraet.

Kleppisius, Gregorius (1617). *Proteus Poeticus Tot Formis Quot Anni jam à Nato Christo Numerantur, MDCXVII Conspiciendus.* Hoc est; *Versus Unicus, In Splendidißimo Dresdae Conventu, Divi Mathhiae, Romanorum Imperatoris &c. Dn. Ferdinandi, Bojohemorum Regis &c. Dn. Maximiliani, Archiducis Austriae &c. Summi Honoris, & Memoriae gratiâ, Millies, Sexenties, & Decies Septies Variatus.* Leipzig: Rehfeld and Grossius and Lanckish.

Lactantius (2003). *Divine Institutes,* (ed.) Anthony Bowen and Peter Garnsey. Liverpool: Liverpool University Press.

Lactantius (2007). *Divinarum Institutionum Libri Septem,* vol. 2 (Books III and IV), (ed.) Eberhard Heck and Antonie Wlosok. Berlin and New York: De Gruyter.

Lansius, Thomas (1620). *F.A.D.W. Consultatio de principatu inter provincias Europae.* Tübingen: Wildius.

Lantz, Johannes (1616). *Institutionum Arithmeticarum Libri Quatuor.* Munich: Henricus.

Laurenberg, Johannes (1640). *Ocium Soranum, Sive Epigrammata, Continentia varias Historias, & res scitu jucundas, ex Graecis Latinisque Scriptoribus depromptas, & exercitationibus Arithmeticis accommodatas.* Copenhagen: Joachim Moltken.

Lavinheta, Bernard de (1523). *Explanatio compendiosaque applicatio artis Raymundi Lulli.* Lyons [s.n.]. Reprinted (1977) (ed.) Erhard-Wolfram Platzeck. Hildesheim: Gerstenberg Verlag.

Lavinheta, Bernard de (1612). *Opera Omnia Quibus tradidit Artis Raymundi Lullii Compendiosam Explicationem, et Eiusdem applicationem ad I. Logica, II. Rhetorica, III. Physica, IV. Mathematica, V. Mechanica, VI. Medica, VII. Metaphysica, VIII. Theologica, IX. Ethica, X. Iuridica, XI. Problematica,* (ed.) Johannes Henricus Alstedius. Cologne: Lazarus Zetner.

Leichner, Eccardus (1652). *De Philosophia Scholaruvm Emendatione Jsagogicon.* Johann Bickner and Paul Michaelis.

Leichner, Eccardus (1664). *Schediasmatum De Principiis Medicis: Sive, De Apodicticâ Scholarum Medicarum Emendatione,* Πεντας *Prima.* Erfurt: Paul Michaels heirs.

Leichner, Eccardus (1669). *Apodictischer Prüfe-Spiegel Wissen- und Gewissenhafter Liebhaber des Christlichen Schul- und allgemeinen Wol-Wesens: Worinne zugleich eine Summarische Abbildung Wahrer und Irriger Logica; auch Physica, Metaphysica und Ethica; Nebst gewiehriger Anzeige wie leicht-müglich die Apodictische Emendation seye.* Erfurt: Kirscher and Paul Michaels heirs.

Leurechon, Jean (1622). *Selectae Propositiones in Tota Sparsim Mathematica Pulcherrimae. Quas in Solemni Festo Sanctorum Ignatii et Xaverii et Anniversaria Collegii Mussipontani Celebritate Literaria Propugnabunt Mathematicarum Auditores.* Pont-à-Mousson: Sébastien Cramoisy.

Lullius, Raymundus (1598). *Opera Quae ad adinventam ab ipso Artem Universalem, Scientiarum Artiumque Omnium Brevi compendio, firmaque memoria appraehendendarum, locupletissimaque vel oratione ex tempore pertractandarum, pertinent. Ut et in eandem quorundam interpretum scripti commentarii.* Strasbourg: Zetzner heirs.

Magnen, Jean Chrisostome (1643). *Democritus Reviviscens: Sive Vita et Philosophia Democriti.* Leiden: Adriaen Wyngaerden.

Maternus, Firmicus (1992). *Mathesis*, vol. 1. Paris: Les Belles Lettres.

Menochio, Giovanni Stefano (1615). *De Arbitrariis Iudicum Quaestionibus et Causis, Libri duo: Quibvs tota fere iuris pars, quae a iudicum arbitrio et potestate pendet, perquam doctè, late, & eleganter pertractantur, magno iusdicentium, docentium & discentium commodo.* Cologne: Antonius Hieratus.

Mersenne, Marin (1625). *La vérité des sciences contre les sceptiques ou Pyrrhoniens.* Paris: T. du Bray. Reprinted (1969). Stuttgart-Bad Cannstatt: Frommann-Holzboog.

Neperus, Ioannes (1617). *Rabdologiae, seu Numerationis per Virgulas libri duo.* Edinburgh: Andreas Hart. Reprinted (1966). Osnabrück: Otto Zeller.

Ocellus Lucanus (1661). *Peri tês tou Pantos Phuseôs / De universi natura*, (ed.) Carolus Emmanuel Vizzanius. Amsterdam: J. Blaeu.

Ocellus Lucanus (1831). *On the Nature of the Universe*, (ed.) Thomas Taylor. London: Bohn, Bohn, and Rodd; (1976) (ed.) Manly P. Hall. Los Angeles, CA: The Philosophical Research Society.

Optatianus Porphyrius (1595). *Panegyricus dictus Constantino Augusto. Ex codice manuscripto Paulli Velseri patricij Aug. Vindelicorum.* Augsburg: Pinus.

Optatianus Porphyrius (1973). *Carmina*, (ed.) G. Polara, in 2 vols. Turin: G.B. Paravia.

Ovid (1984). *Metamorphoses Books IX–XV*, (ed.) Frank Justus Miller and G. P. Goold. Cambridge, MA, and London: Harvard University Press.

Pascal, Blaise (1665). *Traité du triangle arithmétique avec quelques autres petits traitez sur la mesme matière.* Paris: Guillaume Desprez, in Pascal, Blaise (1908). *Oeuvres* vol 3. L. Brunschvicg and P. Boutroux (eds). Paris: Hachette, 433–593.

Patrizi, Francesco (1593). Franciscus Patritius, *Nova de Universis Philosophia Libris quiquaginta comprehensa. In qua Aristotelica methodus non per motum, sed per lucem, & lumina ad primam causam ascenditur. Deinde nova quadam, ac peculiari methodus tota in contemplationem venit diuinitas. Postremo methodus Platonica rerum vniversitas à conditore Deo deducitur.* Venice: Robertus Meiettus.

Pelletier, J. (1557). Iacobus Peletarius Cenomanus, *In Euclidis Elementa Geometrica Demonstrationum Libri sex.* Lyon: Tornaesius and Gazeius.

Plato (1602). *Opera Omnia Quae Exstant*, Marsilio Ficino Interprete. Frankfurt: Cl. Marnius and Ioh. Abrius heirs.

Plato (1992). *Theaetetus*, (ed.) Bernard Williams, M. J. Levett, and Myles Burnyeat. Indianapolis and Cambridge: Hackett.

Puteanus, Erycius (1602). *Musathena, sive Notarum Heptas, Ad Harmonicae lectionis Novum & Facilem usum*. Hanover: Wechalianus, Marnius, and Aubrius heirs.

Puteanus, Erycius (1612–3). *Epistolarum Atticarum Promulsis. Centuria I. & innovata; Epistolarum Fercula secunda. Centuria II; Epistolarum Bellaria. Centuria III. & Nova; Epistolarum Apophoreta. Centuria IV & Recens; Epistolarum Reliquae. Centuria V & Postrema*. Leuven: Jean-Christophe Flavius.

Puteanus, Erycius (1617). *Pietatis thaumata in Bernardi Bauhusii è Societate Jesu Proteum Parthenium, unius libri Versum, vnius Versus Librum, Stellarum numero, sive formis M. XXII. variatum*. Antwerp: Plantin-Moretus.

Puteanus, Erycius (1643). *De Anagrammatismo, Quae Cabalae Pars est, Diatriba, Amoenitatis caussâ scripta, utilitatis edita*. Brussels: Mommaert.

Puteanus, Erycius (1647). *Epistolarum Apparatus*. Amsterdam: Jodocus Janssonius.

Rabanus Maurus (1997). *In honorem Sanctae Crucis*, (ed.) M. Perrin, Corpus Christianorum Continuatio Mediaevalis C. Turnhout: Brepols.

Ramus, Petrus (1543). *Dialecticae institutiones*. Paris: Iacobus Bogardus. Reprinted (1964), Stuttgart-Bad Canstatt: Fromann-Holzboog.

Ramus, Petrus (1556). *Dialecticae libri duo*. Paris: Andreas Wechelus.

Ramus, Petrus (1581). *Scholarvm Dialecticarum, seu animadversionum in Organum Aristotelis, libri xx*, (ed.) Johannes Piscator. Frankfurt: Andreas Wechelus.

Raue, Johannes (1638). *Prior Fundamentalis Controversia pro Logica Novissima*. Rostock: Kilius.

Regius, Johannes (1615). *Commentariorum ac Disputationum Logicarum Libri V: In Quibus & Theoremata Logica accuratè explicantur & quaecunque uspiam inter Logicos tàm veteres, quàm novitios motae extant controversiae, perspicuè & solidè discutiuntur, atque determinantur*. Wittenberg: C. Bergerus and Z. Schurer.

Reimarus, Henricus (1619). *Proteus Poeticus: Hoc est Precatiuncula Metrica Novem Verborum, Annum Christi continentium. Quae, instar Proteï, in bis mille, centum & quinquaginta formas conversae atque hoc turbulento seculo tàm supremis quàm infimis, senibus quàm juvenibus, pro salute & conservanda pace in Choro, Foro & Toro, ad Emanuelem, pacis principem fundenda, offertur*. Hamburg: Mose.

Riccioli, Giovan Battista SJ (1651). *Almagestum novum astronomiam veterum novamque complectens observationibvs aliorum, et propriis novisque theorematibus, problematibus, ac tabulis promotam, in tres tomos distributam quorum argumentum sequens pagina explicabit*. Bologna: Benatius.

Saccheri, Girolamo (1701). *Logica demonstrativa*. Pavia: Carlo Francesco Magri heirs.

Savonarola, Girolamo (1982). *Compendium logicae*, in *Scritti filosofici*, vol. 1. Rome: Angelo Belardetti Editore, 1–208.

Scaliger, Julius Caesar (1540). *De Causis linguae Latinae libri tredecim*. Lyon: S. Gryphius.

Scaliger, Julius Caesar (1561). *Poetices libri septem*. Lyon: Antonius Vincentius. Reprinted (1994) *Poetices libri septem: Sieben Bücher über die Dichtkunst*, Book 1, Ch. 30, (eds) Luc Deitz, Gregor Vogt-Spira, and Manfred Fuhrmann. Stuttgart-Bad Cannstatt: Frommann-Holzboog.

Scaliger, Josephus Justus (1606a). *Animadversiones in Chronologia Eusebii*, in Scaliger, Josephus Justus (1606b).

Scaliger, Josephus Justus (1606b). *Thesaurus Temporum. Eusebii Pamphili Caesareae Palaestinae Episcopi Chronicorum Canonum omnimodae historiae libri duo, interprete Hieronymo, ex fide vetustissimorum Codicum castigati. Item auctores omnes derelicta ab Eusebio, & Hieronymo continuantes*[etc.]. Leiden: Thomas Basson.

Schooten, Frans van (1649). *Geometria, à Renato Des Cartes Anno 1637, Gallicè edita*. Leiden: I. Maire.

Schooten, Frans van (1651). *Principia matheseos universalis seu introductio ad geometriae methodum Renati Des Cartes*. Leiden: Elsevier.

Schott, Caspar SJ (1662). *Physica Curiosa sive Mirabilia Naturae et Artis, libris XII Comprehensa*. Nuremberg: J. A. Endter.

Schott, Caspar SJ (1664). *Technica Curiosa, sive Mirabilia Artis, Libris XII Comprehensa; Quibus varia Experimenta, variaque Technasmata Pneumatica, Hydraulica, Hydrotechnica, Mechanica, Graphica, Cyclometrica, Chronometrica, Automatica, Cabalistica, aliaque Artis arcana ac miracula, rara, curiosa, ingeniosa, magnamque partem nova & antehac inaudita, eruditi Orbis utilitati, delectationi, disceptationique proponuntur*. Nuremberg: Endter/Hertz. Reprinted (1977). Hildesheim and New York: Georg Olms.

Schwenter, Daniel (1636). *Deliciae Physico-Mathematicae oder Mathematische und Philosophische Erquickstunden*. Nuremberg: I. Dümler. [see also: Harsdörffer (1990)].

Schwenter, Daniel (1991). *Deliciae Physico-Mathematicae oder Mathematische und Philosophische Erquickstunden*. Frankfurt: Keip.

Selenus, Gustav (1624). *Cryptomenytices et Cryptographiae Libri IX. In quibus & planißima Stenographiae à Johanne Trithemio, Abbate Spenheymensi & Herbipolensi; admirandi ingenij Viro, magicè & aenigmaticè olim conscriptae, Enodatio traditur*. Lüneburg: [Stern].

Sextus Empiricus (1960). *Against the Physicists*, in *Sextus Empiricus in Four Volumes*, vol. III, *Against the Physicists/ Against the Ethicists*, (ed.) R. G. Bury, Loeb Classical Library, vol. 268. Cambridge, MA, and London: Harvard Unversity Press and Heinemann.

Sturm, Johann Christopher (1661). *Universalia Euclidea, hoc est Liber Quintus Euclidis universalissimis inque omnium entium genere veris demonstrationibus confirmatus*. The Hague: Adrianus Vlacq.

Thomasius, Jakob (1678). *Erotemata Logica pro incipientibus. Accessit pro adultis Processsus disputandi* (...). Leipzig: Frommannus and Colerus. Reprinted in Thomasius (2003).

Thomasius, Jakob (2003). *Gesammelte Schriften*, (ed.) Walter Sparn, vol. 2. Hildesheim, Zürich, and New York: Olms-Weidmann.

Trittenheim, Johann von (1518). Johannes Trithemius, *Poligraphiae Libri sex*. Basel: J. Haselberg.

Ursinus, Benjamin (1623). *Rhabdologia Neperiana. Das ist / Newe / und sehr leichte art durch etliche Stäbichen allerhand Zahlen ohne mühe / und hergegen gar gewiß / zu Multipliciren und zu dividiren, auch die Regulam Detri, und beyderley ins gemein ubliche Radices zu extrahirn: ohne allen brauch des sonsten ob- und nützlichen Ein mahl Eins / Alß in dem man sich leichtlich verstossen kan / Erstlich erfunden durch einen vornehmen Schottländischen Freyherrn Herrn Johannem Neperum Herrn zu Merchiston. rc. Anitzo aber auffskürtzeste / alß immer müglich gewesen / nach vorhergehenden genugsamen Probstücken ins Deutsche ubergesetzt.* Berlin: G. Rungen.

Valla, Lorenzo (1540). *Pro se et contra Calumniatores, ad Eugenium iiii. Pont. Max. Apologia* in *Opera*. Basel: Henricus Petrus.

Valla, Lorenzo (1977). *On Pleasure / De voluptate*, (eds) A. Kent Hieatt and Maristella Lorch. New York: Abaris.

Valla, Lorenzo (2004). *Von der Lust oder Vom wahren Guten / De voluptate sive De vero bono* (ed.) Peter Michael Schenkel. Munich: Fink.

Velserus: *see* Welser.

Viète, François (1591). *In Artem Analyticem Isagogen*. Tournon: Mettayer.

Voetius, Gisbertus (1655). *Disputationes Theologicae Selectae*, vol. 2. Utrecht: Johannes à Waesberge.

Vossius, Johannes Gerardus (1623). *De historicis Graecis libri quatuor*. Leiden: Jean Maire.

Vossius, Johannes Gerardus (1654). *De veterum poetarum temporibus libri duo, qui sunt de poetis Graecis et Latinis*. Amsterdam: Johannes Blaeu.

Wallis, John (1699). *Opera*, vol. 3, *Miscellanea*. Oxford: Sheldonian Theatre.

Ward, Seth (1654). *Vindiciae Academiarum*. Oxford: Robinson.

Wecker, Johann Jacob (1582). *De Secretis Libri XVII. Ex variis authoribus collecti, methodiceque digesti*. Basel: [s.n.].

Weigel, Valentin (2003). *Selected Spiritual Writings* (eds) Andrew Weeks and R. Emmet McLaughlin. New York and Mahwah, NJ: Paulist Press.

Weigel, Valentin (1996–2015). *Sämtliche Schriften*, Neue Edition, (ed.) Horst Pfefferl. Stuttgart-Bad Canstatt: Frommann-Holzboog.

Welser, Paul (1595). Paullus Velserus, *Publilii Optatiani Porphyrii Panegyricus Dictus Constantino Augusto. Ex codice manuscripto*. Augsburg: Willer.

Welser, Markus (1682). Marcus Velserus, *Opera Historica et Philologica, Sacra et Profana. In quibus Historia Boica, Res Augustanae, Conversio & Passio SS. Martyrum, Afrae, Hilariae, Dignae, Eunommiae, Eutropiae, Vitae S. Udalrici, & S. Severini, Narratio eorum, quae contingerunt Apollonio Tyrio, Tabulae Peutingerianae integrae, Epistolae ad Viros Illustres Latinae Italiaeque, & Proteus satyra continentur. Accessit P. Optatiani Porphyrii Panegyricus, Constantino M. missus, ex optimo Codice à Paullo Velsero divulgatus, unà cum Spicilegio Critico Christiani Daumii*[etc.] (= *Opera in Unum Collecta*). Nuremberg: Wolfgang Mauritius and Johann Andreae heirs.

# C Secondary Sources

Acerbi, Fabio (2010). *Il silenzio delle sirene. La matematica greca antica*. Rome: Carocci.

Amunátegui, Giovanni Iommi (2015). 'À propos d'une nouvelle interprétation du *caput variationis* de Leibniz', *Studia Leibnitiana*, 47: 103–7.

Angelelli, Ignatio (1990). 'On Johannes Raue's Logic', in Ingrid Marchlewitz and Albert Heinekamp (eds), *Leibniz' Auseinandersetzung mit Vorgängers und Zeitgenossen*, Studia Leibnitiana Supplementa 27. Stuttgart: Franz Steiner.

Antognazza, Maria Rosa (2001). '*Debilissimae Entitates*? Bisterfeld and Leibniz's Ontology of Relations', *The Leibniz Review* 1: 1–22.

Antognazza, Maria Rosa (2009). *Leibniz: An Intellectual Biography*. Cambridge: Cambridge University Press.

Antognazza, Maria Rosa, and Howard Hotson (1999). *Alsted and Leibniz on God, the Magistrate and the Millennium*. Texts edited with introduction and commentary by Antognazza M. R. and Hotson H. Wiesbaden: Harrassowitz Verlag.

Arthur, Richard T. W. (2004). 'The Enigma of Leibniz's Atomism', in Daniel Garber and Steven Nadler (eds), *Oxford Studies in Early Modern Philosophy* Volume 1. Oxford: Oxford University Press, 183–227.

Barnes, T. D. (1975). 'Publilius Optatianus Porfyrius', *The American Journal of Philology* 96: 173–86.

Beaney, M. (2016). 'Analysis', in The Stanford Encyclopedia of Philosophy, (ed.) Edward N. Zalta (Summer 2016 Edition).

Beeley, Philip (2014). 'Review of N. Rescher's *Leibniz and Cryptography*', *The Leibniz Review* 214: 111–22.

Beer, Susanna de, Karl A. E. Enenkel, and David Rijser (2009). *The Neo-Latin Epigram: A Learned and Witty Genre*. Leuven: Leuven University Press.

Blok, Josine H, and André, P. M. H. Lardinois (eds) (2006). *Solon of Athens. New Historical and Philological Approaches*. Leiden: Brill.

Bonner, Anthony (2007). *The Art and Logic of Ramon Llull: A User's Guide*. Leiden and Boston: Brill.

Bonner, Anthony, and Eve Bonner (eds) (1985). *Ramón Llull, Doctor Illuminatus: A Ramón Llull Reader*.Princeton, NJ: Princeton University Press.

Brann, Noel L. (1981). *The Abbot Trithemius (1462–1516): The Renaissance of Monastic Humanism*. Leiden: Brill.

Brann, Noel L. (1999). *Trithemius and Magical Theology: A Chapter in the Controversy over Occult Studies in Early Modern Europe*. Albany: State University of New York Press.

Brickman, Benjamin (1941). *An Introduction to Francesco Patrizi's Nova de universis Philosophia*. New York: Diss. Columbia Univ.

Brugmans, H. (1974). 'Boxhorn (Marcus Zuerius)', in NNBW 6, 178–80.

Castañeda, H.-N. (1990). 'Leibniz's complete propositional logic', *Topoi* 9: 15–28.

Coudert, Allison P. (1995). *Leibniz and the Kabbala*. Dordrecht: Kluwer.

Coudert, Allison P., Richard H. Popkin, and Gordon M. Weiner (eds) (1998). *Leibniz, Mysticism and Religion*.Dordrecht: Kluwer.

Couturat, Louis (1901). *La logique de Leibniz, d'après des documents inédits*. Paris: Presses Universitaires de France.

Cramer, Florian (2003). 'Poetische Weisheitskunst: Quirinus Kuhlmanns *XLI. Libes-Kuß*' in Renate Lachmann and Stefan Rieger (eds), *Text und Wissen: Technologische und anthropologische Aspekte*: Tübingen: Gunter Narr, 213–26.

De Risi, Vincenzo (2016). 'The Development of Euclidean Axiomatics. The Systems of Principles and the Foundations of Mathematics in Editions of the Elements in the Early Modern Age', *Archive of History of Exact Sciences*, 70: 591–676.

Donalson, Malcom Drew (1996). *A Translation of Jerome's* Chronicon *with Historical Commentary*. Lewiston: Mellen University Press.

Dürr, Karl (1955). 'Die Syllogistik des Johannes Hospinianus', *Synthese* 9: 272–84.

Dürr, Karl (1949). 'Leibniz' Forschungen im Gebiet der Syllogistik', in E. Hochstetter (ed.), *Leibniz zu seinem 300 Geburtstag 1646–1946*. Berlin, Lieferung 5.

Edwards, J. S. (2005). 'The *Carmina* of Publilius Optatianus Porphyrius', in Carl Deroux (ed.), *Studies in Latin Literature and Roman History*, vol. 12, Collection Latomus, vol. 287. Brussels: Latomus, 447–66.

Eis, Elko (1965). 'Zur Rezeption der kanonischen Verwandschaftsbäume Johannes Andreae's: Untersuchungen und Texte', Dissertation, Heidelberg Univeristy.

Findlen, Paula (2004). *The Last Man Who Knew Everything... Or did he? Athanasius Kircher, S. J. (1602–80) and His World*', in Paula Findlen (ed.), *Athanasius Kircher: The Last Man Who Knew Everything*. New York and London: Routledge.

Garber, Daniel (2009). *Leibniz: Body, Substance, Monad*. Oxford and New York: Oxford University Press.

Glidden, Hope H. (1987). 'Polygraphia and the Renaissance Sign: The Case of Trithemius', *Neophilologus* 71: 183–95.

Göncz Zoltán (2013). *Bach's Testament: On the Philosophical and Theological Background of* The Art of Fugue. Lanham, MD: The Scarecrow Press.

Gooch, Paul W. (1983). 'Aristotle and the Happy Dead', *Classical Philology* 78: 112–6.

Häberlein, Mark, Johannes Burckhardt, Judith Holuba, and Theresia Hörmann (eds) (2002). *Die Welser: Neue Forschungen zur Geschichte und Kultur des oberdeutschen Handelshauses*, Colloquia Augustana 16. Berlin: Akademie Verlag.

Hall, Trevor Henry (1969). *Mathematical Recreations: An Exercise in Seventeenth-Century Bibliography*. Leeds: Leeds University School of English.

Heeffer, Albrecht (2004). '*Récréations Mathématiques* (1624): A Study on its Authorship, Sources and Influence'. University of Ghent. http://logica.ugent.be/albrecht/thesis/Etten-intro.pdf (revised 7 October 2004).

Hotson, Howard (2000). *Johann Heinrich Alsted 1588–1638: Between Renaissance, Reformation, and Universal Reform*. Oxford: Clarendon Press.

Hudson, Hoyt Hopewell (1947). *The Epigram in the English Renaissance*. Princeton: Princeton University Press, reprinted (1966), New York: Octagon Books.

Johnson, Christine R. (2008). *The German Discovery of the World: Renaissance Encounters with the Strange and Marvelous*. Charlottesville: University of Virginia Press.

Jolley, Nicholas (ed.) (1995). *The Cambridge Companion to Leibniz*. Cambridge: Cambridge University Press.

Kahn, Charles H. (2001). *Pythagoras and the Pythagoreans: A Brief History*. Indianapolis and Cambridge: Hackett.

Keynes, John Neville (1906). *Studies and Exercises in Formal Logic*. London: MacMillan and Co.

Knecht, Herbert H. (1981). *La logique chez Leibniz: Essai sur le rationalisme baroque*. Lausanne: L'Âge de l'homme.

Knobloch, Eberhard (1973). *Die Mathematischen Studien von G.W. Leibniz zur Kombinatorik*, Studia Leibnitiana Suppl. XI. Wiesbaden: Franz Steiner Verlag.

Knobloch, Eberhard (1974). 'The Mathematical Studies of G. W. Leibniz on Combinatorics', *Historia Mathematica*, 1: 409–30.

Knobloch, Eberhard (2012). 'The Notion of Variation in Leibniz', in Jed Buchwald (ed.), *A Master of Science History, Essays in Honor of Charles Coulston Gillispie*. Dordrecht and New York: Springer, 235–51.

Korcik, Antoni (1955). 'Teoria sylogizmu Hospiniana i Leibniza (The Theory of Syllogism according to Hospinianus and Leibniz)', Polish with English summary, *Roczniki filozoficzne*4: 51–70.

Kügel, Paul, and Wilhelm Studemunde (eds) (1877). *Gai Institutiones ad codicis Veronensis Apographum Studemundianum in usum scholarum*. Berlin: Weidmann.

Lagerlund, Henrik (2000). *Modal Syllogistics in the Middle Ages*. Leiden: Brill.

Lenzen, Wolfgang (2004). 'Leibniz's logic', in D. M. Gabbay and J. Woods (eds), *The Rise of Modern Logic: From Leibniz to Frege*. *Handbook of the History of Logic*, vol. 3. Amsterdam: Elsevier, 1–83.

Loemker, Leroy E. (1961). 'Leibniz and the Herborn Encyclopedists', *Journal of the History of Ideas*, 22: 323–38.

Lombraña, Julián Velarde (1987). 'Proyectos de lengua universal ideados por españoles', *Quadernos de pensamiento*7–8: 7–78.

Lovejoy, Arthur (1936). *The Great Chain of Being: A Study of the History of an Idea*. Cambridge, MA: Harvard University Press.

Łukasiewicz, Jan (1957). *Aristotle's Syllogistic from the Standpoint of Modern Logic*. Oxford: Oxford University Press.

Malink Marko-Vasudevan Anubav (2016). 'The Logic of Leibniz's Generales inquisitiones de analysi notionum et veritatum', *The Review of Symbolic Logic*, 9: 686–751.

Masaracchia, Agostino (1958). *Solone*. Florence: La Nuova Italia.

Mercer, Christia (2000). *Leibniz's Metaphysics: Its Origins and Development*. Cambridge: Cambridge University Press.

Morison, Ben (2008). 'Logic', in R. J. Hankinson (ed.), *The Cambridge Companion to Galen*. Cambridge: Cambridge University Press.

Mugnai, Massimo (1973). 'Der Begriff der Harmonie als metaphysische Grundlage der Logik und Kombinatorik bei Johann Heinrich Bisterfeld und Leibniz', *Studia Leibnitiana* 5: 43–73.

Mugnai, Massimo (2010). 'Logic and Mathematics in the Seventeenth Century', *History and Philosophy of Logic* 31, 297–314.

North, John (1994). *The Fontana History of Astronomy and Cosmology*. London: Fontana Press.

Ong, Walter J. (SJ) (2004). *Ramus, Method and the Decay of Dialogue. From the Art of Discourse to the Art of Dialogue*. Chicago: The University of Chicago Press.

Pater, Wilhelmus Antonius de, SJ (1965). *Les Topiques d'Aristote et la dialectique platonicienne: La méthodologie de la définition*. Fribourg: St. Paul.

Pritzl, Kurt (1983). 'Aristotle and Happiness after Death: Nicomachean Ethics 1. 10–11', *Classical Philology* 78: 101–11.

Rescher, Nicholas (1966). *Galen and the Syllogism*. Pittsburgh: University of Pittsburgh Press.

Rescher, Nicholas (1991). *G.W. Leibniz's* Monadology: *An Edition for Students*. London: Routledge.

Rescher, Nicholas (2012). *Leibniz and Cryptography: An Account of the Occasion of the Initial Exhibition of the Reconstruction of Leibniz's Cipher Machine*. Pittsburgh: University of Pittsburgh Library System.

Rescher, Nicholas (2014). 'Leibniz's Machina Deciphratoria: A Seventeenth-Century Enigma Machine', in *Cryptologia* 38: 103–15.

Roncaglia, Gino (1996). *Palaestra Rationis. Discussioni sulla natura della copula e modalità nella filosofia 'scolastica' tedesca del XVII secolo.* Florence: Leo D. Olschki.

Rossi, Paolo (2006). *Logic and the Art of Memory: the Quest for a Universal Language.* New York and London: Continuum.

Rutherford, Donald (1995). 'Philosophy and Language in Leibniz', in Jolley (1995), 224–69.

Rutherford, Donald (1998). 'Leibniz and Mysticism', in Coudert (1998), 22–46.

Schmidt, Justus (1970). *Johann Kepler: sein Leben in Bildern und eigenen Berichter.* Linz: R. Trauner.

Schuchard, Marsha Keith (1998). 'Leibniz, Benzelius, and the Kabbalistic Roots of Swedish Illuminism', in Coudert (1998), 84–106.

Slaughter, Mary M. (1982). *Universal Languages and Scientific Taxonomies in the Seventeenth Century.* Cambridge: Cambridge University Press.

Smith, David Eugene (1951). *History of Modern Mathematics.* New York: Dover.

Swoyer, C. (1994). 'Leibniz's Calculus of Real Addition', *Studia Leibnitiana*, 26: 1–30.

Tarver, Thomas (1997). 'Varro and the Antiquarianism of Philosophy', in Jonathan Barnes and Miriam Griffin (eds), *Philosophia Togata II: Plato and Aristotle at Rome.* Oxford: Clarendon Press, 130–64.

Thomas, Ivo (1957). 'Review of Dürr (1955) and Korcik (1955)', *The Journal of Symbolic Logic* 22: 382.

Weeks, Andrew (2000). *Valentin Weigel (1533–1588): German Religious Dissenter, Speculative Theorist, and Advocate of Tolerance.* New York: State University of New York.

Withbread, Leslie George (ed.) (1971). *Fulgentius the Mythographer.* Ohio: Ohio State University Press.

Yates, Frances A. (1989). *Lull & Bruno.* London: Routledge & Kegan Paul.

Yates, Frances A. (1991). *Giordano Bruno and the Hermetic Tradition.* Chicago: Chicago University Press.

Yates, Frances A. (1999). *Selected Works*, in 10 vols. London: Routledge.

# D. Abbreviations

| | |
|---|---|
| ADB | *Allgemeine Deutsche Biographie Auf Veranlassung Seiner Majestät des Königs von Bayern herausgegeben durch die historische Commission bei der Königl. Akademie der Wissenschaften* Leipzig: Duncker and Humblot, 1875–1912. |
| AT | *Œuvres de Descartes* (1897–1913). (eds) Charles Adam and Paul Tannery. Paris: L. Cerf. Reprinted 1964–71, Paris: Vrin) / 1996 (Paris: Vrin). |
| Chauvin | Stephanus Chauvin (1713). *Lexicon Philosophicum.* Leeuwarden: Franciscus Halma. Reprinted (1967) (ed.) Lutz Geldsetzer. Düsseldorf: Stern-Verlag Janssen & Co. |

*Corpus Iuris*
*Civilis*      *Corpus Iuris Civilis* (1894). vol. 1, (eds) Paul Krueger and Theodor
             Mommsen. Berlin: Weidmann.
*Digesta*      Justinianus, *Digesta*, in *Corpus Iuris Civilis*.
CSM          *The Philosophical Writings of Descartes* (1985–91) in 3 vols, (eds) John
             Cottingham, Robert Stoothoff, Dugald Murdoch, and Anthony
             Kenny. Cambridge: Cambridge University Press.
DBE          *Deutsche Biographische Enzyclopädie* (1995–2003) in 13 vols, (eds)
             Walther Killy and Rudolf Vierhaus. Munich: K.G. Saur.
DDP          *The Dictionary of Seventeenth and Eighteenth-Century Dutch Philo-
             sophers* (2003) in 2 vols, (eds) Wiep van Bunge, Henri Krop, Bart
             Leeuwenburgh, Han van Ruler, Paul Schuurman, and Michiel Wie-
             lema. Bristol: Thoemmes Press.
DSB          *Dictionary of Scientific Biography* (1970–8) in 16 vols, (ed.) Charles
             Coulston Gillispie. New York: Charles Scribner's Sons.
Gaius        *Institutiones ad Codicis Veronensi apographum* (1877). (eds) Paul
             Krüger and Wilhelm Studemund. Berlin: Weidmann.
GB           *Der Briefwechsel von Gottfried Wilhelm Leibniz mit Mathematikern*
             (1899). (ed.) Carl Immanuel Gerhardt. Berlin: Mayer and Müller.
             Reprinted (1962). Hildesheim: Olms.
*Hofmann*      *Register zu Gottfried Wilhelm Leibniz: Mathematische Schriften und
             Der Briefwechsel mit Mathematikern* (1977). (ed.) Joseph Ehrenfried
             Hofmann. Hildesheim and New York: Olms.
*Institutiones* Justinianus, *Digesta*, in *Corpus Iuris Civilis*.
*Institutes*   *The Institutes of Justinian: Text, Translation and Commentary* (1975).
             (ed.) J. A. C. Thomas. Amsterdam and Oxford: North-Holland Pub-
             lishing Company.
*Jöcher*       Christian Gottlieb Jöcher (1751). *Allgemeines Gelehrten-Lexicon.*
             Leipzig: J. F. Gleditschens.
*Jöcher-F*     *Fortzetsung und Ergänzungen zu Christian Gottlieb Jöchers allgemeinem
             Gelehrten-Lexicon*, (eds) Johann Christoph Adelung and Heinrich
             Wilhelm Rotermund. Leipzig and Bremen: J. F. Gleditschen,
             G. Jöngen, J. G. Heyse, Selbstverlag der Deutschen Gesellschaft,
             1784–1897. Reprinted (1960–1). Hildesheim: Olms.
NDB          *Neue Deutsche Biographie* (1953 ff.) Herausgegeben van der Histor-
             ischen Kommission bei der Bayerischen Akademie der Wissenschaften.
             Berlin: Duncker and Humblot.
NNBW         *Nieuw Nederlands Biografisch Woordenboek* (1911–37) in 10 vols,
             (eds) P. C. Molhuysen, P. J. Blok, and Fr. K. H. Kossmann. Leiden:
             A. Sijthoff. Reprinted (1974). Amsterdam: Israel.
*Zedler*       *Grosses vollständiges Universal Lexicon aller Wissenschaften und Künste,
             welche bisshero durch menschlichen Verstand und Witz erfunden und
             verbessert worden* (1730–54). Halle – Leipzig: J. H. Zedler. / *Grosses
             Vollständiges Universal-Lexikon* (1961–4). Graz: Akademische Druck
             und Verlagsanstalt.

# Index of Names